NEUROMETHODS

Series Editor
Wolfgang Walz
University of Saskatchewan
Saskatoon, SK, Canada

For further volumes:
http://www.springer.com/series/7657

Animal Models of Spinal Cord Repair

Edited by

Håkan Aldskogius

Department of Neuroscience, Biomedical Center,
Uppsala University, Uppsala, Sweden

Humana Press

Editor
Håkan Aldskogius
Department of Neuroscience
Biomedical Center
Uppsala University
Uppsala, Sweden

ISSN 0893-2336 ISSN 1940-6045 (electronic)
ISBN 978-1-4939-6180-1 ISBN 978-1-62703-197-4 (eBook)
DOI 10.1007/978-1-62703-197-4
Springer New York Heidelberg Dordrecht London

Cover illustration: Gustaf Retzius & Axel Key, Studien in der Anatomie des Nervensystems und des Bindegewebes I. Stockholm, Norstedt & Söner, 1875. Tafel I, Fig. 7. Courtesy of the Hagströmer Medico-Historical Library, Karolinska Institutet, Stockholm.

Printed on acid-free paper

Humana Press is a brand of Springer
Springer is part of Springer Science+Business Media (www.springer.com)

Preface to the Series

Under the guidance of its founders Alan Boulton and Glen Baker, the Neuromethods series by Humana Press has been very successful since the first volume appeared in 1985. In about 17 years, 37 volumes have been published. In 2006, Springer Science + Business Media made a renewed commitment to this series. The new program will focus on methods that are either unique to the nervous system and excitable cells or which need special consideration to be applied to the neurosciences. The program will strike a balance between recent and exciting developments such as those concerning new animal models of disease, imaging, in vivo methods, and more established techniques. These include immunocytochemistry and electrophysiological technologies. New trainees in neurosciences still need a sound footing in these older methods in order to apply a critical approach to their results. The careful application of methods is probably the most important step in the process of scientific inquiry. In the past, new methodologies led the way in developing new disciplines in the biological and medical sciences. For example, Physiology emerged out of Anatomy in the nineteenth century by harnessing new methods based on the newly discovered phenomenon of electricity. Nowadays, the relationships between disciplines and methods are more complex. Methods are now widely shared between disciplines and research areas. New developments in electronic publishing also make it possible for scientists to download chapters or protocols selectively within a very short time of encountering them. This new approach has been taken into account in the design of individual volumes and chapters in this series.

Saskatoon, SK, Canada *Wolfgang Walz*

Preface

The development of treatment strategies that can help patients with spinal cord injury to regain lost functions and an improved quality of life is a major medical challenge. Extensive experimental research at the molecular, cellular, and systems level has opened up new and promising avenues towards meeting this challenge. Translation of some of this research to clinical trials for spinal cord injury is already underway, and the results of these trials can be expected to, first of all, provide answers as to the feasibility and safety of the interventions. Indication of potential efficacy may also emerge, although this information is likely to have to wait until sufficient data have been collected and analyzed from ongoing and forthcoming trials. The continued rapid progress in basic spinal cord injury research will most likely also provide foundations for routes to spinal cord repair which are not yet envisioned. In addition, and perhaps most importantly, clinical trials will raise issues which will need to be dealt with by refined experimental research before subsequent possible translations into modified clinical management strategies can be considered. Experimental spinal cord research has to meet these challenges by resolving fundamental problems, establishing a basis for possible novel treatment strategies of spinal cord injury, and motivating their clinical translation.

To place animal models of spinal cord repair in its "human" context, the book features thorough overviews of the anatomy, physiology, and pathophysiology of clinical spinal cord injuries (Chaps. 1, 2). With this background, the reader is subsequently introduced to a broad range of experimental models for research on spinal cord injury, how they have contributed to our current state of knowledge, and what their advantages are in the further advancement towards spinal cord repair. The information presented is intended to guide the implementation of animal models for spinal cord repair, as well as to raise the awareness of the relevance of experimental models which may not be in the current mainstream of this research. There are numerous striking examples on how the choice of a specific, and not always the most obvious, experimental model has been instrumental in the progress of biomedical research. Studies on less-complex systems of spinal cord injury, such as in vitro (Chap. 3) and lamprey (Chap. 4), provide insights into basic mechanisms of neurodegeneration and regeneration, which offer clues to how the injured mammalian spinal cord injury can be repaired. In the other end, a broader use of larger mammalian species, particularly for preclinical assessments, could be of great predictive value and thereby accelerate the development of clinically useful treatment options (Chaps. 7, 8, 10). In this context, veterinarian medicine can also make a significant contribution as a clinical research setting in the care of pets which have suffered accidental spinal cord injury where the pathology is likely to mimic many aspects of human spinal cord injury (Chap. 9).

The rat has been by far the most popular species for in vivo experimentation with mouse now catching up as a result of the development of the transgenic technology, and the preferences for these species are likely to prevail for a foreseeable future. Rat and mouse offer considerable practical advantages for spinal cord injury research. Several well-characterized spinal cord injury models and the principles for their outcome assessment are described for these species (Chaps. 5, 13), including a recently developed mouse injury model, which circumvents the need for laminectomy (Chap. 6). Animal models of spinal cord ischemia, a condition of increasing clinical importance and also highly relevant for spinal cord pathophysiology in general, are extensively covered in Chaps. 6 and 11. In the context of assessment tools, an update is provided of the increasingly important role of magnetic resonance imaging (MRI) for monitoring and assessing pathophysiology and repair of experimental spinal cord injury (Chap. 12).

The final chapter (Chap. 14) places the current state of experimental research in spinal cord repair in a clinical perspective. Major progress has been made over the last decades in the clinical management of patients with spinal cord injury, but despite promising experimental data, no effective clinically applicable therapies exist that are able to significantly counteract secondary injury processes and repair functional circuitry. The gap from the many times promising outcome of experimental research to successful clinical translation is still large. However, through the background information and hands-on methods descriptions, as well as the basic and clinical issues presented in this book, it is hoped to stimulate and guide researchers with different backgrounds towards the development of improved strategies for functionally relevant repair of the injured human spinal cord.

Uppsala, Sweden *Håkan Aldskogius*

Contents

Contributors

ELISABET ÅKESSON • *Division of Neurodegeneration, Department of Neurobiology, Care Sciences, and Society, Karolinska Institutet, Stockholm, Sweden; R&D Unit, Stockholms Sjukhem, Stockholm, Sweden*

HÅKAN ALDSKOGIUS • *Department of Neuroscience, Biomedical Center, Uppsala University, Uppsala, Sweden*

LEIF ANDERBERG • *Department of Clinical Science, Neurosurgery, Lund University, Lund, Sweden*

HAMDY AWAD • *Department of Anesthesiology, The Ohio State University Medical Center, Columbus, OH, USA*

SHALA GHADERI BERNTSSON • *Department of Neuroscience, Neurology, Uppsala University, Uppsala, Sweden*

CINZIA CALZAROSSA • *Division of Neurodegeneration, Department of Neurobiology, Care Sciences, and Society, Karolinska Institutet, Stockholm, Sweden*

CORINNA DARIAN-SMITH • *Department of Comparative Medicine, School of Medicine, Stanford University, Stanford, CA, USA*

HAYTHAM ELGHARABLY • *Department of Anesthesiology, The Ohio State University Medical Center, Columbus, OH, USA*

PAULA J. FOSTER • *Robarts Research Institute, Western University, London, ON, CanadaThe Department of Medical Biophysics, Schulich School of Medicine and Dentistry, Western University, London, ON, Canada*

ROBIN J.M. FRANKLIN • *Department of Veterinary Medicine, University of Cambridge, Cambridge, UK*

ALAIN FRIGON • *Faculty of Medicine and Health Sciences, Department of Physiology and Biophysics, Université de Sherbrooke, Sherbrooke, QC, Canada*

JÁN GÁLIK • *Institute of Neurobiology, Slovak Academy of Sciences, Košice, Slovak Republic*

LAURA E. GONZALEZ-LARA • *Robarts Research Institute, Western University, London, ON, Canada; The Department of Medical Biophysics, Schulich School of Medicine and Dentistry, Western University, London, ON, Canada*

NICOLAS GRANGER • *School of Veterinary Sciences, University of Bristol, Langford, UK*

PIERRE A. GUERTIN • *Faculty of Medicine, Department of Psychiatry and Neurosciences, Laval University, Quebec City, QC, Canada*

ANDERS HOLTZ • *Department of Neuroscience, Neurosurgery, Uppsala University, Uppsala, Sweden*

FREDA JAWAN • *Robarts Research Institute, Western University, London, ON, Canada; The Department of Medical Biophysics, Schulich School of Medicine and Dentistry, Western University, London, ON, Canada*

NICK D. JEFFERY • *Department of Veterinary Clinical Sciences, College of Veterinary Medicine, Iowa State University, Ames, IA, USA*

RONALD C. KIM • *Department of Pathology and Laboratory Medicine, University of California Irvine, Douglas Medical Center, Orange, CA, USA*
ELENA N. KOZLOVA • *Department of Neuroscience, Uppsala University Biomedical Center, Uppsala, Sweden*
NADEŽDA LUKÁČOVÁ • *Institute of Neurobiology, Slovak Academy of Sciences, Košice, Slovak Republic*
ANDREW D. MCCLELLAN • *Division of Biological Sciences, and Interdisciplinary Neuroscience Program, University of Missouri, Columbia, MO, USA*
MIRANDA MLADINIC • *Neuroscience Department, International School for Advanced Studies (SISSA), Trieste, Italy; SPINAL (Spinal Person Injury Neurorehabilitation Applied Laboratory), Istituto di Medicina Fisica e Riabilitazione, Udine, Italy*
ANDREA NISTRI • *Neurobiology Sector, International School for Advanced Studies (SISSA), Trieste, Italy; SPINAL (Spinal Person Injury Neurorehabilitation Applied Laboratory), Istituto di Medicina Fisica e Riabilitazione, Udine, Italy*
JAROSLAV PAVEL • *Institute of Neurobiology, Slovak Academy of Sciences, Košice, Slovak Republic*
PHILLIP POPOVICH • *Department of Neuroscience, Center for Brain and Spinal Cord Repair, The Ohio State University, Columbus, OH, USA*
ERIK SUNDSTRÖM • *Division of Neurodegeneration, Department of Neurobiology, Care Sciences, and Society, Karolinska Institutet, Stockholm, Sweden; R&D Unit, Stockholms Sjukhem, Stockholm, Sweden*
GIULIANO TACCOLA • *Neurobiology Sector, International School for Advanced Studies (SISSA), Trieste, Italy; SPINAL (Spinal Person Injury Neurorehabilitation Applied Laboratory), Istituto di Medicina Fisica e Riabilitazione, Udine, Italy*

Chapter 1

The Spinal Cord: Functional Organization, Diseases, and Dysfunctions

Pierre A. Guertin

Abstract

The spinal cord constitutes a key structure of the central nervous system. It is involved in the transmission of signals between the brain and the rest of the body through numerous ascending and descending pathways. It also contains relatively simple reflex arcs as well as more complex neuronal circuits that can control central pattern generated-motor behaviors. This section of the book essentially provides a summary description of the functional organization, diseases, and dysfunctions of the spinal cord. It is intended to be a comprehensive review of background knowledge that may be helpful for understanding the chapters that compose the present book on animal models of spinal cord injury.

Key words: Spinal cord organization, Spinal pathologies, Central pattern generators, Locomotion, Ejaculation, Micturition, Review

1. The Functional Organization

1.1. Gross Anatomy

The spinal cord constitutes the most caudally located structure of the central nervous system (CNS). It is essentially a long and relatively thin neural structure that extends from the base of the skull (the brain, i.e., specifically the medulla) to the first lumbar vertebra (between the first and second vertebrae). Thus, although it is contained within the vertebral column (the protective bone structure), the spinal cord does not extend the entire length of that column. It varies in length (on average, between 43 and 45 cm) and size (e.g., cervical and lumbar enlargements) with a shape that is compressed dorsoventrally, giving it an elliptical form. The spinal cord possesses grooves dorsally and ventrally called the posterior median sulcus and the anterior median fissure, respectively.

In human, the spinal cord is divided into 31 different segments (species-specific variations exist)—8 cervical, 12 thoracic, 5 lumbar, 5 sacral, and 1 coccygeal segments. Every segment is associated

Håkan Aldskogius (ed.), *Animal Models of Spinal Cord Repair*, Neuromethods, vol. 76,
DOI 10.1007/978-1-62703-197-4_1, © Springer Science+Business Media, LLC 2013

with a pair (right and left) of spinal nerves. The spinal nerves comprise the sensory nerve roots (composed of 6–8 rootlets), which enter the spinal cord at each level, and the motor roots (also with 6–8 rootlets), which emerge from the cord at each level. In rostral parts, the spinal nerves exit directly from (just above from C1 to C7 or below from C8 and lower) the vertebra associated with the corresponding spinal cord segment. However, in caudal parts of the spinal cord, the spinal nerves travel further down the column before exiting.

In cross section, the spinal cord displays two types of tissue, the white and the gray matters (see also Sect. 1.2). Peripherally located, the former contains white matter tracts (ascending and descending myelinated fibers) containing both sensory and motor axons. Centrally located, the latter is characterized by its butterfly-shape that contains gray matter cells (unmyelinated). In the center of the gray matter, a region called the central canal contains cerebrospinal fluids as the ventricules in the brain to which it is connected. The spinal cord is protected by three surrounding layers of tissue, called the meninges—the dura which is the most peripheral layer separated from the vertebrae by the epidural space, the arachnoid which is the middle protective layer, and the pia mater, the innermost protective layer. The latter is relatively thin and tightly associated with the surface of the spinal cord. The spinal cord is supplied by a vast system of blood vessels comprising mainly the anterior spinal artery, the bilateral sulcal branches, the bilateral posterior spinal arteries, the pial arterial plexus as well as the anterior/posterior spinal veins, the anterior/posterior sulcal veins, and the pial venous plexus.

1.1.1. Myotomes

Each spinal cord segment and pair of nerves are associated with the innervation of different sets of muscles so that each muscle (e.g., soleus of the left and right legs) is bilaterally innervated by a corresponding pair of nerve roots. By definition, a myotome is defined as a group of muscles supplied by a specific pair of nerves and spinal cord segment. Cervical segments innervate muscles involved in respiration, head, neck, and arm movements. Thoracic segments provide motor control of the finger, chest, back, and abdominal muscles, whereas the lumbar and sacral segments are associated with the control of muscles involved in locomotion, micturition, bowel, and reproductive functions.

Main myotomes:

C1–C4	Skull, neck, upper thoracic area
C3–C5	Diaphragm (lower thoracic area)
C5	Shoulder
C5–C7	Elbow
C6	Wrist

C7–T1	Finger
C8	Thumb
T1, T12	Chest wall and abdominal
L2–L4	Hip
L3	Knee
L4–S1	Ankle
L5, S1	Toe
S2–S5	Bladder, bowel and sex organs

1.1.2. Dermatomes

A comparable organization exists for the sensory system where each part of the body is associated with a specific pair of nerves and its spinal cord segment. Cervical segments receive sensory signals from the head (skull), neck, arms, and fingers. Thoracic segments receive inputs from the chest and abdominal area, whereas the lumbar and cervical segments integrate sensory information from the leg, foot, anal, and sex organ areas.

Main dermatomes:

C2	Skull
C3	Rostral neck
C4	Caudal neck
C5	Clavicle
C5–C7	Lateral parts of upper limb
C6	Thumb
C7	2nd and 3rd digit
C8	4th and 5th digit
T4	Nipple (T4)
T5	Mammary fold
T6–T7	Xiphoid process
T10	Umbilicus
T12	Pubic bone
T12–L1	Inguinal or groin
L1–L4	Anterior and inner surfaces of leg
L4	Knee
L4–S1	Foot
S1	Foot and little toe
S2–S3	Sex organs
S2–S4	Perineum

1.2. White Matter

1.2.1. Ascending Somatosensory Tracts

The white matter is divided into the dorsal, lateral, and ventral columns. The first two convey ascending sensory signals, whereas the ventral columns are associated mainly with descending motor signals. Each column is composed of bundles of axons, also called tracts. In the dorsal column, signals associated with touch, pressure, proprioception, and vibration travel from the spinal cord via relatively large myelinated fibers in the gracile fasciculus or the cuneate fusciculus (i.e., depending upon the level of entry) to different areas of the brain. Then, synapses and decussation occur in the medulla prior to being redirected (signals) toward the thalamus and eventually the sensory cortex.

In contrast, the lateral or so-called anterolateral columns convey information about pain and thermal sensation. Synapses and decussation occurs at the spinal cord level (close to the entry of the corresponding primary afferents) prior to ascend to the brain via a number of tracts (see section "Anterolateral (Ventrolateral) System").

Dorsal Columns

The gracile fasciculus, often referred to as the posterior column-mediated lemniscus system or the tract of Goll (Swiss neuroanatomist), travels dorsomedially in the white matter and carries information about fine touch, vibration, and conscious proprioception from the lower part of the body (T6–S5) to the medulla oblongata (nucleus gracilis).

The cuneate fasciculus, also called the tract of Burdach, is located in the dorsal column just laterally to the gracile fasciculus. The cuneate fasciculus carries to the brain information about touch, vibration, and proprioception originating from the upper part of the body (C1–T6, e.g., arms).

Main dorsal column tracts:

- Gracile fasciculus
- Cuneate fasciculus

Anterolateral (Ventrolateral) System

Each tract composing the anterolateral system or columns mediates pain and temperature information that originates in different spinal cord regions and ascends to terminate in different areas of the brain. This posterolateral tract (Lissauer's tract) is a relatively small ascending pathway located next to the cuneate fasciculus. Lissauer's tract was described by the German neurologist Heinrich Lissauer in the late nineteenth century. It contains information about pain (non-discriminative) and temperature that typically enters in the gray matter one or two segments rostrally for synapsing onto the secondary sensory neuron and decussating contralaterally prior to penetrate again the white matter as part of the ascending spinothalamic tract in direction for the thalamus.

The dorsal spinocerebellar tract, also called the posterior spinocerebellar or Flechsig's tract, carries information about proprioception after a first synapse in the Clarke's nucleus. Unlike the ventral spinocerebellar tract, the Flechsig's tract does not mainly decussate in the spinal cord or elsewhere. The ventral spinocerebellar tract, also called the anterior spinocerebellar or Gowers's tract (described by Sir William Richard Gowers), conveys proprioceptive information (muscles, joints, etc.) to the cerebellum via one synapse (a primary and a secondary neurons). Decussation generally occurs in the spinal cord as well as in the brainstem (double-cross).

The spinomesencephalic pathway travels in the anterolateral system and ends in the mesencephalic region. The dorsal (posterior or lateral) spinothalamic tract is a pathway mainly composed of sparsely myelinated axons and slow-conducting, unmyelinated axons carrying to the thalamus pain, temperature, information from the skin and other organs. The neospinothalamic tract is one of two subclasses of dorsal spinothalamic tracts. It specifically carries fast pain signals (sparsely myelinated fibers). It decussates in the spinal cord and terminates in most cases in the thalamus (some terminates also in the brainstem, see subclasses below). In turn, the paleospinothalamic tract specifically conveys slow pain signals via unmyelinated C fibers to the brain (most terminate in the brainstem, whereas some terminate in the thalamus).

The ventral spinothalamic tract conveys information to the thalamus from the skin and other organs (itch and crude touch) after synapsing and decussating in the spinal cord and again in the brain (double-cross). The spinotectal pathway conveys information about pain and temperature to the inferior and superior colliculi. It is positioned next to the dorsal spinothalamic tract and decussates at the spinal cord level. The spino-olivary tract is more ventrally located than most of the other anterolateral tracts, and conveys information about muscles, tendons and skin to the olivary nucleus. The spinoreticular tract is located just beside the spinothalamic tract, and carries deep and chronic pain signals to the reticular formation. The spinovestibular tract is located near to the posterior spinocerebellar tract and carries information to the lateral, medial and spinal vestibular nuclei. The interfascicular tract, also called the comma tract of Schultze (discovered by Friedrich Schultze, a German neurologist), is located near the tract of Goll and conveys information about pain and temperature. The spinocortical tract is a bundle of ascending fibers traveling within the descending corticospinal tract conveying information to the cerebral cortex. It originates from all levels of the spinal cord, decussates in the brainstem and terminates in the deep layers of the cerebral cortex.

Main anterolateral system tracts:

- Lissauer's tract
- Dorsal spinocerebellar tract

- Ventral spinocerebellar tract
- Spinomesencephalic tract
- Dorsal (postero or lateral) spinothalamic tract
- Neospinothalamic tract
- Paleospinothalamic tract
- Ventral (antero) spinothalamic tract
- Spinotectal tract
- Spino-olivary tract
- Spinoreticular tract
- Spinovestibular tract
- Interfascicular tract
- Spinocortical tract

1.2.2. Descending Motor Tracts

The descending motor system can be divided into the pyramidal and extrapyramidal tracts. The corticospinal tracts also called the pyramidal tracts contain about one million axons on each side that are involved essentially in skilled movements. In contrast, the extrapyramidal tracts are mainly involved in the control of complex behaviors such as postural control and locomotion.

Pyramidal Tracts

This ensemble of axons originates from the cerebral cortex and from some brainstem motor nuclei. It constitutes the most direct descending motor pathway (e.g., some monosynaptic connections) between the motor cortex (Brodmann's areas 1, 2, 3, 4, and 6) and the final common motor pathway, namely motoneurons in the spinal cord located at all segmental levels. Although often referred to as upper motor neurons, axons composing the pyramidal tracks have their cell bodies in the motor cortex and should accordingly and more appropriately be called corticospinal neurons. The term upper motor neuron may be misleading because they are not really motor neurons although their activation produces movements and, inversely, movement disorders occur when they are injured. About, 80–90% of these corticospinal axons are known to decussate to the contralateral side at the pyramid level in the medulla oblongata (also called crossed lateral pyramidal tract). From there, they constitute the lateral corticospinal tract that sends input to spinal motoneurons (ventral horn area). The remaining 10–20% descends ipsilaterally as part of the ventral corticospinal tract and decussate in the spinal cord prior to synapsing with the corresponding motoneurons (also called uncrossed ventral corticospinal tract). Some of the corticospinal neurons form instead the corticobulbar tract that sends input to brainstem motoneurons involved in the control of face, head and neck muscles. An injury at this level is generally associated with the pyramidal tract syndrome characterized by spasticity, paralysis and loss of skilled movements.

Main pyramidal tracts:

- Crossed pyramidal tract (large lateral corticospinal tract)
- Direct pyramidal tract (uncrossed ventral corticospinal tract)
- Corticobulbar tract

Extrapyramidal Tracts

Cell bodies of the extrapyramidal tracts are mainly found in subcortical nuclei in the pons (reticular formation) and medulla oblongata. They send their axons in the spinal cord to motoneurons at all segmental levels. Specifically, the red, vestibular, tectum (superior colliculus) and reticular formation nuclei have cell bodies sending their axons through their respective descending extrapyramidal tract—the rubrospinal, vestibulospinal, tectospinal, and reticulospinal tracts. The former tract descends laterally in the spinal cord, whereas the other three tracts travel ventrally (anterior column). The rubrospinal tract receives inputs from the motor and premotor areas 4 and 6. It sends projections to contralateral (decussation in the brain) motoneurons of the upper spinal cord segments for the control of both fine skilled movements and large powerful ones. It is generally considered to facilitate flexion and inhibit extension specifically in the upper extremities contralaterally. The tectospinal tract sends its axons in premotor lamina of the cervical spinal cord (VI–VIII) and is involved in neck and head motor control. In contrast with the other two tracks, the reticulospinal track sends projections to all levels of the spinal cord. It is involved in the control of autonomic functions (cardiovascular and respiratory functions, blood pressure, micturition, etc.). The vestibulospinal tract sends axons that travel either laterally or ventrally on the ipsilateral side of the spinal cord to premotor laminae VII and VIII. It is known to control muscular contraction levels specifically in extensors. The olivospinal tract originates in the inferior olivary nucleus and innervates contralaterally segments in the spinal cord. Lesioned extrapyramidal tracts are associated with the extrapyramidal syndrome characterized by involuntary movements, muscular rigidity, and immobility without paralysis.

Main extrapyramidal tracts:

- Rubrospinal tract
- Tectospinal tract
- Reticulospinal tract
- Vestibulospinal tract
- Olivospinal tract

1.3. Gray Matter

As described above, a part of the spinal cord (i.e., white matter) serves as a relay for descending (motor control) and ascending (sensory-related) information between rostral and caudal structures. In recent years, spinal cord physiologists have begun unravelling the existence of other roles of the spinal cord associated mainly with neurons in the gray matter. Some of these neurons form local spinal reflex pathways (so-called reflex arcs), whereas others are part of more complex circuits involved in specific functions. As of today, it is unclear how many different populations (classes) of neurons exist in the spinal cord. However, based on some estimates, it may well exist between one and two billion spinal cord neurons.

1.3.1. Simple Reflex Pathways

A reflex may be defined as a rapid, predictable, repeatable, stereotyped and involuntary motor response or movement induced by a specific stimulus. Most reflexes are mediated locally (within one or two segments) by relatively simple neural pathways (arcs) in the spinal cord—involving generally either one (monosynaptic), two (disynaptic) or more (polysynaptic) synapses and neurons involving interneurons and motoneurons. Note that partially "involuntary controlled" motor behaviors such as locomotion should not be considered as reflexes (i.e., not induced stereotypically and specifically by a corresponding peripheral afferent input). There are two main types of reflexes—autonomic (related with inner organs, eyes, blood vessels, etc.) and somatic (related with skeletal muscle responses). This said, among all reflexes, it is the somatic reflex system mediated in the spinal cord that has been most extensively studied (mainly in cats). Here is a list of spinal reflexes and corresponding reflex arcs for which a detailed description has been provided.

Ia Reflex Pathway

The best known somatic reflex pathway is undoubtedly the Ia monosynaptic reflex arc which is also considered the simplest reflex. It mediates primary afferent (Ia) inputs originating from muscle spindles typically activated by muscle stretching (e.g., also called tendon jerk, tendon tap, or myotatic reflex). It is generally considered to play a role in tonus and postural adjustments. The Ia afferent input enters the spinal cord (dorsally via the corresponding spinal nerve) and establishes monosynaptic connections with homonymous alpha-motoneurons (e.g., soleus Ia afferent input sent to soleus motoneurons) in the gray matter (ventral horn area). This excitatory reflex is known to increase homonymous muscle contraction in response to muscle elongation. In addition to monosynaptic connections with the corresponding homonymous muscle, Ia afferents possess branches that also establish monosynaptic connections with synergistic alpha-motoneurons (e.g., soleus Ia afferent with gastrocnemius motoneurons).

Although generally considered as stereotyped responses, spinal reflexes can nonetheless be modulated. For instance, recent evidence has shown that activity levels in the Ia monosynaptic arc can vary depending upon circumstances, phases, and tasks—e.g., amplitudes of the H-reflex (experimentally induced myotatic reflex) decrease during walking compared with standing quietly. Also, during walking, H-reflex amplitudes are smaller during the swing phase (flexion) than the extension phase (extension). An assessment of Ia reflex pathway function can provide clinically valuable information on neuropathological conditions. A reflex hammer stimulation eliciting an exaggerated (hyperreflexia) or diminished (hyporeflexia) monosynaptic reflex is generally considered as a sign of CNS or PNS damage, respectively.

Ib Reflex Pathways

Also known as the inverse myotatic reflex or autogenic inhibitory reflex pathway, this reflex arc is associated with peripheral afferent inputs from the group Ib afferent fibers (Golgi tendon organs). It has been described initially as inhibiting the homonymous and synergistic alpha-motoneurons at rest although it may remain quiescent in other conditions such as during locomotion. Inputs from the Ib afferents enter the spinal cord and inhibit disynaptically (two synapses involving one inhibitory neuron called the Ib interneuron) alpha-motoneurons. This inhibitory reflex was originally believed to act as a protective mechanism against excessive muscle contraction. However, this view has been challenged since in normal individuals (e.g., not spinal cord-injured), Golgi tendon organs and Ib reflex pathways have been found to mediate inputs throughout a wide range of muscle activity and load. As with the Ia reflex, the Ib autogenic inhibitory reflex has been found to undergo extensive task-dependent modulation. Actually, during locomotor activity, it was found to completely disappear. It is replaced by another Ib afferent-induced reflex pathway that provides extensive excitation of homonymous and synergistic alpha-motoneurons (at all joints of the lower extremities). The candidate excitatory interneurons mediating this disynaptic reflex (and possibly also locomotor-dependent Ia afferent-induced excitation of extensors) is located in lumbar enlargement segments (premotor interneuron in lamina VII) and rhythmically active during locomotion.

Although the role of Ib autogenic inhibition remains largely unclear in resting conditions, the role of the Ib disynaptic excitatory pathway was proposed, during locomotion, to significantly enhance muscular contraction of extensors during the stance phase (20–50% contribution to final force generation). No clinically induced reflex movement has been associated with stimulation of Golgi tendon organs. However, the clasp knife reflex that was thought originally to be mediated only by autogenic inhibition is now believed to be the result of an interaction between Ia monosynaptic excitatory and Ib autogenic inhibitory actions on alpha-motoneurons in spastic patients (e.g., with injured descending tracts).

Flexion Reflex Pathways

Another well-described spinal reflex arc is the flexion reflex afferent (FRA) pathway activated by relatively high threshold fibers (e.g., associated with cutaneous nociceptor A or C fibers, group II, III, and IV muscle afferent fibers, etc.). On the ipsilateral side, it involves at least two interneurons (three or more synapses) in several segments of the spinal cord and alpha-motoneurons of several flexor muscles. It is also called the nocifensive reflex or withdrawal reflex pathway. Specifically, it comprises at least two excitatory interneurons (three synapses) for flexor motoneuron activation and two interneurons (one excitatory and one inhibitory) for extensor motoneuron inhibition. If the stimulus is strong (intense) enough, it may be accompanied by an activation of an accompanying reflex pathway, the crossed-extension reflex arc that produces an extension of the contralateral limb. Contralaterally, additional interneurons (and decussation) are involved for producing the crossed-extension reflex—two excitatory interneurons for contralateral extensor motoneuron activation and two interneurons (one inhibitory and one excitatory) for contralateral flexor muscle inhibition. These reflex pathways are thought to play a role in withdrawal of a limb (unilateral flexion and contralateral extension) from a painful stimulus.

Task-dependent modulatory responses have also been found during locomotor activity since a long-lasting burst of activity is unravelled ipsilaterally during pharmacologically induced fictive locomotion (long-lasting FRA response). The corresponding locomotor-related interneurons have even been proposed to be part per se of the Central Pattern Generator (CPG) for locomotion (specifically part of the flexor portion of a half-center organized-like network) since FRA stimulation under experimental conditions was shown to reset the step cycle to flexion.

Clinically, this reflex pathway, in resting conditions, may be investigated by induction of the Babinski sign (i.e., tongue depressor-induced plantar extension). Although normally found in infants, it is an indication of neurological problems in adults (e.g., spinal pyramidal tracts-induced injury caused by trauma or tumor). The long-lasting FRA response as well as myoclonus can also be uncovered following FRA stimulation in spinal cord-injured patients.

Reciprocal Inhibitory Reflex Pathway

Unlike the other classical reflexes describe above, this reflex pathway is not specifically associated with one type of peripheral afferent input. It has been found in association with the Ia monosynaptic reflex and the Ib autogenic inhibitory reflex, etc. The same principle remains in all cases which is that during a stereotypic response in the main muscle group, the group of antagonist muscles will display the opposite motor response in order to facilitate movement execution. For instance, Ia monosynaptic excitation of soleus motoneurons (extensor muscle) will be accompanied of reciprocal disynaptic inhibition of tibialis anterior motoneurons (flexor muscle) mediated by one inhibitory interneuron (that directly receives input from the primary afferent fiber). A comparable pathway exists for Ib autogenic inhibition which, in parallel, through synaptic connections with excitatory interneurons, will activate the antagonistic populations of motoneurons. Clinically, dysfunctional reciprocal inhibitory reflex pathways have been associated with increased co-contraction. It has been reported in different pathological conditions such as in Parkinson's disease patients or in patients displaying spasticity.

1.3.2. Complex Spinal Circuits

The spinal cord has often been considered as a simple relay between brain cells and effective organs (muscles, skin, etc.). However, in the last 40 years, spinal cord physiologists have clearly begun to unravel the existence of "command centers" involved significantly in the control and modulation of several relatively complex functions (largely under involuntary control—i.e., not entirely depending upon corticospinal or other supraspinal inputs). With at least 1–2 billion neurons, the spinal cord is now recognized as a key structure of the CNS. In fact, it has been clearly shown to control both simple motor acts (as briefly described above) such as reflexes (e.g., monosynaptic excitatory, reciprocal inhibitory, flexion and crossed-extension reflexes) as well as more complex motor behaviors such as locomotion, micturition and ejaculation (Fig.1). Note that, although involuntary controlled in many ways, these spinal-mediated functions are not reflexes per se but instead, complex motor behaviors that may be induced by different methods (central activation in the spinal cord with epidural stimulation or pharmacological activation, unlike reflexes) and modulated by descending and peripheral afferent input.

Central Pattern Generator for Locomotion

This spinal cord circuitry can be defined as a neural network that can endogenously (i.e., independently of rhythmic sensory or brain descending input) produce the basic commands underlying rhythm and pattern motor output sent to peripheral nerves and muscles for locomotion. The pioneer work of Sherrington's and Graham-Brown on spinal cord-transected and de-afferented cats and dogs has provided evidence suggesting that the basic neural commands for locomotion (e.g., walking, running) were generated, for most part, in the spinal cord (i.e., in the absence of descending and peripheral afferent inputs). Since then, no consensus has been reached as to how this CPG is structurally organized. Several models have been proposed such as the half-center model, the Miller and Scott model, the Ring model, the Flexor Burst Generator model, and the Unit Burst Generator model. Nonetheless, in recent years, some populations of neurons have been proposed as constitutive candidate elements of this "black box." Experiments using activity-dependent labeling and/or selective lesions have shown locomotor-like active neurons in the intermediate zone of the lumbar spinal cord (e.g., L1–L2, Fig. 1). Electrical stimulation of comparable segments in rodents and humans has revealed key CPG elements in upper lumbar cord segments.

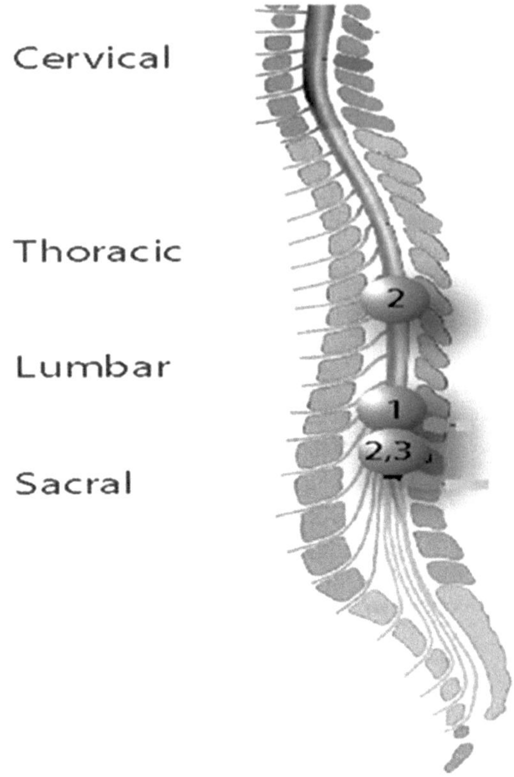

Fig. 1. Key central pattern generators in the spinal cord. (1) Central Pattern Generator for Locomotion with key elements in L1–L2 (humans, mice). (2) Spinal Center for Micturition presumably located in two key areas—low thoracic and mid-lumbar L3–L4 (cats, rats). (3) Spinal Generator for Ejaculation located in L3–L4 (rats).

Using molecular and genetic techniques, neuroscientists have very recently identified several populations of lumbar cord neurons candidates that may play different roles in locomotor rhythm and pattern generation. For instance, V0 interneurons were associated with left-right alternation, V1 inhibitory interneurons with fast locomotor rhythms, Chx10-expressing neurons (V2a glutamatergic and V2b GABAergic interneurons) with frequency, amplitude and bilateral coordination, V3 interneurons with robust and balanced rhythm generation, EphA4 receptor-expressing neurons with bilateral alternation, and HB9 neurons with excitation and locomotor rhythm generation.

Electrophysiological and pharmacological approaches have also been used to identify a few distinct populations of CPG neuron candidates in the intermediate zone (lamina VII, lumbar cord area). These interneurons are typically active only during locomotion (silent and nonresponsive in absence of locomotion) and responsive either to group I afferent, group II afferent or FRA stimulation. Finally, agonists, antagonists, and knockout animal models have been used to identify at least some of the transmembranal receptor families (and subclasses) associated with CPG activity. Among them, serotonin (5-HT) 1A, 5-HT2A, 5-HT7, dopamine (DA) I, and glutamate (e.g., n-methyl-d-aspartate), gamma-aminobutyric acid (GABA) A, receptors were shown to be involved in potent activation of the CPG for locomotion.

Spinal Center for Micturition

As for locomotion, micturition can be controlled for its activation by descending input from the brain (e.g., pontine micturition center) and spinobulbospinal pathways although compelling evidence suggests that the main neural commands are generated in the spinal cord by a CPG network for micturition. Micturition essentially depends, for the storage and periodic elimination of urine, on the coordinated activity of different smooth and striated muscles including the urinary bladder, bladder neck, urethra and internal urethral sphincter. For instance, for the storage of urine, spinal commands are sent to contract the internal urethral sphincter and to relax the detrusor muscle. Inversely, during elimination, spinal commands are sent to contract the detrusor muscle and to relax the internal urethral sphincter (much like in flexors and extensors during locomotion). As mentioned above, coordinated activity between these muscles is largely mediated by a neural circuit located in the spinal cord that was identified mainly with labeling techniques. Spinal interneurons retrogradely labeled by injection of pseudorabies virus into the urinary bladder were found in regions receiving afferent input from the bladder. A comparable distribution was shown following injections of virus into the urethra or the external urethral sphincter, indicating a prominent overlap of the interneuronal pathways controlling the various target muscles of the urinary system. In addition, spinal interneurons (located near the dorsal commissure, superficial dorsal horn, and sacral parasympathic nucleus) involved in processing afferent input from the lower urinary tract have been identified by c fos expression following noxious or non-noxious stimulation of the bladder and urethra. Some of these interneurons send long projections to the brain, whereas others make local connections in the spinal cord.

Electrical stimulation of the spinal cord above the sacral segments can elicit coordinated micturition motor response in decerebrate cats. Specifically, the superficial aspect of the dorsolateral funiculus (DLF) within the lower thoracic (T9–T13) segments can produce, once stimulated electrically, fictive micturition-like response—increasing firing activity in nerve innervating the bladder coordinated with decreased firing activity in the external urethral sphincter branch of the pudendal nerve. These data have provided compelling evidence suggesting the existence of a CPG for micturition in various segmental areas including thoracolumbar (T8–9) and lumbosacral (L3–L4, L6–S1) segments of the spinal cord. This is also strongly supported by experiments conducted in de-afferented animals by Dr. Schefchyk and colleagues showing reflex-independent-induced micturition following pontine micturition center stimulation. Using various pharmacological approaches, a number of neurotransmitters and receptor systems in the spinal cord have recently been proposed to play different roles in micturition (e.g., 5-HT, dopamine, muscarinic, GABA, opiod, glutamatergic, purinergic, neuropeptides, etc.). Recently, 5-HT1A receptors have been associated with activity in an apparently distinct spinal network for urine storage—referred to as the Spinal Urine Storage Reflex Inhibitory Center or SUSRIC.

Spinal Generator for Ejaculation

Until recently, ejaculation was considered as a rather simple sequence of spinal reflexes induced mainly by sensory genital stimulation facilitated in intact subjects by so-called psychogenic descending inputs from the medial preoptic area, hypothalamus, and paragigantocellularis area. Ejaculation comprises two main phases—the emissive phase and the expulsive phase. The former was shown to involve parasympathetic responses including secretion of seminal fluids from epithelial cells and accessory sex glands, as well as sympathetic responses that move the seminal fluids to the urethra (i.e., contraction of seminal vesicles, prostate, vas deferens, and coagulant glands) and closure of the bladder neck (that is the emissive phase). In turn, the expulsive phase (the final phase of ejaculation) is defined as the forceful ejection of semen from the urethral meatus caused by rhythmic contractions of genital muscles surrounding the genital tract (e.g., urethralis, bulbospongiosus, and ischiocavernous muscles supported by the anal and urethral sphincters). Thus, ejaculation clearly depends upon coordinated responses involving autonomic and somatic events involving, in the latter case, pudendal motoneurons located medially in the spinal nucleus of the bulbocavernous as well as pudendal motoneurons located laterally for those innervating the bulbospongiosus, ischiocavernous, external anal, and urethral muscles and sphincters.

A recent breakthrough has drastically changed our view of the neurobiology of sexual function. In 2002, Truitt and Coolen have provided clear evidence showing that ejaculation depends upon a neural network in the spinal cord (lumbar segments L3 and L4) referred to as the Spinal Generator for Ejaculation or SGE (Fig. 1). It is defined as a circuit capable of producing self-sustained rhythmic output to pudendal motoneurons. It contains lumbar spinothalamic neurons also called LSt cells that project to forebrain, that project to pudendal motoneurons (in the Onuf's nucleus in men) and that receive input from sex organs via the pudendal and dorsal nerve of the penis. Moreover, ejaculation is completely lost in animals undergoing LSt cell-lesioned procedures. LSt cells are found in the vicinity of the central canal (lamina X and medial portion of lamina VII), and most of them specifically contain galanin, CCK, enkephalin, and NK-1 receptors. Electrical stimulation of the pudendal or dorsal nerve of the penis nerves was shown to elicit ejaculatory responses in low-thoracic-transected rats. Comparable data were found in SCI men supporting the existence of LSt cells (or at least of a SGE) in mammals including men.

Other Spinal Cord Networks

Although less studied, some evidence suggests the existence of at least a few other behavior-controlling networks of neurons in the spinal cord. For instance, the control of defecation or scratching was shown to be partially controlled by the spinal cord. Additional rhythm generators are likely to be identified in the next few years since behaviors including the restless legs syndrome, shivering, and nocturnal myoclonus may possibly have also their own spinal "command centers."

2. Diseases and Dysfunctions

Spinal cord injury (SCI) can occur accidentally (i.e., trauma caused by motor vehicle accident, fall, gunshot, etc.) or be caused by a disease (i.e., non-traumatic SCI associated with diseases such as multiple sclerosis, cancer, etc.). The incidence and prevalence of traumatic and non-traumatic spinal cord injury (TSCI and NTSCI) remain incompletely assessed worldwide. In the USA, an extensive study recently sponsored by the Christopher and Dana Reeve Foundation has reported incidence values (geographic-specific differences) around 40 cases per million population and prevalence values in the range of 1,275,000 TSCI patients (0.4% of the US population) which is nearly five times greater than previously assessed some years ago by the University of Alabama' National Spinal Cord Injury Statistical Center. Different incidence values for TSCI have been reported in a few other countries such as Germany (66 cases per million population), France (12.7 cases per million population), and Canada (37–46 cases per million population). Estimated incidences of NTSCI have rarely been studied. This said, numbers as high as 80 cases per million population in the USA have been proposed.

An increasingly large literature is available on dysfunctions and medical complications associated with TSCI. This is in contrast with NTSCI that remains poorly described etiologically, functionally and clinically although a few comparative studies have reported some pathophysiological differences between TSCI and NTSCI patients.

2.1. Traumatic Spinal Cord Injury

TSCI is associated with a sudden and generally irreversible loss of sensory and voluntary motor control caused by an accident. The type of accidents associated with TSCI is increasingly recognized to be geographically and sociodemographically dependent.

For instance, in Canada, TSCI patients are injured mainly due to vehicle accidents (36.7% for cars, 8.4% for motorcycles and 7.4% for recreational vehicles), falls (15.8%), and accidents at work (11.6%). In contrast, in the USA (data from the Christopher and Dana Reeve Foundation-sponsored report), TSCI cases are caused mainly by accidents at work (28%), vehicle accidents (24%), and falls (9%). In countries such as India, falls (e.g., from a tree) constitute the main cause of TSCI (above 50% of all cases). Higher risks of trauma are generally found during adolescence and early adulthood males and females with a male to female ratio of 4:1. Injuries at the cervical level lead to tetraplegia (also called quadriplegia), whereas those at the thoracic level or lower cause paraplegia. In the USA (statistics from the NSCISC), incomplete tetraplegia is the most frequent neurologic category of TSCI (30.1%), followed by complete paraplegia (25.6%), complete quadriplegia (20.4%), and incomplete paraplegia (18.5%).

Despite injury level (e.g., cervical or thoracic) or extent (e.g., complete or incomplete), a spinal shock and neurogenic shock normally occur immediately after trauma. Spinal shocks are usually temporary, lasting only for 24–48 h. It is characterized as the temporary absence of sensory and motor functions (even in incompletely injured patients). A neurogenic shock may last for weeks and can lead to a loss of muscle tone below the injured site. At the cellular level, the initial injury results into spinal cord deformation, laceration, crush and persistent post-injury cord compression occurring within seconds to minutes post-accident. This leads to immediate cell death, axonal disruption, vascular and metabolic changes, which have subsequent effects or so-called secondary injury processes occurring within a few minutes to a few weeks of injury. They are associated still with an incompletely understood cascade of biochemical events (e.g., free radicals, glutamate toxicity, inflammatory reactions, etc.) leading mainly to apoptosis (programmed cell death), greater injury and increased functional loss.

At the systemic level, the state of paralysis is associated with the progressive development of health problems or so-called secondary complications including muscle wasting, osteoporosis, cardiovascular problems, obesity, type II diabetes, immune deficiencies, bladder and bowel problems, sexual dysfunction, spasticity, autonomic dysreflexia, skin ulcers, urinary tract infections, septicemia, anemia, deep vein thrombosis, depression, etc.

Main causes of TSCI:

- Car
- Motorcycle and recreational vehicles
- Fall
- Injury at work
- Pool accident

- Sports
- Act of violence—e.g., gun shot

2.2. Non-traumatic Spinal Cord Injury

Non-traumatic Spinal Cord Injury (NTSCI) is any injury or damage to the spinal cord that is not associated or caused by a trauma. The main causes are cancer, infection, vascular problems, and neurological problems. Specifically, in the USA, stroke (29%) and multiple sclerosis (17%) (although discrepancies were reported in specific geographical areas, e.g., Virginia with 53% of all NTSCI caused by stenosis) constitute the main causes, whereas in Canada (Quebec), the main causes of NTSCI are stenosis-associated myelopathies (34.5%). As mentioned earlier, incidence and prevalence numbers have rarely been examined for NTSCI (in the USA or elsewhere). Some articles have reported numbers as high as 80 cases per million population or prevalence values 4–5 times greater in NTSCI than TSCI. The incidence (and prevalence) obviously depends upon each specific disease causing NTSCI. Some differences in age, gender, extent and level of injury or associated conditions (so-called secondary complications) have been found between TSCI and NTSCI patients. For instance, higher risks of TSCI were reported during adolescence and early adulthood males and females with a male-to-female ratio of 4:1, whereas higher risks of NTSCI were reported in older adults with a more or less even male-to-female ratio (1:1).

Main causes of NTSCI:

- Cancer (e.g., spinal cord, vertebrae, myeloma)
- Multiple sclerosis
- Vascular causes (e.g., stenosis)
- Syringomyelia
- Amyotrophic lateral sclerosis
- Friendrich's ataxia
- Spina bifida
- Arthritis (vertebrae)
- Osteoporosis (vertebrae)
- Hernia
- Myelitis

In Canada (Quebec), essentially no completely injured patients were found among NTSCI patients, whereas completely injured patients constituted 37.8% of all TSCI patients. In contrast, in India (Karnataka), 64% were reported as being incompletely injured and 36% as complete NTSCI patients. NTSCI attributed to cancer is commonly found (most important cause in the USA). In fact, among all cancer patients, approximately 70% have metastatic disease at death with 40% of them who will develop spine cancer and spinal cord compression (5–10% of all cancer patients). Spinal tumors can origin in different tissues within the spinal cord including gliomas (associated with glial cells, i.e., astrocytomas, oligodendrogliomas, ependymomas). Spinal cord tumors are much less common than brain tumors (cancerous or noncancerous) and may be originally of non-CNS origin (so-called secondary spinal tumors or of metastatic origin). Specifically, vertebral hemangioma within the spinal column is also relatively common. Most often benign, vascular tumors rarely cause neurological loss or spinal cord compression. Bone malignancies account for merely about 1.5% of all cancers, with a small percentage of these tumors developing in the cervical spine.

Multiple sclerosis (MS) is an incompletely understood disease generally believed to result of some combination of genetic, environmental and infectious factors. In multiple sclerosis patients, myelin sheaths around axons of the descending and ascending tracts in the spinal cord are damaged (immune system-related), leading to demyelination and scars and progressive loss of motor and sensory functions (much like in TSCI patients). It is often associated with relapses (sudden loss of motor control followed by a period of spontaneous functional recovery in 90% of the cases) although slow and continuously progressing cases also exist (progressive form in approximately 10% of the cases). Secondary complications found in TSCI patients have also been found in MS patients specifically towards the end of progression. The onset usually occurs in young adults (more frequent in females than males). The incidence is higher in Northern countries (North America) ranging between 20 and 100 cases per million population per year. In the USA, approximately 390,000 patients are currently living with MS.

Amyotrophic lateral sclerosis (ALS), also called Lou Gehring's disease in North American or Maladie de Charcot in France, is a type of motoneuron disease that may affect both corticospinal neurons (motor cortex) and/or spinal motoneurons. These neurons specifically die progressively for unclear reasons (5–10% are genetically inherited cases associated in some cases with SOD1 gene mutations, whereas most patients have no known cause although an association is believed with free radicals, mitochondrial problems, glutamate toxicity, axonal transport defects, environmental toxins, etc.) leading to voluntary and involuntary motor control losses (cognitive functions are typically spared). In most cases, it begins between 40 and 60 years of age in muscles of one leg that will display weaknesses and locomotor function problems. In cases where it begins in upper extremities, problems with fine movements such as buttoning a shirt, writing, or turning a key may be found. However, regardless of where it begins and how fast it progresses, muscle weakness and atrophy normally increase until it reaches motoneurons that control respiration (phrenic motoneurons).

Unlike most other cases of SCI (traumatic and non-traumatic) who can live many years (and decades) post-injury onset, ALS patients do not generally survive more than 3 years after diagnosis. Between 10 and 20 cases per million population per year are found in the USA. Currently, approximately 30,000 patients live with ALS (probably between 0.3 and 0.5 million patients worldwide).

Stenosis is one of the most important indirect causes of NTSCI. It is essentially a spinally located blood vessel flow problem (typically a reduction due to a narrowing) induced originally by atherosclerosis, diabetes, birth defect problems, inflammation, neoplasm (cancer), etc. Consequently, the results are complex and may be associated with mixed spinal cord compression and vascular-related necrosis. As with other SCI cases, it is thus expected to be associated with comparable functional loss problems and the potential development of secondary complications as with most types of SCI (e.g., neurogenic bladder, neurogenic bowel, urinary tract infection, pressure ulcers, neuropathic pain, etc.)

2.3. Diagnosis

For both TSCI and NTSCI patients, medical attention is immediately required when a SCI is suspected. Along obvious radiological evaluation (x-ray, myelography, CT scan, and/or MRI), diagnostic tests conducted by physicians should normally include an assessment of functions below suspected or confirmed injury level. This may be performed using the ASIA (American Spinal Injury Association) Impairment Scale (AIS) where both muscle contraction and sensation are graded (motor 0–5 and sensory 0–2). Other types of classification or functional tests such as the International Standards for Neurological Classification in Spinal Cord Injury (ISCSCI), Berg Balance Scale (equilibrium), Walking Index for Spinal Cord Injury (WISCI) or Functional Independence Measure (FIM) may also be used.

Motor and sensory functions (ASIA Impairment Scale—AIS):

- A (Complete)—No motor or sensory function is preserved below injury level (including the sacral segments S4–S5).
- B (Motor complete, sensory incomplete)—Partially preserved sensory function but not motor function below injury level (including the sacral segments S4–S5).
- C (Motor and sensory incomplete)—Sensory and motor functions are both partially preserved below injury level, and more than half of key muscles below the neurological level have a muscle grade less than 3.
- D (Motor and sensory Incomplete)—Sensory and motor functions are both partially preserved below injury level, and at least half of key muscles below the neurological level have a muscle grade of 3 or more.
- E (Normal)—Motor and sensory functions are normal.

Examination obviously depends on whether or not the cause of injury is known (specifically for NTSCI). To diagnose NTSCI, many neural functions may be tested including tonus (spasticity and contractures) and peak muscle contraction, reflexes, as well as sensation (including pain). The areas of weakness and loss of sensation can give clues about lesion level. Temperature, pulse rate, breathing rate, and blood pressure should be checked as well as signs of spine deformity, skin integrity, pressure points, infection, and pressure ulcers. Lower extremity muscles should be closely examined (doppler) for deep vein thrombosis. It is also recommended to analyze urine specimens for culture as urinary infection is very common. Blood samples should be analyzed to look for any signs of infection (raised white cell count) or anemia (reduced hematocrit levels).

Incomplete SCI-specific classification:

- Anterior Cord Syndrome—partial or complete loss of ability to sense pain, temperature, and touch below the level of injury.
- Central Cord Syndrome—Loss of arm function and some leg, bowel, and bladder control may be preserved..
- Posterior Cord Syndrome—Muscle strength, pain, and temperature sensation is preserved.
- Brown-Sequard Syndrome—Unilateral injury associated with remaining pain and temperature sensation but impaired motor control ipsilaterally only.
- Cauda equine lesion—Partial or complete loss of motor control and sensation below lumbar level.

The effects of an incomplete injury depends upon spinal cord region-specific aspects (medio-lateral or ventro-dorsal localization). Accordingly, five main classes exist to characterize incomplete SCI (ASIA B-D): anterior cord syndrome, central cord syndrome, posterior cord syndrome, Brown-Sequard's syndrome, and cauda equina lesion.

For MS patients, disability is normally measured also with the disease-specific Expanded Disability Status Scale (EDSS), which ranges from 0 (no disability) to 10 (death from multiple sclerosis). 0–4: associated with specific neurological impairments and disability; 4–9.5: associated mainly with reducing levels of mobility and upper-limb and bulbar function. Generally, a lesion of the reflex arc or the corresponding afferent fibers (PNS) is associated with reduced corresponding reflex responses. Comparable reflex amplitude reduction may be found despite the PNS level affected by the pathology. For instance, the H-reflex (Ia monosynaptic pathway) response will normally decrease in patients with amyotrophic lateral sclerosis (motoneuronal cell death) or with Guillain–Barré syndrome (degradation of motor unit axon terminals).

Most (although not all) descending tracts display tonic inhibitory control over spinal reflexes below injury level. Consequently, an injury caused by trauma (traumatic spinal cord injury) or disease (multiple sclerosis or cancer) should normally facilitate and thus increase (hyperreflexia) spinal reflexes (e.g., Ia monosynaptic stretch reflex). This said, during the acute phase, as mentioned earlier, a lesion of descending tracts may be associated with an initial stage of spinal shock or hyporeflexia. In extreme cases such as in complete spinal cord injury (e.g., transection), weeks of hypoactivity may even be found. In contrast with most TSCI cases where hyperreflexia, hypertonia and muscle weakness or paralysis are found, motor neuronal damage results in its own characteristic pattern of deficits. Rather than an entire side of deficits, there is a pattern relating to the myotome affected by the damage. Problems such as tumors that would affect only specific pools of motoneurons (ventral spinal cord compression) are characterized by muscle weakness, hypotonia, hyporeflexia, and muscle atrophy.

3. Concluding Remarks

In many medical textbooks, the spinal cord is still considered as a rather simple relay between the brain and effective organs. However, modern experimental approaches and recent advances in spinal cord research have begun to unravel its role also as a significant "command center" involved both in mediating simple reflexes and more complex motor functions. Although the properties of most of its one to two billion neurons remain to be fully characterized, the role of the spinal cord in producing locomotion, micturition, ejaculation and other key functions is progressively being described. It is increasingly clear also that patients with traumatic (accident) and non-traumatic (disease such as spinal tumors) spinal cord injuries share many similarities pathophysiologically, functionally and clinically. Diagnosis and acute care intervention are also comparable in many ways (x-rays and magnetic resonance imaging, functional tests, surgical stabilization, etc.). No significant cure has been found for traumatic or non-traumatic SCI patients, although some cell therapy-based or drug candidates are currently in development to repair (or reduce secondary cell death) spinal cord tissue. This said, several treatment candidates actively being developed by scientists (industry or academia-based) are also designed as palliative & chronic treatments aim at treating (preventing or reducing) secondary health problems (so-called secondary complications) often experienced by both groups of patients (traumatic and non-traumatic).

All in all, there is a critical need for the development of both acute and chronic treatments for SCI patients. Basic research in animal models of SCI and scientifically based therapeutic approaches are critically important to increase the success rate of innovative technologies that reach regulatory approval and commercialization. The different chapters of this book describe a variety of animal models and their utilities as research tools to further understand spinal cord functions and dysfunctions and to discover new drug candidates for SCI.

Recommended Reading

1. Agarwal P, Upadhyay P, Raja K (2007) A demographic profile of traumatic and non-traumatic spinal injury cases: a hospital-based study from India. Spinal Cord 45:597–602
2. Andersson KE (2002) Bladder activation: afferent mechanisms. Urology 59:43–50
3. Andersson KE (2006) Treatment-resistant detrusor overactivity—underlying pharmacology and potential mechanisms. Int J Clin Pract Suppl 151:8–16
4. Barnes DG, Shaw PJ, Timoney AG, Tsokos N (1993) Management of the neuropathic bladder by surapubic catherisation. Br J Urol 72:169–172
5. Chang HY, Cheng CL, Chen JJ, de Groat WC (2007) Serotonergic drugs and spinal cord transections indicate that different spinal circuits are involved in external urethral sphincter activity in rats. Am J Physiol Renal Physiol 292:F1044–F1053
6. Cheng CL, de Groat WC (2010) Role of 5-HT1A receptors in control of lower urinary tract function in anesthetized rats. Am J Physiol Renal Physiol 298:F771–F778
7. Fowler CJ (1999) Neurological disorders of micturition and their treatment. Brain 122: 1213–1231
8. Guertin PA (2009) The mammalian central pattern generator for locomotion. Brain Res Rev 62:45–56
9. Guertin PA (2009) Recovery of locomotor function with combinatory drug treatments designed to synergistically activate specific neuronal networks. Curr Med Chem 16:1366–1371
10. Guertin PA, Steuer I (2009) Key central pattern generators of the spinal cord. J Neurosci Res 87:2399–2405
11. Guertin PA, Ung RV, Rouleau P (2010) Oral administration of a tri-therapy for central pattern generator activation in paraplegic mice: proof-of-concept of efficacy. Biotechnol J 5:421–426
12. Gupta A, Taly AB, Srivastava A, Murali T (2009) Non-traumatic spinal cord lesions: epidemiology, complications, neurological and functional outcome of rehabilitation. Spinal Cord 47:307–311
13. Gupta R, Bathen ME, Smith JS, Levi AD, Bhatia NN, Steward O (2010) Advances in the management of spinal cord injury. J Am Acad Orthop Surg 18:210–2122
14. Henneman E (1974) Spinal reflexes and the control of movement. In: Mountcastle VB (ed) Medical physiology, vol 1, 13th edn. Mosby, St Louis, MO
15. Jackson AB, Dijkers M, DeVivo MJ, Poczatek RB (2004) A demographic profile of new traumatic spinal cord injuries: change and stability over 30 years. Arch Phys Med Rehabil 85:1740–1748
16. Kalat JW (1998) Biological psychology, 6th edn. Brookes/Cole, Pacific Grove, CA
17. Kandel ER, Schwartz JH, Jessell TM (2000) Principles of neural science, 4th edn. McGraw-Hill, New York
18. Karicheti V, Langdale CL, Ukai M, Thor KB (2010) Characterization of a "Spinal—Urine Storage Reflex—Inhibitory Center" (SUSRIC) and its regulation by serotonin 5-HT1A receptors in female cats. Am J Physiol Regul Integr Comp Physiol 298:R1198–R1208
19. Koning W, Frowein RA (1989) Incidence of spinal cord injury in the Federal Republic of Germany. Neurosurg Rev 12:562–566
20. Matthews PBC (1972) Mammalian muscle receptors and their central actions. Williams and Wilkins, Baltimore, MD
21. Matthews PBC (1991) The human stretch reflex and the motor cortex. Trends Neurosci 14:87–90
22. McKenna KE, Nadelhaft I (1986) The organization of the pudendal nerve in the male and female rat. J Comp Neurol 248:532–549
23. Miller H, Simpson CA, Yeates WK (1965) Bladder dysfunction in multiple sclerosis. Br Med J 1:1265–1269
24. Minaire P, Castanier M, Girard R, Berard E, Deidier C, Bourret J (1978) Epidemiology of spinal cord injury in the Rhone-Alps region. France 1970–1975. Paraplegia 16:76–87

25. Newman HF, Reiss H, Northup JD (1982) Physical basis of emission, ejaculation and orgasm in the male. Urology 19:341–350
26. Netter FH (2006) Atlas of human anatomy. Saunders, Philadelphia, PA
27. Noreau L, Proulx P, Gagnon L, Drolet M, Laramée MT (2000) Secondary impairments after spinal cord injury. A population-based study. Am J Phys Med Rehabil 79:526–535
28. Maria Patestas, Leslie P Gartner Chapter 10: Ascending sensory pathways. In: A textbook of neuroanatomy. A textbook of Neuroanatomy John Wiley & Sons, Blackwell Publishing 2006 464 pages ISBN 978-104051-0340-4
29. Pearson KG (2008) Role of sensory feedback in the control of stance duration in walking cats. Brain Res Rev 57:222–227
30. Pickett W, Simpson K, Walker J, Brison RJ (2003) Traumatic spinal cord injury in Ontario, Canada. J Trauma 55:1070–1076
31. Prochazka A (1996) Proprioceptive feedback and movement regulation. In: Rowell L, Sheperd JT (eds) Handbook of physiology: regulation and integration of multiple systems. Am Physiol Soc, New York, pp 89–127
32. Remy-Neris O, Barbeau H, Daniel O, Boiteau F, Bussel B (1999) Effects of intrathecal clonidine injection on spinal reflexes and human locomotion in incomplete paraplegic subjects. Exp Brain Res 129:433–440
33. Rouleau P, Ayoub E, Guertin P (2011) Traumatic and non-traumatic spinal cord-injured patients in Quebec, Canada. Part 1—epidemiological, clinical and functional characteristics. Open Epidemiol J 4:133–139
34. Schmidt MH, Klimo P, Vrionis FD (2005) Metastatic spinal cord compression. J Natl Compr Canc Netw 3:711–719
35. Shafik A, El-Sibai O (2000) Mechanism of ejection during ejaculation: identification of an urethrocavernosus reflex. Arch Androl 44:77–83
36. Shefchyk SJ (1998) Modulation of excitatory perineal reflexes and sacral striated sphincter motoneurons during micturition in the cat. In: Rudomin P, Romo R, Mendell LM (eds) Presynaptic inhibition and neural control. Oxford University Press, Oxford, pp 398–406
37. Shefchyk SJ (2006) Spinal mechanisms contributing to urethral striated sphincter control during continence and micturition: "how good things might go bad". Prog Brain Res 152:85–95
38. Stein RB, Capaday C (1988) The modulation of human reflexes during functional motor tasks. Trends Neurosci 11:328–332
39. Swartz MH (1998) Textbook of physical diagnosis: history and examination, 3rd edn. WB Saunders Co, Philadelphia
40. Truitt WA, Coolen LM (2002) Identification of a potential ejaculation generator in the spinal cord. Science 297:1566–1569
41. Waindberg M, Barbeau H, Gauthier S (1990) The effects of cyproheptadine on locomotion and on spasticity in patients with spinal cord injuries. J Neurol Neurosurg Psychiatry 53: 754–763
42. Yoshimura N, Chancellor MB (2003) Neurophysiology of lower urinary tract function and dysfunction. Rev Urol 5:S3–S10

Chapter 2

The Pathological Findings in Traumatic Injury to the Human Spinal Cord

Ronald C. Kim

Abstract

The anatomical pattern of damage following upon traumatic injury to the human spinal cord is dependent upon the manner in which the injury is sustained. Hyperflexion typically results in anterior spinal cord injury, hyperextension in central spinal cord injury, stab wounds in hemisection injury, and complete crush in total spinal cord injury. Secondary lesions may appear hours to years later and may result in serious additional disability. The presence of congenital spinal stenosis, cervical spondylosis, or ossification of the posterior longitudinal ligament may greatly enhance the likelihood of damage following blunt injuries to the spine.

Key words: Human spinal cord trauma, Traumatic paraplegia, Traumatic quadriplegia, Spinal cord injury, Spinal cord pathology, Human spinal cord injury

1. Introduction

Although methods traditionally employed to produce experimental spinal cord injury, such as the weight-drop technique, the clip compression technique, or the extradural balloon compression method (1), have no counterpart in humans, for purposes of reproducibility it is essential to be able to standardize the pattern and severity of injury in experimental paradigms, despite the fact that no two instances of human spinal cord trauma show precisely the same pattern of damage.

Epidemiologically, the manner in which spinal cord injury occurs in man varies according to the population surveyed. In most large peacetime series, motor vehicle accidents account for 40–50%, violence (gunshot wounds or stab wounds) for 7–23%, falls for 12–20%, and injuries sustained during sports or recreational

Håkan Aldskogius (ed.), *Animal Models of Spinal Cord Repair*, Neuromethods, vol. 76,
DOI 10.1007/978-1-62703-197-4_2,

activities (particularly diving) for up to 24% (2, 3). How the injury is sustained may be an important determinant of the anatomical pattern of damage within the spinal cord.

2. Patterns of Injury

To the clinician caring for the patient who has suffered from traumatic spinal cord injury the neurologic deficits that emerge after the initial period of spinal shock has elapsed will provide important insights into the configuration of spinal cord lesions (4). The *anterior spinal cord syndrome*, which typically follows a *hyperflexion injury* (such as that frequently observed in the thoracolumbar region after a motor vehicle accident), is characterized by hypesthesia (resulting from damage to the anterior spinothalamic tracts), hypalgesia (resulting from damage to the lateral spinothalamic tracts), and spastic weakness (resulting from damage to the lateral corticospinal tracts), with relative sparing of posterior column (touch/proprioceptive) function (Fig. 1). The *central spinal cord syndrome*, which is ordinarily associated with a *hyperextension injury*, such as that observed in most diving accidents, is usually characterized, depending on the severity of the injury, by variable loss of posterior column function and spastic weakness that is more pronounced in the upper than in the lower extremities (Fig. 2). The *Brown-Séquard syndrome*, which is usually the result of a *penetrating injury* such as a stab wound (5), is due to incomplete hemisection of the spinal cord and manifests clinically as contralateral loss of pain and temperature perception and ipsilateral spastic

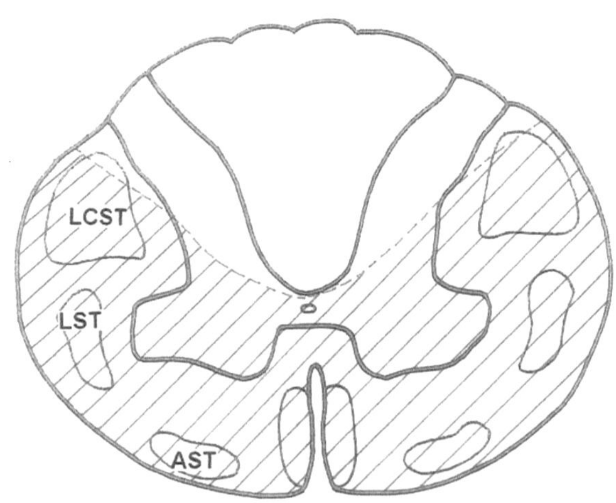

Fig. 1. Hyperflexion injury is typically associated with an *anterior spinal cord syndrome*, with spastic paraplegia and hypalgesia and sparing of posterior column sensation (*LCST* lateral corticospinal tract, *LST* lateral spinothalamic tract, *AST* anterior spinothalamic tract).

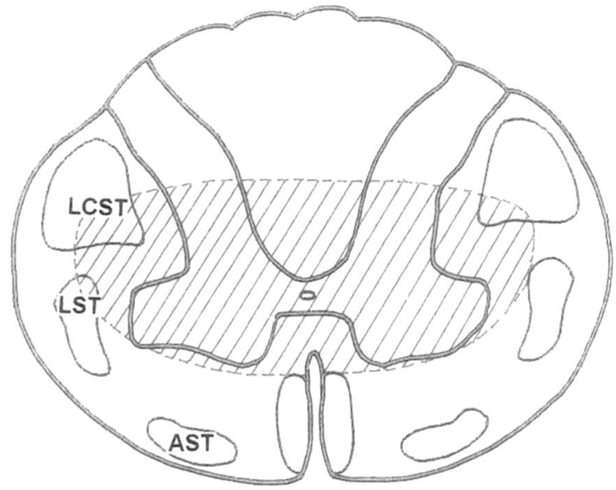

Fig. 2. Hyperextension injury leads to a *central spinal cord syndrome*, which is characterized, in cervical injuries, by spastic weakness that is more prominent in the upper than in the lower extremities and by hypalgesia with variable loss of posterior column sensation.

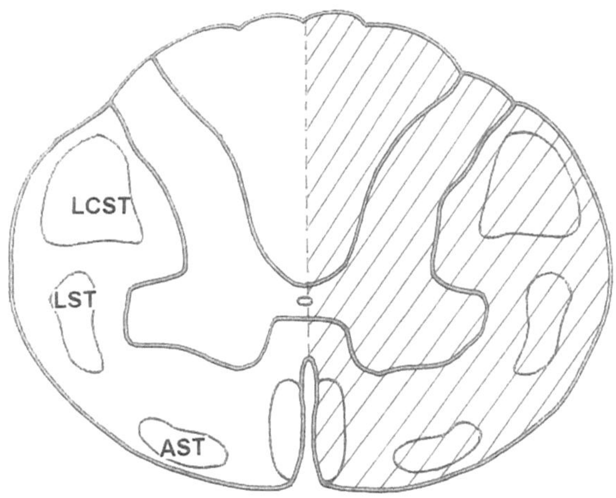

Fig. 3. The *Brown-Séquard syndrome*, i.e., ipsilateral weakness and loss of posterior column sensation with contralateral hypalgesia, is usually the result of a penetrating injury such as a stab wound.

weakness and loss of proprioception (Fig. 3). The *complete spinal cord syndrome*, which may follow upon a *complete crush injury* to the spinal cord, is characterized by complete, permanent spastic paralysis and loss of all sensory function below the level of the injury (Fig. 4).

2.1. Penetrating Injuries

Penetrating injuries are those in which the integrity of the dura and leptomeninges are breached by the injuring agent which, in peacetime civilian life within the United States, is most frequently a bullet or a sharp object such as a knife blade. The resulting neurological

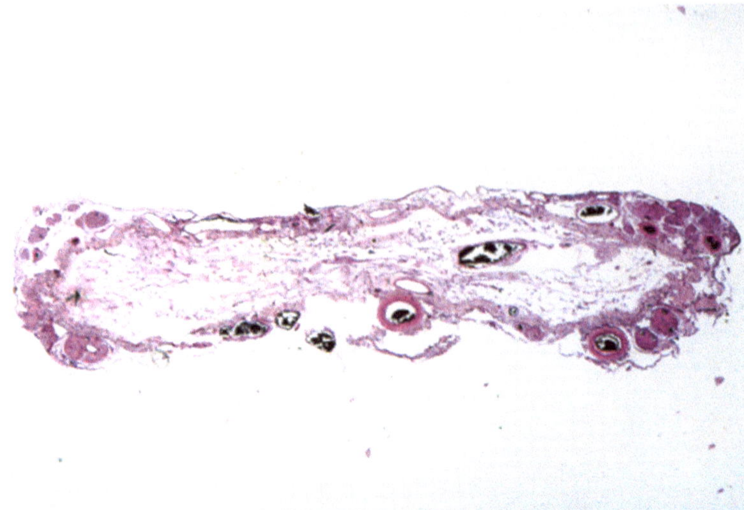

Fig. 4. Spinal cord at T10 in a man with a *complete spinal cord syndrome*, 39 years after a shell-fragment injury to the spine. Only the leptomeninges and leptomeningeal blood vessels are visible; no spinal cord parenchyma remains (hematoxylin and eosin stain).

deficits will depend heavily upon the extent and level of the injury. Missiles may penetrate the spinal cord directly or injure it by driving bone fragments into it. Closed injuries may also breach the dura and leptomeninges by producing vertebral fractures and driving bone directly into the spinal cord parenchyma.

2.2. Nonpenetrating Injuries

Acutely fatal injuries tend to occur at high cervical spinal cord levels and may show very little in the way of dramatic pathological findings, as the continuity of the vertebral column and spinal cord is typically restored immediately following the injury (6). Damage to the vertebral column should alert the examiner to the possibility of structural damage to the spinal cord proper. Surprisingly little may be seen (7). Epidural, subdural, or subarachnoid hemorrhage or damage to external blood vessels such as the vertebral or anterior spinal arteries is seldom observed. The main finding may only be extravasation of erythrocytes into the spinal cord parenchyma, which may evolve, on rare occasion, into frank hemorrhage (hematomyelia). Histologically, recognition of the presence of structural damage to the tissue itself requires that the victim survive for at least several hours after the injury.

3. Primary Spinal Cord Lesions

In those instances in which the injury is not immediately lethal and in which the afflicted individual survives for more than a few hours, histological abnormalities are more readily observed. For purposes of this discussion it is convenient to separate lesions into two categories, namely, those that develop at the time of impact (primary lesions) and those that develop afterward (secondary lesions).

For primary lesions, some investigators use the same terms as those used for traumatic brain injuries, i.e., concussion, contusion, laceration, and compression (8). Not all of these terms are in common usage, however, because of differences among investigators in the way they are defined. The word *concussion*, for example, is applied by some to transitory spinal cord dysfunction in the absence of structural damage and by others to injuries in which damage may be moderately severe (8, 9). Given the limited opportunity to examine human material in such circumstances, it would probably wiser to restrict its usage to those clinical situations in which, following traumatically induced spinal cord dysfunction, there is rapid and complete neurological recovery (10).

The word *contusion*, by analogy with its use in association with craniocerebral injuries, implies that the spinal cord has impacted against the bony wall of the spinal canal, a situation that seldom if ever prevails in actuality. Injury following upon contact between the spinal cord and its bony encasement is virtually always due to bony displacement (e.g., fracture or dislocation), in which case the term *compression* or *crush injury* is more appropriate.

Spinal cord *laceration* is seldom observed in the absence of a penetrating injury and, when it occurs, is most often the result of bone fragments being driven into the spinal cord following a vertebral fracture.

Spinal cord *compression* is generally due to fracture/dislocation or subluxation within the bony vertebral column. Vertebral body continuity may be restored completely immediately following the injury. When acute it may lead to the development of a cylindrical core of necrosis within the ventralmost portion of the posterior column that tapers, in conical fashion, above and below the level at which compression has been maximal (9, 11–13), much in the manner in which a tube of toothpaste has been squeezed (14) (Fig. 5). This type of lesion has sometimes been referred to as "pencil necrosis" (15, 16). With very severe injuries the spinal cord may be transected or completely crushed, although this is rare. Usually, some residual parenchyma can be observed traversing the site of maximal injury (3, 17).

4. Secondary Spinal Cord Lesions

Provided that the spinal cord is not transected or completely crushed, the earliest change that is visible by light microscopy is the presence of pericapillary extravasation of erythrocytes and serum constituents (6). These hemorrhages may become more numerous over the next few hours, but frank hematomyelia is uncommon.

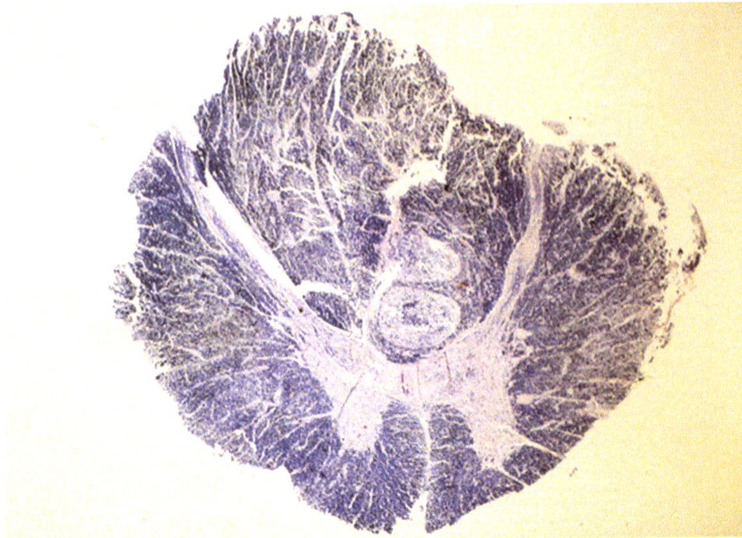

Fig. 5. Acute spinal cord compression. Note the "cores" of displaced tissue within the ventral portions of the posterior columns (Klüver–Barrera stain).

Swollen, injured axons may become visible as early as 30 min after the injury, as evidenced by immunoreactivity for β-amyloid precursor protein, owing either to axonal retraction bulb formation after transaction or to impairment of axoplasmic flow through otherwise anatomically intact axons (18, 19). Between 8 and 24 h post-injury small numbers of neutrophils begin to appear and the white matter becomes progressively more edematous, as evidenced by pallor of myelin staining. Macrophages appear at 24–48 h and steadily increase in numbers thereafter, and GFAP-positive reactive astrocytes appear at 48–72 h. At 24–48 h there is extensive coagulative necrosis with prominent neuronal ischemic cell change. Following this there is tissue breakdown progressing to cavitation. Neovascularization begins at approximately 1 week. In both human and experimental material there is evidence to suggest that, in the brain (and, presumably, also in the spinal cord), there is prolonged activation of resident microglial cells that may persist for months to years after injury (20, 21).

Although experimental studies have demonstrated a number of physiological and biochemical abnormalities, including impairment of microcirculatory perfusion (22, 23) and of autoregulation of blood flow (24), release of vasoactive amines and of excitotoxic neurotransmitters (25–27), Ca++ and K+ ion shifts (28, 29), and generation of oxygen-free radicals (30, 31), such abnormalities have not, for obvious reasons, been documented in human spinal cord-injured subjects.

Traumatic demyelination, i.e., loss of myelin with relative axonal sparing, though well studied in experimental models of spinal cord injury (3–34), does not appear to be as prominent in human

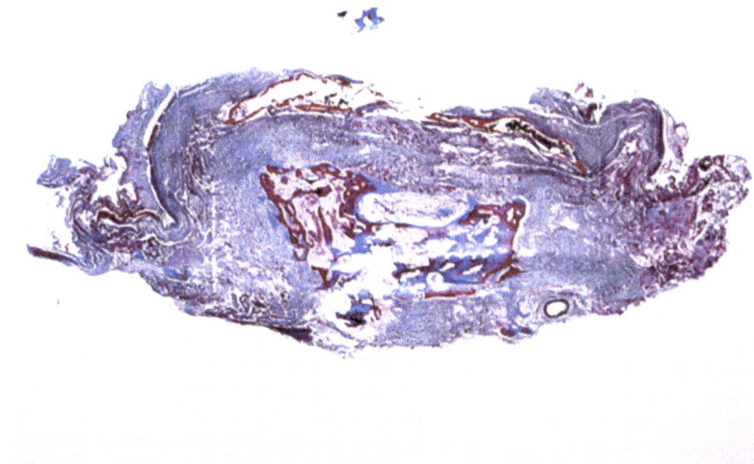

Fig. 6. Spinal cord at T10 ten years after a fall, showing complete replacement of the spinal cord by collagenous connective tissue (*blue*). Heterotopic bone formation is present within the central portion of the scar (Masson trichrome stain).

material (3, 18). When it occurs, it typically does so in close relation to the site of primary impact, and then only around isolated axons. Evidence of oligodendroglial apoptosis has been described in human material (35). Remyelination of central axons, when observed, appears to be achieved by Schwann cells rather than by oligodendrocytes, a process sometimes referred to as "schwannosis" (3, 18, 36).

Collagenous scar formation occurs whenever the pia-arachnoid is breached. If the damage to the spinal cord parenchyma is extensive, the collagen may encroach upon what remains of the parenchyma to such an extent that it appears to replace it completely (Fig. 6). The collagen is presumably elaborated by arachnoidal cells and is limited to the subarachnoid compartment, as evidenced by the presence of a layer of arachnoidal cells between the mass of collagen and the dura (37) (Fig. 7). A recent study in experimental animals suggests that pericytes may also contribute to the appearance of collagen (38). The presence of scar tissue tethers the spinal cord and renders it less mobile within the spinal canal, making it more vulnerable to complications such as posttraumatic syringomyelia (see below).

Traumatic neuromata are present in great abundance where damage to spinal cord parenchyma is maximal (Fig. 8). These neuromata are presumed to represent regenerating sprouts, mainly from the centrally directed neurites of dorsal root ganglion cells (39–41).

Wallerian degeneration of ascending tracts above and descending tracts below the level of injury represents degeneration of those axonal

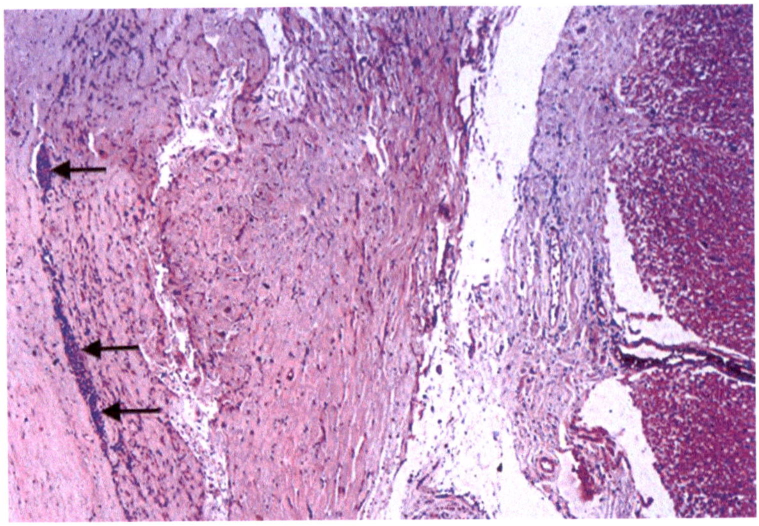

Fig. 7. Spinal cord at C6 in a patient rendered quadriplegic 9 years previously in a motor vehicle accident. Dura is on the *left* and spinal cord on the *right*, with collagenous connective scar tissue in between. Note the layer of arachnoidal cells (*arrows*) between the scar tissue and the dura, thereby localizing the collagen within the subarachnoid compartment (hematoxylin and eosin stain).

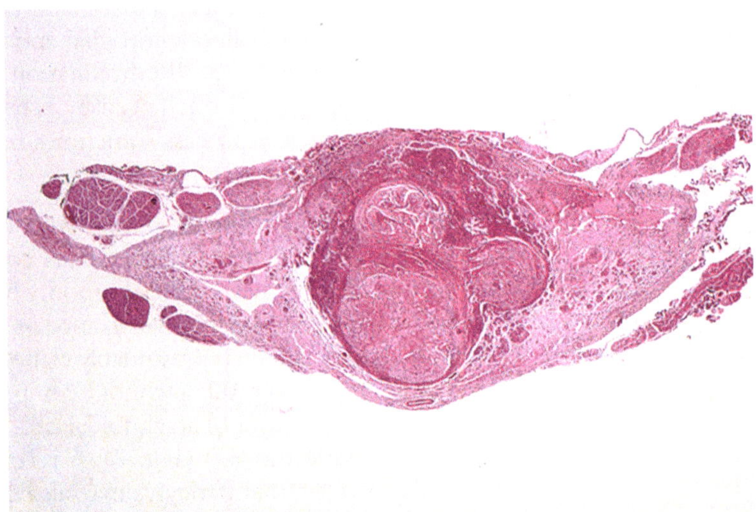

Fig. 8. Spinal cord at C6 in a patient rendered quadriplegic 45 years earlier in a motor vehicle accident. Note the presence of numerous traumatic neuromas (hematoxylin and eosin stain).

segments distal to the site of transaction. The evolution of this process, which may take place over a period of years, is much more attenuated than it is in the peripheral nervous system. Marchi staining permits detection as early as 10 days after injury, but this is not commonly done nowadays. CD68-immunoreactive macrophages are seen 1–4 months after injury (42, 43) and are accompanied by reactive

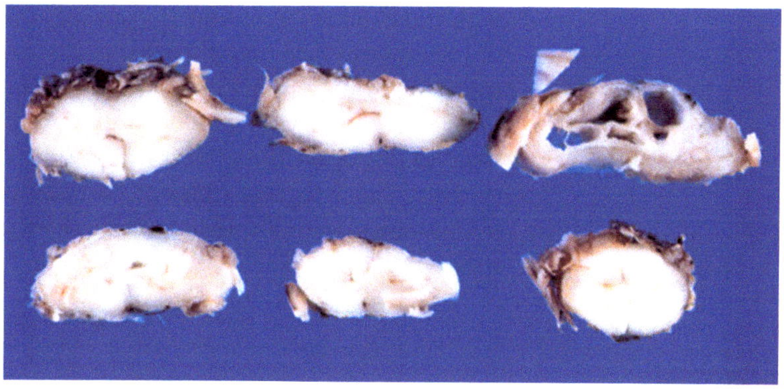

Fig. 9. Spinal cord at C3, C4, and C5 (*top row*) and at C6, T4, and T10 (*bottom row*) in a C4/C5 quadriplegic who sustained his injury 37 years earlier in a motor vehicle accident. A syringomyelic cavity extended from the level of the injury down to T10.

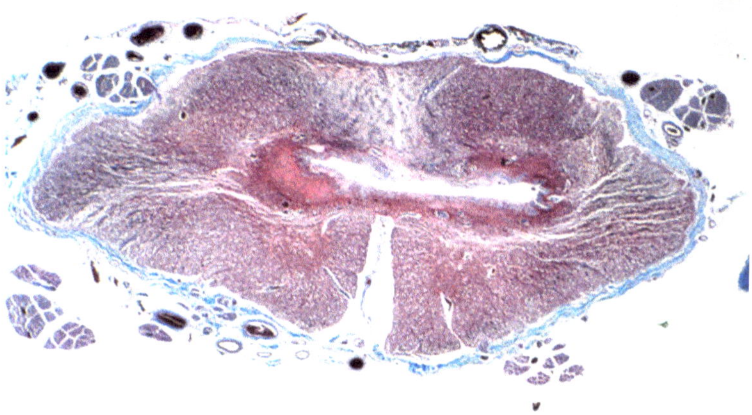

Fig. 10. Spinal cord at T11 showing a syringomyelic cavity in a man who became quadriplegic after a motor vehicle accident 43 years earlier (Masson trichrome stain).

astrocytosis (44). Pallor of myelin staining of affected tracts may not become pronounced until 6 weeks to 2 months have elapsed.

Delayed posttraumatic syringomyelia is of singular clinical importance, as it may increase significantly the extent of neurologic impairment. The condition is characterized by the development, after an interval of several years, of one or more syringomyelic cavities extending either upward or downward for variable distances from the original site of injury (45) (Fig. 9). The consequences can be dire. If, for example, the cavity extends upward from a thoracolumbar lesion, a paraplegic patient may be rendered quadriplegic. The frequency with which this complication develops is not known but is probably over 20% (46). The syringomyelic cavity, which contains no lining epithelium and which is bordered by a zone of isomorphic gliosis (Fig. 10), has no direct communication with the central canal, fourth ventricle, subarachnoid compartment, or, necessarily, cavities formed at the original site of injury. The mechanism by which posttraumatic syringomyelia develops is not known, but the view has been expressed that progressive tearing may follow upon the effects of episodic elevation of venous back-pressure (such as that induced by coughing, sneezing, or Valsalva maneuvers) upon a spinal cord that is tethered by adhesive arachnoidopathy (37, 45).

5. Cervical Spondylosis

Patients with cervical spondylosis are predisposed to develop damage to the spinal cord after traumatic injury that may be relatively minor (37). Increasing age is accompanied by progressive dessication and disruption of the integrity of intervertebral discs, particularly at the C5–C6 and C6–C7 interspaces, where spine mobility is greatest (47–49). The resulting narrowing of disc spaces leads to direct contact between adjoining vertebral bodies and the formation of bone spurs (osteophytes) which, if directed posteriorly, will narrow the spinal canal (50, 51). Such narrowing would predispose the afflicted individual to parenchymal damage to the spinal cord following upon a hyperextension or hyperflexion injury that would ordinarily be considered to be relatively minor (51). The pattern of spinal cord damage may vary according to the location of the bone spurs (52). Classically, with centrally positioned spondylotic protrusions, parenchymal damage, which is usually the result of hyperextension, is characterized by the presence of an ovoid spinal cord contour and of a "butterfly" pattern of tissue necrosis that affects the lateral columns, the lateral spinal gray matter, and the ventral portions of the posterior columns (50, 51, 53–56). The manner in which this pattern of damage develops is unknown, although a variety of vascular mechanisms have been proposed (57–59). Quencer and colleagues (60) have suggested, in a correlative MRI pathological study, that hyperextension-associated buckling of the ligamenta flava into a spinal canal that has already been narrowed by cervical spondylosis may lead to axonal disruption at the sites indicated above.

6. Concluding Remarks

The anatomical distribution of the lesions within the traumatized spinal cord will be determined by the manner in which the injury is sustained and will, in turn, determine the resulting signs and symptoms. The primary lesions that develop at the time of injury will be followed by secondary lesions that may appear hours or years later, with potentially major clinical consequences. Predisposing factors such as congenital narrowing of the spinal canal and, particularly, cervical spondylosis and, among East Asian subjects, ossification of the posterior longitudinal ligament (61) render some individuals more susceptible than others to the development of parenchymal damage to the spinal cord following upon blunt injuries to the spine.

References

1. Khan M, Griebel R (1983) Acute spinal cord injury in the rat: comparison of three experimental techniques. Can J Neurol Sci 10(2):161–165
2. Sekhon LHS, Fehlings MG (2001) Epidemiology, demographics, and pathophysiology of acute spinal cord injury. Spine 26(Suppl 24):S2–S12
3. Norenberg MD, Smith J, Marcillo A (2004) The pathology of human spinal cord injury: defining the problems. J Neurotrauma 21(4):429–440
4. Tator CH (1996) Classification of spinal cord injury based on neurological presentation. In: Narayan R, Wilberger JE Jr, Povlishock JT (eds) Neurotrauma. McGraw-Hill, New York, pp 1059–1073
5. Lipschitz R (1976) Stab wounds of the spinal cord. In: Vinken PJ, Bruyn GW (eds) Handbook of clinical neurology, vol 25. North-Holland, Amsterdam, pp 197–207
6. Kakulas BA (1984) Pathology of spinal injuries. CNS Trauma 1(2):117–129
7. Tator CH, Koyanagi I (1997) Vascular mechanisms in the pathophysiology of human spinal cord injury. J Neurosurg 86(3):483–492
8. Bailey FW (1971) Trauma of the spinal cord. In: Minckler J (ed) Pathology of the nervous system, vol 2. McGraw-Hill, New York, pp 1765–1774
9. Jellinger K (1976) Neuropathology of cord injuries. In: Vinken PJ, Bruyn GW (eds) Handbook of clinical neurology, vol 25. North-Holland, Amsterdam, pp 43–121
10. Zwimpfer TJ, Bernstein M (1990) Spinal cord concussion. J Neurosurg 72(6):894–900
11. Davison C (1945) General pathology of injuries of the spinal cord. Res Publ Assoc Nerv Ment Dis 24:151–187
12. Jellinger K (1976) Traumatic vascular disease of the spinal cord. In: Vinken PJ, Bruyn GW (eds) Handbook of clinical neurology, vol 12. North-Holland, Amsterdam, pp 556–630
13. Balentine JD (1988) Impact injuries of the spine and spinal cord. In: Leestma J (ed) Forensic neuropathology. McGraw-Hill, New York, pp 254–275
14. McVeigh JF (1923) Experimental cord crushes with special reference to the mechanical factors involved and subsequent changes in the areas of the cord affected. Arch Surg 7(3):573–600
15. Hashizume Y, Iijima S, Kishimoto H, Hirano A (1983) Pencil-shaped softening of the spinal cord: pathologic study in 12 autopsy cases. Acta Neuropathol 61(3–4):219–224
16. Ito T, Oyanagi K, Wakabayashi K, Ikuta F (1997) Traumatic spinal cord injury: a neuropathological study on the longitudinal spreading of the lesions. Acta Neuropathol 93(1):13–18
17. Bunge RP, Puckett WR, Becerra JL, Marcillo A, Quencer RM (1993) Observations on the pathology of human spinal cord injury: a review and classification of 22 new cases with details from a case of chronic cord compression with extensive focal demyelination. Adv Neurol 59:75–89
18. Kakulas BA (1999) A review of the neuropathology of human spinal cord injury with emphasis on special features. J Spinal Cord Med 119(2):117–124
19. Cornish R, Blumbergs PC, Manavis J, Scott G, Jones NR, Reilly PL (2000) Topography and severity of axonal injury in human spinal cord trauma using amyloid precursor protein as a marker of axonal injury. Spine 25(2):1227–1233
20. Nagamoto-Combs K, McNeal DW, Morecraft RJ, Combs CK (2007) Prolonged microgliosis in the rhesus monkey central nervous system after traumatic brain injury. J Neurotrauma 24(11):1719–1742
21. Ramlackhansingh AF, Brooks DJ, Greenwood RJ, Bose SK, Turkheimer FE, Kinnunen KM et al (2011) Inflammation after trauma: microglial activation and traumatic brain injury. Ann Neurol 670(3):374–383

22. Wallace MC, Tator CH, Frazee P (1986) Relationship between posttraumatic ischemia and hemorrhage in the injured rat spinal cord as shown by colloidal carbon angiography. Neurosurgery 18(4):433–439
23. Tator CH, Fehlings MG (1991) Review of the secondary injury theory of acute spinal cord trauma with emphasis on vascular mechanisms. J Neurosurg 75(1):15–26
24. Guha A, Tator CH (1988) Acute cardiovascular effects of experimental spinal cord injury. J Trauma 28(3):481–490
25. Choi DW (1988) Glutamate neurotoxicity and disease of the nervous system. Neuron 1(8):623–634
26. Panter SS, Yum SW, Faden AI (1990) Alteration in extracellular amino acids after traumatic spinal cord injury. Ann Neurol 27(1):96–99
27. Liu D, Thangnipan W, McAdoo DJ (1991) Excitatory amino acids rise to toxic levels upon impact injury to the rat spinal cord. Brain Res 547:344–348
28. Balentine JD, Spector M (1977) Calcification of axons in experimental spinal cord trauma. Ann Neurol 2(6):520–523
29. Young W (1992) Role of calcium in central nervous system injuries. J Neurotrauma 9(Suppl 1):S9–S25
30. Braughler JM, Hall ED (1989) Central nervous system trauma and stroke: biochemical considerations for oxygen radical formation and lipid peroxidation. J Free Rad Biol Med 6(4):289–301
31. Braughler JM, Hall ED (1992) Involvement of lipid peroxidation in CNS injury. J Neurotrauma 9(Suppl 1):S1–S7
32. Blight AR (1985) Delayed demyelination and macrophage invasion: a candidate for secondary cell damage in spinal cord injury. CNS Trauma 2(4):299–315
33. Totoiu MO, Keirstead HS (2005) Spinal cord injury is accompanied by chronic progressive demyelination. J Comp Neurol 486(4):373–383
34. Siegenthaler MM, Tu MK, Keirstead HS (2007) The extent of myelin pathology differs following contusion and transection spinal cord injury. J Neurotrauma 24(10):1631–1646
35. Emery E, Aldana P, Bunge MB, Puckett W, Srinivasan A, Keane RW et al (1998) Apoptosis after traumatic human spinal cord injury. J Neurosurg 89(6):911–920
36. Bruce JH, Norenberg MD, Kraydieh S, Puckett W, Marcillo A, Dietrich D (2000) Schwannosis: role of gliosis and proteoglycan in human spinal cord injury. J Neurotrauma 17(9): 781–788
37. Kim RC (2003) Pathology of non-neoplastic, regional disorders of the spinal cord. In: Nelson JS, Mena H, Parisi JE, Schochet SS Jr (eds) Principles and practice of neuropathology, 2nd edn. Oxford University Press, New York, pp 459–496
38. Göritz C, Dias DO, Tomilin N, Barbacid M, Shupliakov O, Frisén J (2011) A pericyte origin of spinal cord scar tissue. Science 333(6039):238–242
39. Wolman L (1967) Posttraumatic regeneration of nerve fibers in the human spinal cord and its relation to intramedullary neuroma. J Pathol Bacteriol 94(1):123–129
40. Sung JH, Mastri AR, Chen KTK (1981) Aberrant peripheral nerves and neuromas in normal and injured spinal cords. J Neuropathol Exp Neurol 40(5):551–565
41. Kamiya M, Hashizume Y (1989) Pathological studies of aberrant peripheral nerve bundles of spinal cords. Acta Neuropathol 79(1):18–22
42. Schmitt AB, Buss A, Breuer S, Brook GA, Pech K, Martin D et al (2000) Major histocompatibility complex class II expression by activated microglia caudal to lesions of descending tracts in the human spinal cord is not associated with a T cell response. Acta Neuropathol 100(5):528–536
43. Buss A, Brook GA, Kakulas B, Martin D, Franzen R, Schoenen J et al (2004) Gradual loss of myelin and formation of an astrocytic scar during Wallerian degeneration in the human spinal cord. Brain 127(1):34–44
44. Puckett WR, Hiester ED, Norenberg MD, Marcillo AE, Bunge RP (1997) The astroglial response to Wallerian degeneration after spinal cord injury in humans. Exp Neurol 148(2):424–432
45. Barnett HJM, Jousse AT (1976) Posttraumatic syringomyelia (cystic myelopathy). In: Vinken PJ, Bruyn GW (eds) Handbook of clinical neurology, vol 26. North-Holland, Amsterdam, pp 113–157
46. Nielsen OA, Biering-Sørensen F, Bötel H, Gardner BP, Little J, Ohta H et al (1999) Posttraumatic syringomyelia. Spinal Cord 37(10):680–684
47. Püschel J (1930) Der Wassergehalt normaler und degenerierter Zwischenwirbelscheiber. Beitr Pathol Anat 84:123–130
48. Saunders JBdeCM, Inman VT (1939) Intervertebral disc: critical and collective review. Int Abstr Surg 69:14–29
49. Bowden REM (1971) The applied anatomy of the cervical spine. In: Wilkinson M (ed) Cervical spondylosis: its early diagnosis and treatment. Saunders, Philadelphia, pp 10–34

50. Wilkinson M (1960) The morbid anatomy of cervical spondylosis and myelopathy. Brain 83(4):589–617
51. Wilkinson M (1971) Pathology. In: Wilkinson M (ed) Cervical spondylosis: its early diagnosis and treatment. Saunders, Philadelphia, pp 35–58
52. Yu YL, du Boulay GH, Stevens JM, Kendall BE (1986) Computer-assisted myelography in cervical spondylotic myelopathy and radiculopathy: clinical correlations and pathogenetic mechanisms. Brain 109(2):259–278
53. Bedford PD, Bosanquet FD, Russell WR (1952) Degeneration of the spinal cord associated with cervical spondylosis. Lancet 260(6724):55–59
54. Brain WR, Northfield D, Wilkinson M (1952) The neurological manifestations of cervical spondylosis. Brain 75(2):187–225
55. Mair WGP, Druckman R (1953) The pathology of spinal cord lesions and their relation to the clinical features in protrusion of cervical intervertebral discs (a report of four cases). Brain 76(1):70–91
56. Ogino H, Tada K, Okada K, Yonenobu K, Yamamoto T, Ono K et al (1983) Canal diameter, anteroposterior compression ratio, and spondylotic myeloopathy of the cervical spine. Spine 8(1):1–15
57. Taylor AR (1964) Vascular factors in the myelopathy associated with cervical spondylosis. Neurology 14(1):62–68
58. Turnbull IM (1971) Microvasculature of the human spinal cord. J Neurosurg 35(2): 141–147
59. Gooding MR (1974) Pathogenesis of myelopathy in cervical spondylosis. Lancet 304(7890):1180–1181
60. Quencer RM, Bunge RP, Egnor M, Green BA, Puckett W, Naidich TP, Post MJD, Norenberg M (1992) Acute traumatic central cord syndrome: MRI-pathological correlations. Neuroradiology 34(2):85–94
61. Chikuda H, Seichi A, Takeshita K, Matsunaga S, Watanabe M, Nakagawa Y et al (2011) Acute cervical spinal cord injury complicated by preexisting ossification of the posterior longitudinal ligament. Spine 36(18):1453–1458

Chapter 3

Acute Spinal Cord Injury In Vitro: Insight into Basic Mechanisms

Miranda Mladinic, Andrea Nistri, and Giuliano Taccola

Abstract

Spinal injury is produced by traumatic and nontraumatic causes that often induce long-term disability. In order to limit tissue damage and functional impairment, it is important to intervene at an early stage after the primary insult to restrain damage extension. This objective is difficult to achieve since the mechanisms underlying cell death in the spinal cord are complex and incompletely understood. We discuss how to set up a simple in vitro preparation of the neonatal rat spinal cord that retains full activity of locomotor networks and allows detailed investigations into the elementary pathological processes after inducing an experimental injury in vitro. We report different methods for producing an in vitro lesion: one is based on the excitotoxic action of the glutamate analogue kainate; one on the application of dysmetabolic conditions comprising hypoxia, aglycemia, and free oxygen radicals; and finally one due to the combination of both methods. Furthermore, the toxic insult can be applied to the entire in vitro preparation or to a few spinal segments isolated from the rest of the spinal cord by placing transverse barriers that limit toxic solutions to preselected regions only. In all tests, the end point (24 h after the transient application of the experimental insult) is the ability to express locomotor-like rhythmic patterns recorded electrophysiologically from lumbar ventral roots and to correlate this function with cell survival, cell type identification, and demonstration of pathological processes responsible for cell loss. This approach can be useful to investigate the basic mechanisms of action of drugs designed to exert early neuroprotection and the time course of effectiveness against distinct cell types.

Key words: Motoneurons, Central pattern generator, Rat spinal cord, Cell death, Excitotoxicity, Hypoxia, Ischemia

1. Introduction

One of the prime challenges for neuroscientists is to develop effective therapies to treat acute spinal cord injury (SCI) by arresting damage progression through the secondary phase that follows the initial insult (1). Such a phase initiates within hours after the primary lesion and spreads to unaffected spinal areas, thus amplifying

Håkan Aldskogius (ed.), *Animal Models of Spinal Cord Repair*, Neuromethods, vol. 76,
DOI 10.1007/978-1-62703-197-4_3, © Springer Science+Business Media, LLC 2013

the severity of the primary damage (2). It is widely acknowledged that deranged intracellular signal cascades are responsible for such a secondary lesion (3). Recent clinical studies indicate that, in addition to trauma, SCI is often produced by nontraumatic causes like vertebral collapse (cancer, osteoporosis), vascular disorders, and degenerative diseases with progressive stenosis of the vertebral canal. It is currently unclear if a common biochemical process underlies causes so etiologically different (these issues are addressed elsewhere in the present book).

Various studies have considered the possibility that acute trauma by damaging cells and generating a wave of depolarization might induce cell death by excitotoxicity, a mechanism elicited by massive release of the transmitter glutamate with subsequent neuronal overexcitation (4). In the case of nontraumatic origin, hypoxia/dysmetabolic changes/generation of toxic-free radicals are thought to play an essential role in activating the secondary damage (1). Both processes may, however, converge onto common pathways depending on the time dependence of lesion evolution.

Thus, investigations into different aspects of SCI, such as limitation or block of secondary injury cascade, promotion of axon regeneration, replacement of the damaged and lost cells or, at later stages, neurorehabilitation, have led to development of different laboratory animal models (in vivo and in vitro) of SCI. After the first well-documented animal experimentation of SCI described by Allen (5) who used a weight-drop technique on the dog spinal cord, a variety of techniques (weight drop, contusion, aneurysm clip compression, calibrated forceps compression, laceration including complete transection or hemisection, and chemically mediated injury (6)) have been used to induce in vivo damage to spinal cord of vertebrates from lamprey to primates as discussed in detail in the present book. Even though the use of multiple experimental paradigms has advanced our understanding of the pathophysiology of the SCI, no animal model completely addresses all aspects of human SCI. In fact, the limitations of in vivo models include the complex pathology surrounding the injury and the restricted follow-up, which makes unavoidable terminating the experiment at each time point of pathophysiological analysis. Furthermore, experimental animal lesions are produced, for ethical concerns, under general anesthesia, a condition obviously very different from human cases and capable of influencing the lesion outcome, which in the experimental animal is rarely treated in the same way as the acute patient in the intensive care unit.

In vitro models simplify the complexity of injury pathophysiology, thus allowing dissecting out the components of specific injury processes and the identification of affected cells without the interference by general anesthesia or blood pressure changes.

Nonetheless, in vitro models lack certain properties of the in vivo lesion like the effects due to disruption of vascular supply, the role of blood-borne substances and cells, the delayed effects caused by immune reactions, and the lack of behavioral alterations. It seems, therefore, advantageous that in vivo and in vitro data are combined to obtain a more integrated view of the pathological progress.

In vitro models comprise cell cultures (7), organotypic cultures (8), and isolated spinal cord preparations, like, for example, the one of the neonatal opossum that is very useful to study regenerative capacities of the injured axons (9, 10). For the sake of simplicity, the present report will focus on the rat isolated spinal cord preparation that has been recently employed in our laboratory (11, 12).

Isolated spinal cord preparations of rodents have been largely used to study sensory/motor pathways and their integration in controlling major functions like nociception or locomotion (13). Much of the functional circuitry of the CNS can be studied in the isolated spinal cord with the additional advantage that the isolated spinal cord can be perfused with known concentrations of ions, neurotransmitters, and novel therapeutics of potential neuroprotective value (13). Furthermore, these substances can be later washed away to observe any recovery of the circuitry and/or reproducibility of the effects.

The rat in vitro neonatal spinal cord was the first mammalian nervous system preparation (isolated from the brainstem to the cauda equina) characterized by the fact that all the essential network components related to locomotion are retained with functional output investigated by recording either rhythmic bursts from ventral roots (VRs) supplying hindlimbs or the electromyogram from limb muscles (14–16). Chemical substances can be readily applied to activate rhythmic networks that express their function recorded as alternating cyclic discharges from VRs. This pattern is termed fictive locomotion because it contains all the basic characteristics of real locomotion, namely left/right signal alternation at homosegmental level, and homolateral alternation between flexor and extensor motor pools, all occurring with regular periodicity (13). In addition, the intrinsic rhythmicity of spinal networks is readily unveiled by pharmacological block of synaptic inhibition (disinhibited bursting) (17–19), thus providing useful data on the basic excitability of spinal circuits. Of course, this preparation is also suitable to study synaptic transmission at segmental level, and motoneuron/interneuron properties depending on the recording configuration.

The rat spinal cord in vitro has been shown to be particularly advantageous to study SCI. In our laboratory, we have developed novel procedures to mimic metabolic perturbations characteristically activated after SCI, or glutamate-mediated excitotoxicity, and we have applied them to the isolated neonatal spinal cord to induce large or local damage of the spinal cord tissue. Monitoring spinal network electrophysiology has enabled us to follow up functional damage evolution during and after spinal cord injury, and to correlate it with a histological map of damaged cells. This novel approach, therefore, provides a tool to study acute experimental spinal cord lesion with a focus on the vulnerability of locomotor networks. In particular, we have investigated the evolution of early secondary damage by taking, as outcome, the functional operation of locomotor networks in relation to cell survival (11). An additional advantage of this model is the opportunity for ongoing monitoring of fictive locomotion at the site of focal injury vs. unaffected segments, and to compare this phenomenon with sham-operated spinal cords (12). Thus, our preparation is best suited to discover damage mechanisms as it retains cellular connections, networks, and activities, including fictive locomotion (unlike the cell culture models). The use of a neonatal preparation can also shed light on the issue of child spinal injury, a scanty researched field (20, 21). Our model has already been shown to be suitable for preclinical testing of neuroprotective drug activity (22), for understanding the role of the extracellular ion homeostasis in determining network outcome (23), and for revealing the molecular mechanisms involved in cell death after SCI (24, 25).

2. Materials

Kainate is purchased from Ascentis Scientific, while strychnine, bicuculline, NMDA, and serotonin (5HT) are from Tocris. Figure 2b lists the composition of the pathological medium used to produce dysmetabolic conditions in the in vitro spinal cord.

Table 1 lists the source and dilution of antibodies used for the current protocols.

Paraformaldehyde (PFA), bovine serum albumin (BSA), Triton X-100, xylene, 4′,6-diamidino-2-phenylindole dihydrochloride (DAPI), and toluidine blue are purchased from Sigma-Aldrich. Normal goat serum (NGS) is obtained from Chemicon, while Vectashield Mounting Media for fluorescence are from Vector Laboratories.

To analyze and count the cells in spinal cord sections after immunofluorescence staining, we use the following softwares: MetaVue from Molecular Devices (Sunnyvale, CA, USA) and CellCounter software from GlanceVision Technologies (Trieste, Italy).

Table 1
List of commercially available antibodies used in our laboratory for immunostaining of different cells or molecules in rat spinal cord

Antibody	Type	Company (catalog number)	Dilution
Mouse Anti-Neuronal Nuclei (**NeuN**) Monoclonal Antibody	Mouse monoclonal	Chemicon, Millipore (MAB377)	1:50
Neurofilament H Non-Phosphorylated (**SMI 32**) Monoclonal Antibody	Mouse monoclonal	Covance (SMI-32R)	1:200
Monoclonal Anti-Glial Fibrillary Acidic Protein (**GFAP**)	Mouse monoclonal	Sigma (G3893)	1:100
Polyclonal Rabbit Anti-**S100**	Rabbit polyclonal	Dako (Z0311)	1:100
Monoclonal Anti-Oligodendrocyte Marker **O4** Antibody	Mouse monoclonal	R&D Systems (MAB1326)	1:200
Rabbit Anti-**NG2** Chondroitin Sulfate Proteoglycan Polyclonal Antibody	Rabbit polyclonal	Chemicon, Milipore (AB5320)	1:100
Rabbit Anti **Iba1** (Ionized calcium binding adaptor molecule 1) antibody	Rabbit polyclonal	Wako (019-19741)	1:500
Mouse Anti-Oligodendrocytes Monoclonal Antibody **RIP**	Mouse monoclonal	Chemicon, Milipore (MAB1580)	1:1,000
Cleaved Caspase-3 (Asp175) Antibody	Rabbit monoclonal	Cell Signaling Technology (9661)	1:50
Poly(ADP-ribose), mAb (10 H) (**PAR**)	Mouse monoclonal	Alexis, Enzo Life Sciences (ALX-804-220)	1:50
Alexa Fluor 488 or **594** Secondary Antibodies	Goat anti-mouse or goat anti-rabbit	Invitrogen (A 10680, A-11008, A-11005, A1012)	1:500

Bold characters indicate standard abbreviations

3. Methods

3.1. Spinal Cord Preparation

Experiments are performed on thoracolumbar spinal cord preparations isolated from neonatal Wistar rats (0–2 days old) under urethane anesthesia (0.2 ml i.p. of a 10% w/v solution). After decapitation and evisceration, the vertebral column with surrounding tissue is pinned to the Sylgard-coated bottom of a Petri dish at room temperature and the spinal cord is carefully dissected out under microscopic control. During this phase, the tissue is superfused with Krebs solution of the following composition (in mM): NaCl, 113; KCl, 4.5; $MgCl_2 7H_2O$, 1; $CaCl_2$, 2; NaH_2PO_4, 1; $NaHCO_3$, 25; glucose, 11; gassed with 95% O_2 5% CO_2; pH 7.4 at room temperature. Extra care is applied to the isolation of pairs of roots (usually between T10 and S1 segments) that will be used for recording and electrical stimulation. The preparation (with roots attached) is transferred to a recording chamber continuously superfused (7.5 ml/min) with the same Krebs solution. Miniature suction electrodes are used to record from four ventral roots (VRs) and to stimulate dorsal roots (DRs). Figure 1 shows a photograph of the spinal cord setup for recording and stimulating with multiple electrodes, where  and  indicate recording and stimulating, respectively.

3.2. Electrophysiological Recording

To record fictive locomotor rhythms for a long time, the majority of experiments are based on DC-coupled recordings from four lumbar VRs (26) as exemplified in Fig. 1b. Fictive locomotion is typically evoked by continuous bath application of NMDA (5 μM) and 5HT (10 μM). In particular, lumbar (L) 2 (denoting the spinal segment) VRs express mainly flexor motor commands to the hindlimb skeletal muscles, while L5 VRs convey mainly extensor motor signals to the same limbs. When these discharges occur at 2–4/s periodicity and alternate between flexor and extensor motor pools and between left (l) and right (r) sides, the pattern has all the properties typical of fictive locomotion (13, 27) as demonstrated in Fig. 1b. Intracellular recordings from lumbar motoneurons are obtained using sharp electrodes filled with 3 M KCl (30–60 MΩ resistance) or 2 M Kacetate (60–100 MΩ; ref. (26)) connected to an AxoClamp amplifier (0.1 preamplifier gain; Molecular Devices, Sunnyvale, CA, USA) in a current-clamp mode. The motoneuron field is approached by tracking down with the sharp electrode to meet a large negative-going field potential. When the field is inverted because of the very close proximity to a single motoneuron, a brief oscillation of the negative capacitance control or of the electrode clearance switch is adequate to impale a motoneuron. Motoneurons are identified by their all-or-none antidromic action potential induced by weak stimulation of the corresponding VR (usually 0.05 ms pulse width; 1 V intensity at 0.5 Hz). Typical electrophysiological parameters are membrane potential, action potential, and synaptic potential induced by stimulation of the corresponding DR. The input resistance of motoneurons is measured by delivering hyperpolarizing current steps (0.1–0.9 nA, 30–50 ms) through the intracellular electrode: current/voltage curves are constructed within the linear range (from −70 to −60 mV). All data are acquired and analyzed off-line with pCLAMP software (Molecular Devices, Sunnyvale, CA, USA).

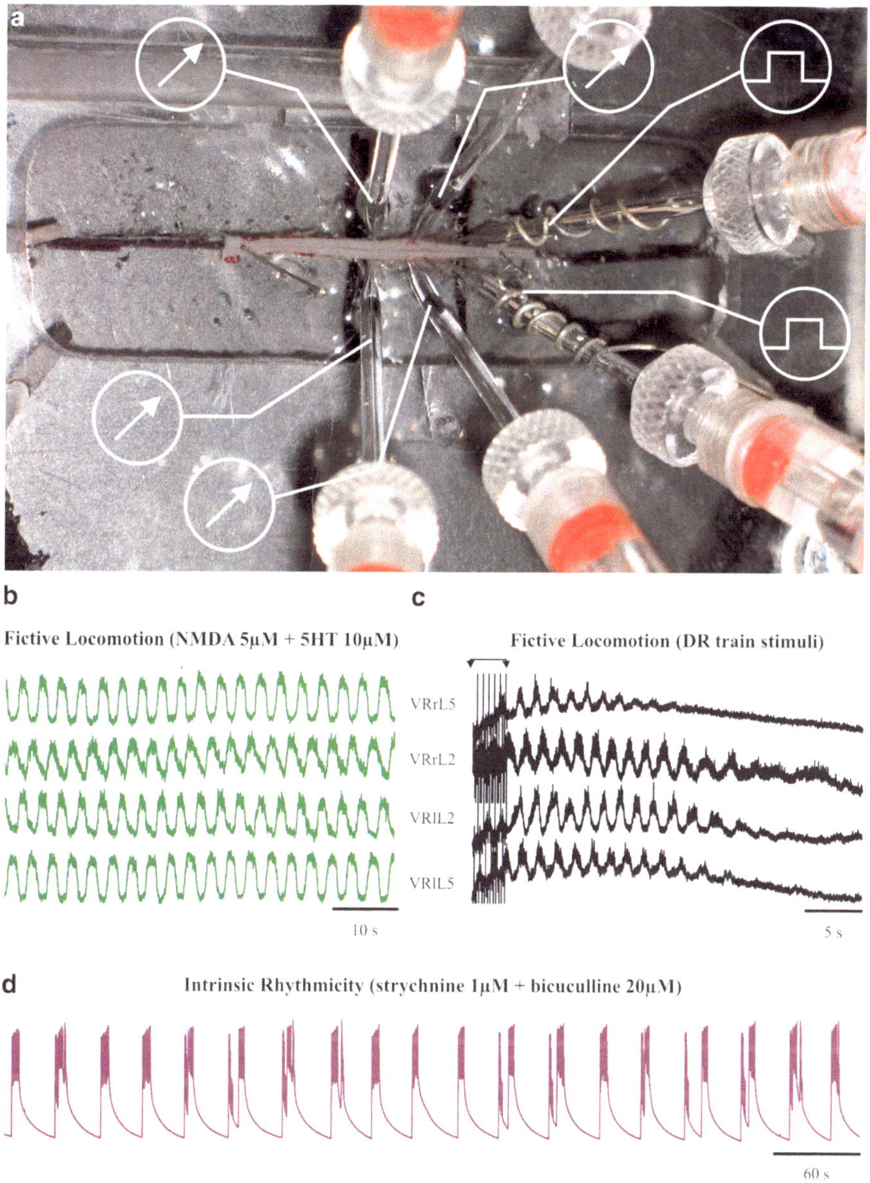

Fig. 1. Electrophysiological setup for studying locomotor network activity. (**a**) Photograph of rat isolated spinal cord preparation pinned to the bottom of a Sylgard-coated Petri dish. An array of miniature suction electrodes is used to stimulate DRs and to record electrophysiological responses from VRs. Recording electrodes are indicated by ↗ while stimulating electrodes are indicated by ⎍. (**b**) Example of chemically evoked fictive locomotion characterized by alternating cyclic oscillations present on four lumbar roots following bath application of NMDA and 5HT. (**c**) Example of a bout of fictive locomotion induced by a train (30 pulses at 10 Hz) applied to one DR and observed as cyclic oscillations on all four VRs. (**d**) Example of spontaneous disinhibited bursting recorded from one VR following application of strychnine (1 μM) and bicuculline (20 μM) (unpublished data by G.T.)

DR electrical stimuli are employed to elicit responses recorded from either a motoneuron or the ipsilateral VR or cumulative depolarization (1–30 V stimulus applied to one DR; 0.1 ms duration; 30 pulse trains at 1–2 Hz). Stimuli are regarded as low T when they do not exceed 2× stimulus strength to evoke a detectable synaptic response from the recorded motoneuron or VR. High T stimuli are larger than 2× T (28). To minimize variability, DR stimulation-induced responses are averages of at least five events, of which the amplitude and area are calculated.

3.3. Induction of Fictive Locomotion

This pattern is normally elicited with two protocols: (1) Trains of DR stimuli (see Sect. 3.2) that evoke cumulative depolarization upon which a bout of alternating oscillations emerges normally for a few tens of seconds (Fig. 1c). (2) Continuous bath application of *N*-methyl-D-aspartate (NMDA; 4 or 5 μM) plus serotonin (5-hydrohytryptamine; 5-HT; 10 μM) (29, 30) as shown in Fig. 1b. This rhythm can be maintained for a long time (tens of min) without fading.

All rhythmic discharges are analyzed for their period defined as the time between the onset of two cycles of oscillatory activity. The regularity of the rhythm is calculated in terms of coefficient of variation (CV; given by standard deviation (SD) mean^{-1}) of the period. Signal correlation between pairs of VRs is expressed with the cross-correlation function (CCF) using pCLAMP software. While CCF > +0.5 indicates synchronicity between two VRs, CCF < −0.5 shows full alternation.

Disinhibited bursting (17, 18, 31) is a highly regular pattern of all-or-none discharges arising from synchronous activation of motor pools in distinct spinal segments. It is characterized by very large depolarization of motoneurons or VRs (lasting 7–10 s) at a low periodicity (1 every 20–30 s). Disinhibited bursting is produced by continuously bath-applied strychnine (1 μM) and bicuculline (10 μM) that unmask the latent excitability of spinal networks in the absence of synaptic inhibition (Fig. 1d). Disinhibited bursting can also be electrically driven by DR stimulation. Burst duration is calculated between the event onset and the start of the rapid decay, while the period is the time interval between the onset of a burst and the next one.

3.4. Experimental Lesion Protocol for the Isolated Spinal Cord

Several clinical studies show that, under the most favorable conditions, the average admission time to hospital emergency care for acute SCI is approximately 3 h (32–34). Nonetheless, stabilization of metabolic conditions is expected to begin earlier at the lesion site (35). Thus, the protocol for in vitro SCI attempts to mimic the early pathophysiological events of SCI. To obtain consistency and to allow cross-sectional study comparison, toxic agents are applied to produce spinal cord damage for 1 h to the whole spinal cord preparation. This approach is in close agreement with the observed excitotoxicity evoked by 1 h microdialysis of glutamate applied to the in vivo rat spinal cord (36). Control preparations are kept for the same time in vitro without any drug or toxic agent administration (sham preparations). The various protocols are schematically summarized in Fig. 2a. Prior to any toxic challenge,

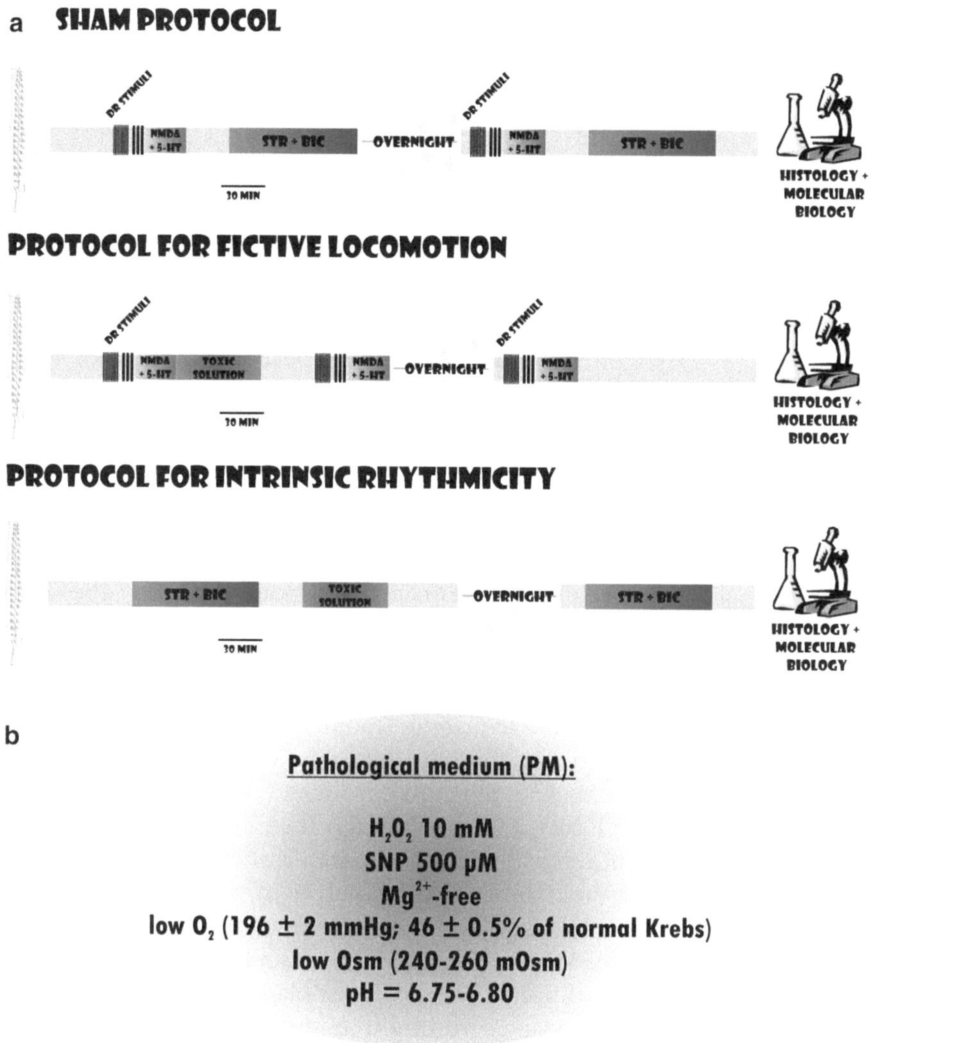

Fig. 2. Scheme of protocols used for SCI experiments. (**a**) The isolated spinal cord is first tested for electrophysiological responses that include DR stimulation, fictive locomotion, or disinhibited bursting (STR + BIC). If using sham conditions, after overnight rest, the preparation is retested 24 h later and then processed for histology (*top row*). *Middle row* shows the protocol for fictive locomotion followed by the application of the toxic solution for 1 h, its washout and retesting of the electrophysiological responses 24 h later. *Bottom row* shows a similar approach in which, instead of fictive locomotion, disinhibited bursting is tested followed by the toxic solution, washout, and retest the day after. (**b**) Recipe for the composition of pathological medium (PM) that mimics the severe metabolic perturbation is thought to be generated in the spinal cord after acute injury. This solution can be applied for 1 h as per protocol in A.

a series of electrophysiological tests (fictive locomotion, DR stimulation, or induction of disinhibited bursting with strychnine and bicuculline) is performed to establish the baseline conditions of network activity. In all protocols, the toxic medium is applied for 1 h only followed by extensive washout with oxygenated standard Krebs solution for 24 h when the outcome of the lesion in terms of morphological and functional features is systematically investigated.

Repeated electrophysiological tests are run during and after the toxic solution washout to monitor any on-going change in network activity before the study end point at 24 h.

It is noteworthy that, on the same preparation, we normally test either fictive locomotion or disinhibited bursting in order to avoid potential interference by the neurochemicals used to elicit such rhythms with the pathological processes. This is particularly important for the application of strychnine and bicuculline, whose effects are very slowly reversible on washout. Sham preparations (24 h in vitro) reproduce standard fictive locomotion or disinhibited bursting indicating no deleterious action of the in vitro maintenance (11).

The present protocol for excitotoxicity relies on the application of kainate, a metabolically stable potent analogue of glutamate. Kainate is not a substrate for the glutamate uptake carriers and has been extensively used to produce excitotoxicity (37–39). In vivo injection of kainate into the rat spinal cord is known to produce extensive damage with motor paralysis (40–42). Thus, to induce excitotoxic damage to spinal networks, kainate is applied for 1 h at 1 mM concentration and then washed out, while its effects are monitored for up to 24 h (Fig. 2a). This concentration of kainate is adequate for rapid and irreversible loss of fictive locomotion. Our studies have also indicated that lower concentrations of kainate (50–500 μM) produce comparable functional damage and cell loss at 24 h, although the rate of damage development is significantly slower (8). Figure 3a shows an example of the relatively rapid onset depolarization of motor pools recorded from four VRs (see Fig. 1a for electrophysiological arrangement) by applying kainate (1 mM). Noisy traces indicate repetitive discharges elicited during the strong depolarization that reaches a plateau usually sustained throughout the kainate application. The functional consequences (24 h later) of 1 h application of this neurotoxic drug on fictive locomotion evoked by NMDA and 5-HT are exemplified in Fig. 3b, in which full suppression of this rhythm is observed. Figure 3e shows that disinhibited bursting persists 24 h later although the burst amplitude is strongly impaired.

A more complex toxic solution (Fig. 2b) is used to mimic the chemical environment thought to be induced at the site of neuronal injury and to contribute to secondary spinal damage. This involves changes due to lack of oxygenation that leads to ionic alterations, pH acidification, edema (with reduced osmolarity), and production of endogenous toxic metabolites. In practice, to achieve this condition, the experimental solution is not gassed with a mixture of O_2/CO_2 because simple arrest of oxygen delivery to a spinal preparation in vitro produces severe hypoxia (43). We have actually measured (with a miniature O_2 electrode) the pO_2 values in the experimental bath and found that they usually fall to 196 ± 2 mm Hg (i.e., $46 \pm 0.5\%$ with respect to normal oxygenated

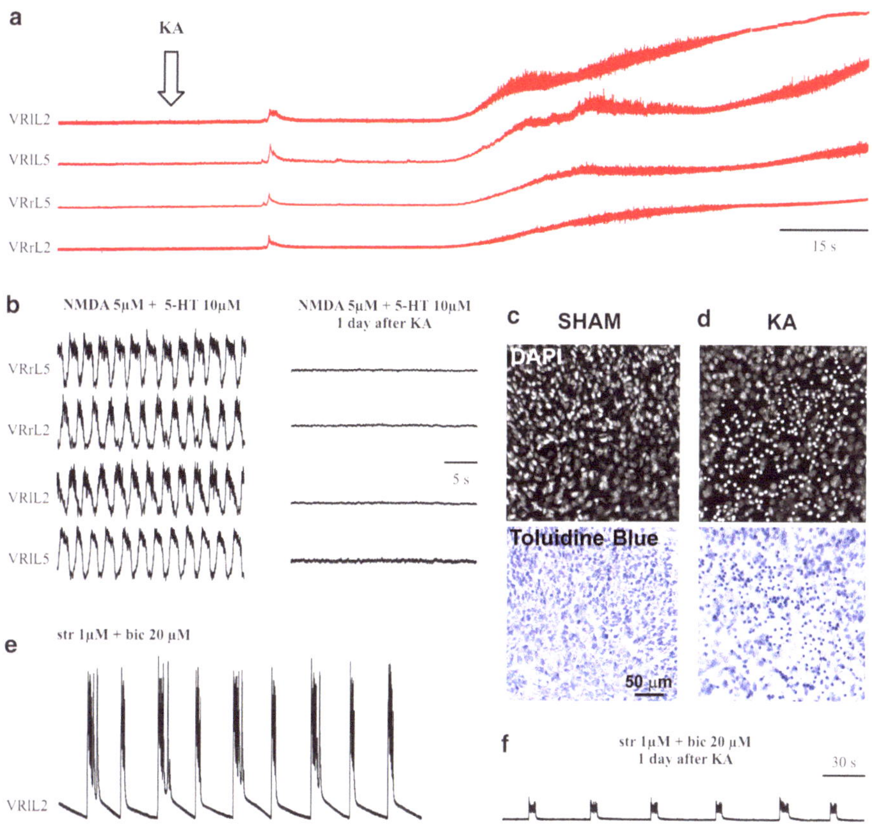

Fig. 3. Effects of kainate on the rat spinal cord in vitro. (**a**) Example of electrophysiological recording from four lumbar VRs. Application of kainate (*arrow*) elicits depolarization with superimposed irregular discharges to reach a steady-state plateau. The kainate application lasts for 1 h and is followed by 24 h wash. (**b**) Chemically evoked fictive locomotion is observed in control condition (sham; *left*) but it is absent 24 h after kainate application (*right*). (**c**) Histological examination of sham spinal tissue in the ventral area shows good cell preservation observed with DAPI (*top*) or toluidine blue (*bottom*) staining. (**d**) 24 h after kainate administration, extensive pyknosis (recognizable by small, round shape of cell nucleus and intense staining) is detected with either dye. (**e**, **f**) Example of disinhibited bursting (sham preparation) with strong regular bursts detected from one VR, a phenomenon greatly attenuated in amplitude after kainate, although burst frequency and regularity are preserved. Unpublished data by M.M. and G.T.

Krebs solution). With this solution, Mg^{2+}, glucose, and $NaHCO_3$ are also omitted, while Hepes is added to reach pH 6.75–6.80 (with 0.1 N NaOH). Osmolarity is 230–240 mOsm, thus recreating the hypoosmotic milieu surrounding an acute lesion (44, 45). H_2O_2 and sodium peroxynitrite (SNP; a NO donor) are also added to produce toxic free radicals. This solution is termed pathological medium (PM; Fig. 2b). Figure 4 exemplifies the effect of PM on the spinal cord in vitro. PM depolarizes motor pools (Fig. 4a) and triggers rapid oscillations recorded from two VRs. After 1 h application, the consequences (24 h later) are less severe on locomotor functions since low amplitude, less regular cycles of fictive locomotion can be observed (Fig. 4b), while disinhibited bursting is very similar to control (Fig. 4c).

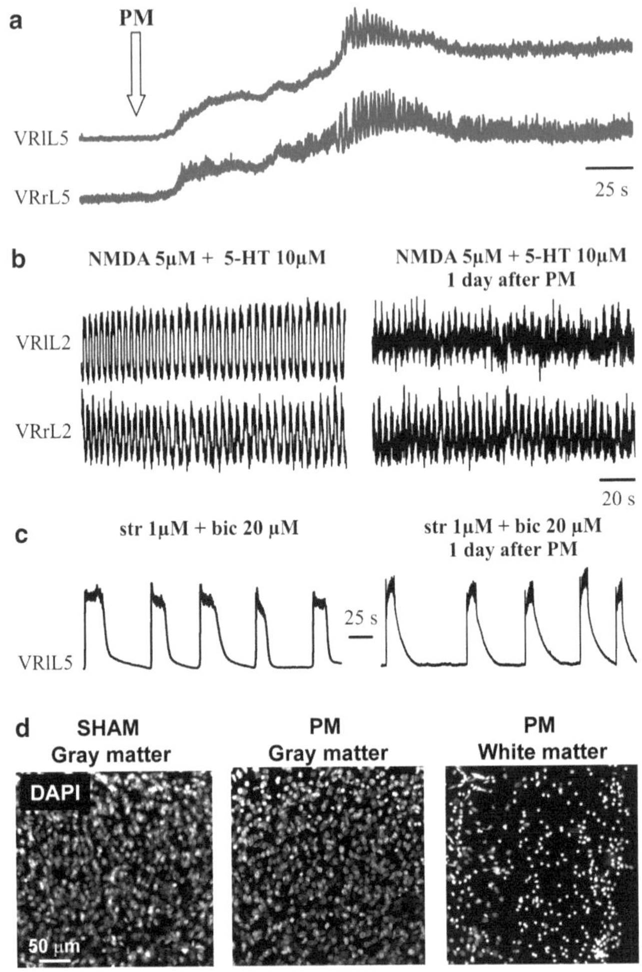

Fig. 4. Effects of pathological medium (PM) on rat spinal cord in vitro. (**a**) Electrophysiological effects of PM applied at the time indicated by the arrow. A slowly developing depolarization with superimposed irregular discharges is observed. This phenomenon is quantitatively less intense than the one detected following application of kainate. (**b**) Chemically evoked fictive locomotion of a sham preparation (*left*) is detectable also 24 h after PM application and washout (*right*). Note, however, that oscillatory cycles are of smaller amplitude and are less regular. (**c**) Example of disinhibited bursting of a sham preparation (*left*) compared with the similar pattern observed 24 h after PM (*right*). Note substantial preservation of this phenomenon. (**d**) Examples of rat neonatal spinal cord sections (30 μm) cut out from sham preparation (dorsal region, *left*) or from preparation formerly exposed to PM (dorsal, *middle*). Note limited number of DAPI-stained pyknotic nuclei. Conversely, the ventrolateral white matter (*right*) shows massive pyknosis of the PM-treated spinal cord preparation. Panels (**a–c**) are modified from ref. (11), while histological pictures are unpublished work by M.M.

An additional protocol includes the combined administration of kainate with the PM to explore the consequences of metabolic disruption plus a primary excitotoxic agent. In this case, the resulting effects are a very severe destruction to the spinal tissue (11).

3.5. Focal Damage to the In Vitro Spinal Cord

The methods described in Sect. 3.4 are helpful to investigate the early evolution of SCI and to provide a substantial amount of tissue suitable for morphological and molecular biology experiments. Nevertheless, as SCI is often patchy and incomplete, it seems important to study whether and when a focal lesion can extend to spared tissue. To achieve this aim, it is very advantageous to partition the rat spinal cord preparation into subcompartments that allow the administration of different solutions to distinct segments of the spinal cord. This approach can be successfully obtained by the use of plastic barriers which do not damage the spinal tissue, yet they restrict the application of substances to preselected areas only.

The use of barriers to investigate the site of action of drugs and the distribution of locomotor networks has been reported before (46–49). Figure 5a shows how this method is applied in our laboratory. It is of paramount importance to establish that barriers themselves do not create tissue mechanical damage with dysfunction of synaptic transmission and impulse conduction. Indeed, at the start of each experiment, it is important to check that the basic

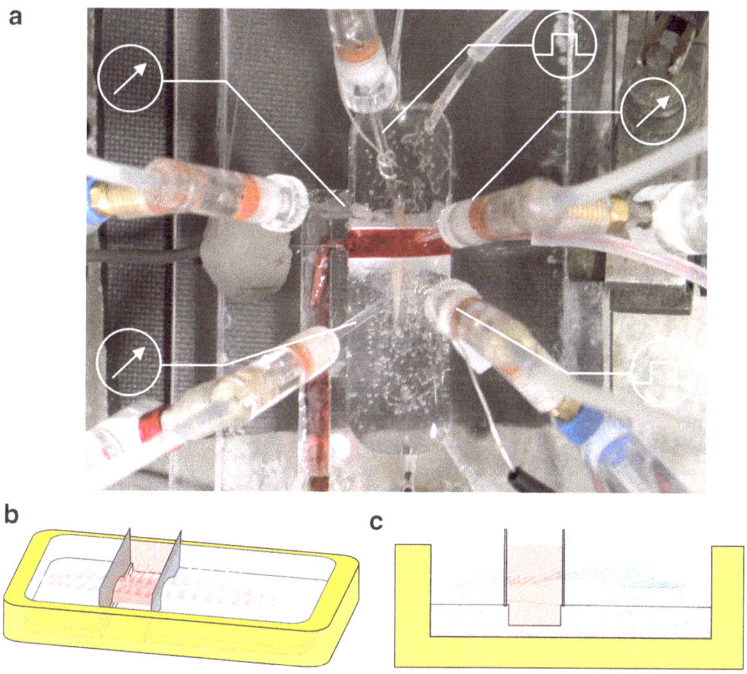

Fig. 5. Transverse barrier arrangement for isolated spinal cord. (**a**) Photograph of rat isolated spinal cord preparation pinned to the bottom of a Sylgard-coated Petri dish and partitioned by two barriers to create a central compartment separately superfused from the top and bottom ones. The central compartment is filled with a phenol red solution to test its leak-proof condition. Note array of miniature suction electrodes to stimulate DRs and to record electrophysiological responses from VRs. Abbreviations are as in Fig. 1a legend. (**b**, **c**) Schematic representation of the transverse barrier arrangement to indicate that the central compartment is fully superfused to avoid a dishomogeneous distribution of the toxic solution.

electrophysiological properties of in vitro spinal cord are not impaired. Hence, we routinely stimulate DRs above and below the barriers to elicit robust responses from VRs below and above, respectively, as a way to confirm full tissue viability. As a norm, we place barriers at the level of the lower thoracic spinal cord to enclose three or four segments down to L1–L2. Figure 5a shows an example of placing such barriers to create an isolated compartment (filled with phenol red solution to check it remains leak proof). One recording electrode is connected to a VR within the central compartment, while other recording or stimulating electrodes are used for VRs and DRs above and below the central compartment. Such responses are present even the day after, with unchanged peak amplitude, even though the late polysynaptic component can be attenuated (12).The schemes in Fig. 5b and c indicate how this arrangement is obtained and the fact that below the segments within the central compartment there is a grove to allow access to the bathing solution. The reason for this device is to avoid applying any toxic substance to just one side of the spinal cord and to restrict the damage to the exposed region only.

Figure 6a shows how, for 1 h of continuous recording, spinal segments above (T6) and below (L5) the central compartment (T13) respond to application of kainate plus PM. T13 produces a large and relatively persistent depolarization (middle trace in Fig. 6a), while the responses of T6 (top trace) is brief and small, and the one from L5 (bottom trace), albeit larger, fades away.

The early effects of kainate plus PM are depicted, on a fast timescale (top four traces), in Fig. 6b showing (different preparation from (a)) irregular oscillations superimposed on the depolarization of four VRs outside the central compartment. The bottom four traces show (on a faster timescale) the irregular nature of discharges recorded from the same four VRs: after a transient bout of few oscillatory cycles, mixed asynchronous activity emerges. This observation indicates that the strong excitation of the networks within the central compartment spreads early and intraspinally to adjacent segments. When we test disinhibited bursting (Fig. 6c), we can observe strong slowing down of this rhythm above the top barrier (T7), disruption within the barriers (T12), and preservation below them (L5) because the latter segment is known to be highly rhythmogenic (13, 18). Figure 3.6d shows that, although fictive locomotion evoked by NMDA (5 μM) and 5HT (10 μM) is absent on all four VRs outside the compartment containing kainate plus PM, it returns 24 h later even if rather irregular. Thus, the adjacent excitation, that had invaded the untreated segments, produces temporary functional inactivation of locomotor networks ("spinal shock") from which subsequent spontaneous recovery is observed.

3.6. Histological Examination of SCI

As a first approach to assess the gross distribution of a lesion under light microscopy, a number of classical histology stainings like

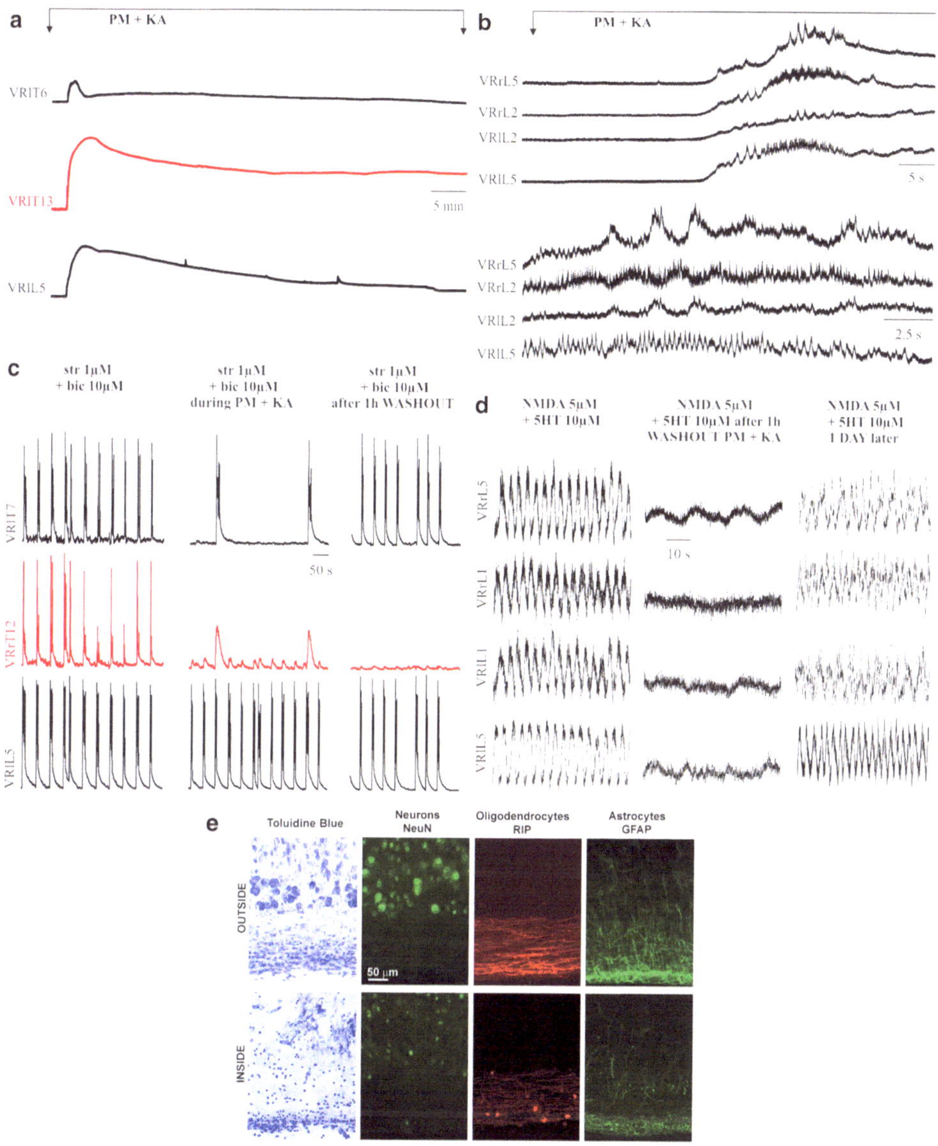

Fig. 6. Remote effects of toxic solution on distant segments of the rat spinal cord with barrier partitions. (**a**) Simultaneous recording from three VRs from the start of PM plus kainate application (*left arrow*) is maintained for 1 h when washout is effected (*right arrow*). The *top* and *bottom* traces are from VRs outside the central compartment to which the toxic solution (PM plus kainate) is applied. Note depolarization recorded from all VRs, although only the middle one is maintained up to the washout time point. (**b**) *Top*, fast-time base records of the same responses shown in (**a**) to indicate the irregular discharges present on four VRs outside the central compartment containing the toxic solution. *Bottom*, on a faster timescale, after a few cyclic oscillations, very asynchronous activity is observed on all VRs. (**c**) *Left*, disinhibited bursting is observed on three VRs prior to the toxic solution administration to the central compartment (*middle trace*). At the end of toxic solution application (PM plus kainate), bursting is slowed down on the top VR, almost blocked on the middle VR (inside the central compartment), and fully present on the bottom VRs. (**d**) Records from four VRs outside the central compartment that was superfused with PM plus kainate. Fictive locomotion evoked by NMDA and 5-HT is lost on all VRs during the washout phase after PM plus kainate. Nonetheless, the rhythm returns 24 h later indicating the transient nature of locomotor network block outside the region immediately treated with the toxic solution. (**e**) Examples of sagittal spinal cord sections (30 μm) cut out from a preparation are locally exposed to the PM plus kainate. The *upper panels* show the spinal cord tissue from lumbar (L3) segment outside the barriers and, thus, not exposed to the toxic medium, while *lower panels* show the lumbar segment (L1) within the two barriers and are exposed to the toxic solution. The sections are stained with toluidine blue (*left panels*) and by immuno-fluorescence with antibodies specifically recognizing different cell types, namely NeuN for neurons, RIP for myelin sheath of oligodendrocytes, and GFAP for astrocytes. Note extensive tissue and cell destruction in the region directly exposed to the toxic medium. a is modified from ref. (12), b, c, d are unpublished data by G.T. and e are unpublished data by M.M.

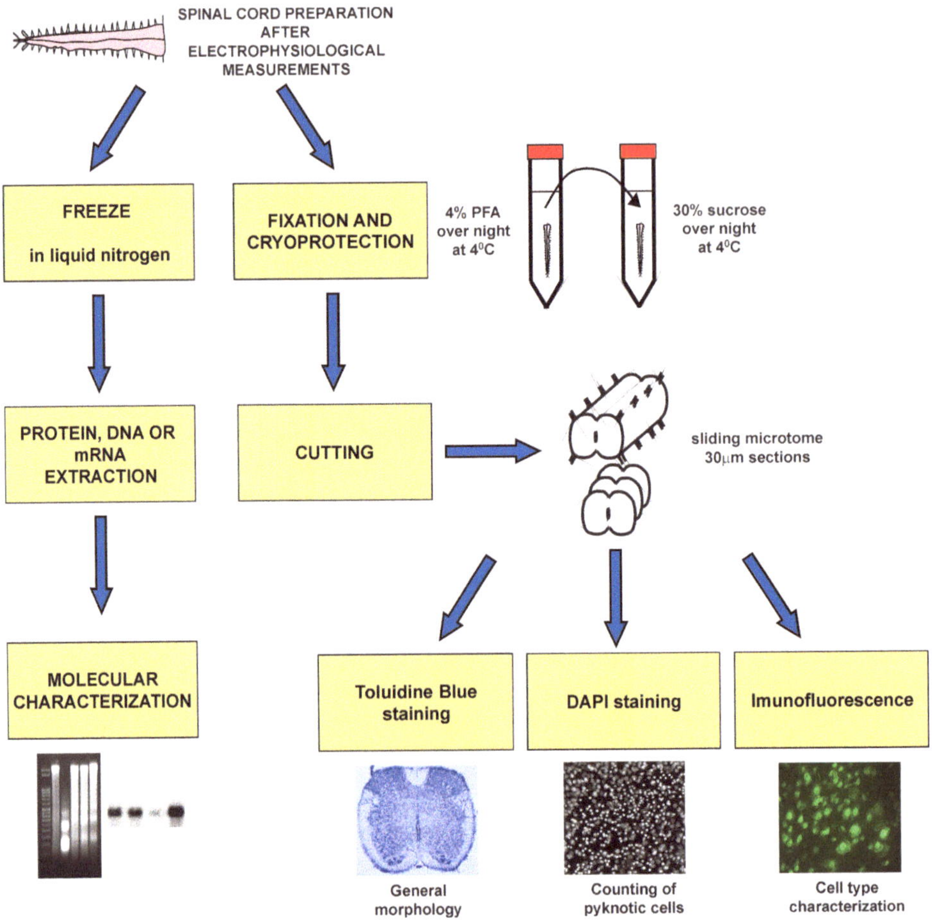

Fig. 7. The flowchart of the procedures for histological and molecular characterization of spinal cord preparations. Spinal cord preparations (either treated with toxic solution or in sham condition), after 24 h in vitro and electrophysiology measurements, are frozen in liquid nitrogen or fixed in 4% paraformaldehyde (PFA). The frozen tissue is stored at −80°C until further use for molecular biology techniques to isolate proteins, mRNA or DNA (*left branch of flow diagram*). For histological characterization, spinal cord preparations are fixed overnight, cryoprotected in 30% sucrose and cut with a sliding microtome (30 μm sections). The sections are then stained with different techniques using toluidine blue to reveal general tissue morphology, fluorescent DAPI to reveal cell nuclei and their pyknosis, immunofluorescence to specifically recognize neuronal or glial cells, or to identify molecules specific for different cell death pathways.

hematoxylin-eosin (50), methylene blue (51), or toluidine blue (52) can be used. The latter is the method employed in our laboratory (see examples in Fig. 3c, d, top for DAPI and bottom for toluidine blue).

After completing the electrophysiological recording one day following the experimental lesion or in sham conditions, neonatal rat spinal cords are fixed in 4% paraformaldehyde for 12–24 h followed by 24 h cryoprotection in 30% sucrose. Figure 7 (left branch) shows a flowchart to indicate the various steps for histological processing. In particular, 30 μm sections are cut with a sliding microtome and stored in phosphate buffer saline (PBS) until use. Our experience shows that fixed spinal cords should not be stored for more than 4 months in order to preserve the anatomical structure. When required, sections are stained in 0.015% solution of toluidine blue for 3 min at room temperature followed by ethanol/xylene cleansing. This staining can supply a good description of the damaged spinal cord region, but it cannot identify (and reliably quantify) dead cells (Figs. 3c, d bottom, 6e, left).

3.7. Quantification of Dead Cells

Electrophysiological experiments of SCI preparations generate a large number of specimens that need reliable assessment of the lesion extent, and the number and type of damaged or lost cells. Hence, it is essential to develop histological protocols that allow the detection and quantification of dead or damaged spinal cord cells in large-scale protocols. For this purpose, there are several techniques suitable for cell culture models, such as flow cytometry fluorescence-activated cell sorting (FACS) based on the propidium iodide (PI) labeling of the dead cells (53): as PI binds to DNA (and becomes fluorescent) only when it can enter membrane-lesioned cells, it is included in immunofluorescence-staining protocols to identify dead cells. Another flow cytometric method is based on annexin V binding, which detects relocation of membrane phosphatidyl serine from the intracellular surface to the extracellular surface (54). Nonetheless, such techniques are not suitable for fixed tissue and do not provide information about the laminar and regional distribution of damage.

Hence, to quantify dead cells on a large-scale manner in spinal cord specimens (in which diverse cell death molecular pathways might have been operating), we have used the simplest benchmark of a dying cell in our experimental conditions, namely the phenomenon of pyknosis. This is a readily observable change in nuclear morphology resulting from chromatin condensation and can take two forms: either nucleolytic (with DNA fragmentation typical of apoptosis) or anucleolytic (without DNA fragmentation; (55)). To observe this change in nuclear morphology, we use DAPI nuclear staining followed by dedicated analysis with a special software ("CellCounter"; GlanceVision Technologies, Trieste, Italy) to identify and count pyknotic nuclei in different regions of the spinal cord.

For DAPI staining, the rat neonatal spinal cord sections (fixed and cut as explained above for the toluidine blue staining procedure) are treated with 1 μg/ml DAPI solution for 20 min to visualize cell nuclei. After brief washing in PBS, the sections are mounted on gelatinized glasses using Vectashield mounting media (Vector Laboratories). One example is shown in Fig. 3c, d (top) in which the pyknosis elicited is observed 24 h after kainate application in comparison with a sham preparation. Likewise, extensive pyknosis is found in the lateral white matter region following application of PM (Fig. 4d, right panel), while the gray matter region is little affected (middle panel of Fig. 4d).

For each experimental group, we analyze three to ten spinal cords. For each spinal cord, three to six different spinal cord thin (30 μm) sections from thoracic (Th) 12 to L1 segments are examined. For each section, four different parts of the spinal cord are investigated: dorsal gray matter (Rexed laminae I–IV), central gray matter (Rexed laminae V–VIII and X), ventral gray matter (Rexed lamina IX), and white matter of the lateral column (11). For each part of the spinal cord, three to seven fields of 280 × 280 μm (gray matter) or 100 × 280 μm (white matter) are counted for pyknotic nuclei. The average values (expressed as percentage of the total number of DAPI-positive cells) of pyknotic nuclei in different parts of the spinal cord are then compared between different experimental groups. In control conditions (24 h), there is no evidence for extensive cell death (in support of the electrophysiological responses recorded under similar conditions) as exemplified in Figs. 3c and 4d (left).

3.8. Immuno-fluorescence Staining

To identify the different neuronal and glial cell types in injured and control neonatal rat spinal cords, we use a range of commercially available antibodies (see Table 1) against distinct cell targets. In particular, neuronal nuclei are labeled with an antibody against NeuN which corresponds to Fox-3 (56) (Fig. 6e, second panels from left). For staining spinal motoneurons (57–59), we employ an antibody against SMI-32, a marker of nonphosphorylated neurofilaments of such cells. An antibody against GFAP is used to recognize glial fibrillary acidic protein, that is an intermediate filament protein characteristic of white matter astrocytes (60) (Fig. 6e, third from left). Furthermore, astrocytes are also recognized by an antibody against S100, a protein belonging to the family of Ca^{2+}-binding proteins present in the cytoplasm and nuclei of these cells (61, 62). An antibody against O4 is employed to recognize a sulfated glycolipid antigen typical of early oligodendroglia (63). Developing and adult oligodendrocyte precursor cells are recognized with an antibody against NG2, a chondroitin sulfate proteoglycan mainly expressed on the surface of these cells (64). An antibody against Iba1 is a standard biomarker for a sulfated glycolipid antigen of microglia (65–67). The myelin sheath of oligodendrocytes (68–70) is labeled with an antibody against RIP (Fig. 6e, rightmost panels).

Immunohistochemical labeling is performed by using a free-floating method with spinal cord sections processed as previously described for the toluidine blue staining procedure (Sect. 3.6). The sections are incubated in 5% NGS/5% BSA/0.3% Triton for 1 h at room temperature, followed by overnight incubation at 4°C in 1% NGS/1% BSA/0.1% Triton X-100 solution containing selected primary antibodies (Table 1). The primary antibodies are visualized using secondary fluorescent anti-mouse Alexa Fluor 488 or 594 antibodies (Table 1).

Quantification of immunofluorescence intensity (gray level intensity expressed in arbitrary units, AU) is performed with MetaVue imaging software (Molecular Devices, Sunnyvale, CA, USA) using the densitometry function to calculate mean signal intensity for areas of interest (640 × 480 μm) reactive to specific antibodies. The values are averaged from at least three different fields in each one of three different sections from three different spinal cords.

3.9. Counting Neurons and Motoneurons

Once a cell becomes pyknotic, it is difficult to identify its type because at this stage a variety of proteins have already been lost. Thus, we calculate the percentage values of dead neurons and motoneurons by comparing the number of NeuN or SMI32 positive cells left in lesioned spinal cords versus those detected in sham preparations run alongside the lesioned ones. A similar strategy is used to quantify the number of dead glial cells.

In accordance with this approach, cells are counted in spinal segments exposed to toxic medium as well as in the same segments of sham preparations. For each experimental group, five different sections from at least three spinal cords are analyzed. The NeuN-positive cells (neurons) are counted in a 300 × 280 μm area of the dorsal part of the gray matter (Rexed laminae I–IV), in a 400 × 350 μm area of the central part of the gray matter (Rexed laminae V–VIII and X), and in a 350 × 250 μm area of the ventral part of the gray matter (Rexed lamina IX). Motoneurons are counted in the ventral horn area.

3.10. Quantification of Microglial Cells

Iba1-positive cells are counted in 30-μm-thick sections of lumbar segments (L2–L4) of lesioned spinal cords or sham preparations. We classify activated microglia on the basis of specific cell morphology (71, 72) that comprises round-shaped inclusions around the pyknotic nucleus (labeled with DAPI) and amoeboid cell shape with few short processes. Following the application of in vitro lesion, microglia can actually surround neurons ("engulfing microglia") observed in each focal plane where neurons are present: this phenomenon is quantified as percentage of total Iba1-positive cells. For this purpose, we use at least six sections from sham preparations and lesioned spinal cords viewed with a Zeiss Plan-NEOFLUAR objective (×20) with a numerical aperture of 0.5.

3.11. Detection of Cell Death Type

The mechanisms responsible for cell death in the central nervous system are complex and involve distinct intracellular signaling pathways (73). Figure 7 (right branch) shows the flowchart used to study cell death mechanisms.

The earliest process is often believed to be necrosis that is readily identified by complete nucleus fragmentation and loss of DAPI staining. With the protocols of lesion described in Sect. 3.4, we have not detected significant necrosis. Conversely, pyknotic cells may have been killed by (or may be dying of) programmed cell death mechanisms like apoptosis which is identified, as a routine, by the marker active caspase-3 (73). Apoptosis can also occur during normal development of the gray and white matter of the spinal cord (74). We have found it to be the prevalent death process in the spinal cord white matter after hypoxic–dysmetabolic insult (24). Immunofluorescence staining for caspase 3 is performed with a selective antibody (Table 1).

Parthanatos is a recently discovered cell death pathway (75, 76) that involves hyperactivity of poly(ADP-ribose) polymerase-1 (PARP-1) and has been considered important to spinal cord injury (77–79). We have found parthanatos strongly responsible for neuronal death in the rat spinal cord in vitro after kainate-mediated excitotoxicity (25), using standard immunostaining protocol and mouse monoclonal antibody against poly ADP ribose (PAR) (Table 1; see refs. (80, 81).

In all cases, the spinal cord tissue is fixed and cut as described in Sect. 3.4 and processed for immunocytochemistry as per Sect. 3.8 followed by DAPI staining, mounted using Vectastain mounting medium (Vector Laboratories, Burlingame, CA, USA) and analyzed with a Zeiss Axioskop2 microscope and MetaVue software (see Sect. 3.8).

4. Conclusions

The isolated spinal cord preparation is a useful in vitro tool to investigate the functional and structural changes induced by an acute lesion to locomotor networks and to screen potential neuroprotective drugs. Although not tested in our laboratory, a similar procedure may be applied, in principle, to the neonatal mouse spinal cord. This approach would open up further possibilities of studying how murine genetic models (82, 83) react to spinal injury, thus expanding our understanding of this important pathological process.

Acknowledgements

This study was supported by grants from the government of the Friuli Venezia Giulia Region and the Italian Ministry for Education and Research (MIUR) under their PRIN program. M.M. is a research biologist of the Local Health Authority (ASS4 MedioFriuli). G.T. gratefully acknowledges the financial support by Vertical Foundation (Rome), Hypobank (Udine), Messina Line (Genova) and the FAIP (Federazione delle Associazioni Italiane Para-Tetraplegici).

References

1. Schwab JM, Brechtel K, Mueller CA, Failli V, Kaps HP, Tuli SK et al (2006) Experimental strategies to promote spinal cord regeneration–an integrative perspective. Prog Neurobiol 78:91–116
2. Mitchell CS, Lee RH (2008) Pathology dynamics predict spinal cord injury therapeutic success. J Neurotrauma 25:1483–1497
3. Lu J, Ashwell KW, Waite P (2000) Advances in secondary spinal cord injury: role of apoptosis. Spine 25:1859–1866
4. Park E, Velumian AA, Fehlings MG (2004) The role of excitotoxicity in secondary mechanisms of spinal cord injury: a review with an emphasis on the implications for white matter degeneration. J Neurotrauma 21:754–774
5. Allen A (1911) Surgery of experimental lesion of spinal cord equivalent to crush injury of fracture dislocation of spinal column: a preliminary report. JAMA 37:878–880
6. Onifer SM, Rabchevsky AG, Scheff SW (2007) Rat models of traumatic spinal cord injury to assess motor recovery. ILAR J 48:385–395
7. Abu-Rub M, McMahon S, Zeugolis DI, Windebank A, Pandit A (2010) Spinal cord injury in vitro: modelling axon growth inhibition. Drug Discov Today 15:436–443
8. Mazzone GL, Margaryan G, Kuzhandaivel A, Ebrahimi Nasrabady S, Mladinic M, Nistri A (2010) Kainate-induced delayed onset of excitotoxicity with functional loss unrelated to the extent of neuronal damage in the in vitro spinal cord. Neuroscience 168:451–462
9. Nicholls J, Saunders N (1996) Regeneration of immature mammalian spinal cord after injury. Trends Neurosci 19:229–234
10. Mladinic M, Muller KJ, Nicholls JG (2009) Central nervous system regeneration: from leech to opossum. J Physiol 587:2775–2782
11. Taccola G, Margaryan G, Mladinic M, Nistri A (2008) Kainate and metabolic perturbation mimicking spinal injury differentially contribute to early damage of locomotor networks in the in vitro neonatal rat spinal cord. Neuroscience 155:538–555
12. Taccola G, Mladinic M, Nistri A (2010) Dynamics of early locomotor network dysfunction following a focal lesion in an in vitro model of spinal injury. Eur J Neurosci 31:60–78
13. Kiehn O (2006) Locomotor circuits in the mammalian spinal cord. Annu Rev Neurosci 29:279–306
14. Arasaki K, Kudo N, Nakanishi T (1984) Firing of spinal motoneurones due to electrical interaction in the rat: an in vitro study. Exp Brain Res 54:437–445
15. Smith JC, Feldman JL, Schmidt BJ (1988) Neural mechanisms generating locomotion studied in mammalian brain stem-spinal cord in vitro. FASEB J 2:2283–2288
16. Clarac F, Pearlstein E, Pflieger JF, Vinay L (2004) The in vitro neonatal rat spinal cord preparation: a new insight into mammalian locomotor mechanisms. J Comp Physiol A Neuroethol Sens Neural Behav Physiol 190: 343–357
17. Bracci E, Ballerini L, Nistri A (1996) Spontaneous rhythmic bursts induced by pharmacological block of inhibition in lumbar motoneurons of the neonatal rat spinal cord. J Neurophysiol 75:640–647
18. Bracci E, Ballerini L, Nistri A (1996) Localization of rhythmogenic networks responsible for spontaneous bursts induced by strychnine and bicuculline in the rat isolated spinal cord. J Neurosci 16:7063–7076
19. Taccola G, Nistri A (2007) Differential modulation by tetraethylammonium of the processes underlying network bursting in the neonatal rat spinal cord in vitro. Neuroscience 146:1906–1917
20. Vitale MG, Goss JM, Matsumoto H, Roye DP Jr (2006) Epidemiology of pediatric spinal cord injury in the United States: years 1997 and 2000. J Pediatr Orthop 26: 745–749
21. Achildi O, Betz RR, Grewal H (2007) Lapbelt injuries and the seatbelt syndrome in pediatric spinal cord injury. J Spinal Cord Med 30 (Suppl 1):S21–S24
22. Margaryan G, Mladinic M, Mattioli C, Nistri A (2009) Extracellular Mg^{2+} enhances the damage to locomotor networks produced by metabolic perturbation mimicking spinal injury in the neonatal rat spinal cord in vitro. Neuroscience 163:669–682
23. Margaryan G, Mattioli C, Mladinic M, Nistri A (2010) Neuroprotection of locomotor networks after experimental injury to the neonatal rat spinal cord in vitro. Neuroscience 165:996–1010
24. Kuzhandaivel A, Margaryan G, Nistri A, Mladinic M (2010) Extensive occurrence of glial apoptosis develops early after hypoxic-dysmetabolic insult to the neonatal rat spinal cord in vitro. Neuroscience 169:325–338
25. Kuzhandaivel A, Nistri A, Mladinic M (2010) Kainate-mediated excitotoxicity induces neuronal death in the rat spinal cord in vitro via a PARP-1 dependent cell death pathway (parthanatos). Cell Mol Neurobiol 30: 1001–1012

26. Taccola G, Nistri A (2006) Fictive locomotor patterns generated by tetraethylammonium application to the neonatal rat spinal cord in vitro. Neuroscience 137:659–670
27. Taccola G, Nistri A (2006) Oscillatory circuits underlying locomotor networks in the rat spinal cord. Crit Rev Neurobiol 18:25–36
28. Marchetti C, Beato M, Nistri A (2001) Alternating rhythmic activity induced by dorsal root stimulation in the neonatal rat spinal cord in vitro. J Physiol 530:105–112
29. Kiehn O, Kjaerulff O (1996) Spatiotemporal characteristics of 5-HT and dopamine-induced rhythmic hindlimb activity in the in vitro neonatal rat. J Neurophysiol 75:1472–1482
30. Butt SJ, Harris-Warrick RM, Kiehn O (2002) Firing properties of identified interneuron populations in the mammalian hindlimb central pattern generator. J Neurosci 22:9961–9971
31. Bracci E, Beato M, Nistri A (1997) Afferent inputs modulate the activity of a rhythmic burst generator in the rat disinhibited spinal cord in vitro. J Neurophysiol 77:3157–3167
32. Bracken MB, Shepard MJ, Collins WF, Holford TR, Young W, Baskin DS et al (1990) A randomized, controlled trial of methylprednisolone or naloxone in the treatment of acute spinal-cord injury. Results of the Second National Acute Spinal Cord Injury Study. N Engl J Med 322:1405–1411
33. Pointillart V, Petitjean ME, Wiart L, Vital JM, Lassié P, Thicoipé M et al (2000) Pharmacological therapy of spinal cord injury during the acute phase. Spinal Cord 38:71–76
34. Hall ED, Springer JE (2004) Neuroprotection and acute spinal cord injury: a reappraisal. NeuroRx 1:80–100
35. Bernhard M, Gries A, Kremer P, Böttiger BW (2005) Spinal cord injury (SCI) prehospital management. Resuscitation 66:127–139
36. Liu D, Xu GY, Pan E, McAdoo DJ (1999) Neurotoxicity of glutamate at the concentration released upon spinal cord injury. Neuroscience 93:1383–1389
37. Alici K, Gloveli T, Weber-Luxenburger G, Motine V, Heinemann U (1996) Comparison of effects induced by toxic applications of kainate and glutamate by glucose deprivation on area CA1 of rat hippocampal slices. Brain Res 738:109–120
38. Hampson DR, Manalo JL (1998) The activation of glutamate receptors by kainic acid and domoic acid. Nat Toxins 6:153–158
39. Wang Q, Yu S, Simonyi A, Sun GY, Sun AY (2005) Kainic acid-mediated excitotoxicity as a model for neurodegeneration. Mol Neurobiol 31:3–16
40. Magnuson DS, Trinder TC, Zhang YP, Burke D, Morassutti DJ, Shields CB (1999) Comparing deficits following excitotoxic and contusion injuries in the thoracic and lumbar spinal cord of the adult rat. Exp Neurol 156:191–204
41. Hadi B, Zhang YP, Burke DA, Shields CB, Magnuson DS (2000) Lasting paraplegia caused by loss of lumbar spinal cord interneurons in rats: no direct correlation with motor neuron loss. J Neurosurg 93:266–275
42. Sun H, Kawahara Y, Ito K, Kanazawa I, Kwak S (2006) Slow and selective death of spinal motor neurons in vivo by intrathecal infusion of kainic acid: implications for AMPA receptor-mediated excitotoxicity in ALS. J Neurochem 98:782–791
43. Carlin KP, Brownstone RM (2006) Rapid pH and PO2 changes in the tissue recording chamber during stoppage of a gas-equilibrated perfusate: effects on calcium currents in ventral horn neurons. Eur J Neurosci 24:1353–1358
44. Dumont RJ, Okonkwo DO, Verma S, Hurlbert RJ, Boulos PT, Ellegala DB et al (2001) Acute spinal cord injury, part I: pathophysiologic mechanisms. Clin Neuropharmacol 24:254–264
45. Norenberg MD, Smith J, Marcillo A (2004) The pathology of human spinal cord injury: defining the problems. J Neurotrauma 21:429–440
46. Matsushima T, Grillner S (1992) Neural mechanisms of intersegmental coordination in lamprey: local excitability changes modify the phase coupling along the spinal cord. J Neurophysiol 67:373–388
47. Cowley KC, Schmidt BJ (1997) Regional distribution of the locomotor pattern-generating network in the neonatal rat spinal cord. J Neurophysiol 77:247–259
48. Beato M, Nistri A (1999) Interaction between disinhibited bursting and fictive locomotor patterns in the rat isolated spinal cord. J Neurophysiol 82:2029–2038
49. Cazalets JR, Bertrand S (2000) Coupling between lumbar and sacral motor networks in the neonatal rat spinal cord. Eur J Neurosci 12:2993–3002
50. Liu F, You SW, Yao LP, Liu HL, Jiao XY, Shi M et al (2006) Secondary degeneration reduced by inosine after spinal cord injury in rats. Spinal Cord 44:421–426
51. Ek CJ, Habgood MD, Callaway JK, Dennis R, Dziegielewska KM, Johansson PA et al (2010)

Spatio-temporal progression of grey and white matter damage following contusion injury in rat spinal cord. PLoS One 5:e12021
52. Tehranipour M, Moghimi A (2010) Neuroprotective effects of testosterone on regenerating spinal cord motoneurons in rats. J Mot Behav 42:151–155
53. Riccardi C, Nicoletti I (2006) Analysis of apoptosis by propidium iodide staining and flow cytometry. Nat Protoc 1:1458–1461
54. Koopman G, Reutelingsperger CP, Kuijten GA, Keehnen RM, Pals ST, van Oers MH (1994) Annexin V for flow cytometric detection of phosphatidylserine expression on B cells undergoing apoptosis. Blood 84:1415–4120
55. Burgoyne LA (1999) Mechanisms of pyknosis: hypercondensation and death. Exp Cell Res 248:214–222
56. Kim KK, Adelstein RS, Kawamoto S (2009) Identification of neuronal nuclei (NeuN) as Fox-3, a new member of the Fox-1 gene family of splicing factors. J Biol Chem 284: 31052–31061
57. Gotow T, Tanaka J (1994) Phosphorylation of neurofilament H subunit as related to arrangement of neurofilaments. J Neurosci Res 37: 691–713
58. Carriedo SG, Yin HZ, Lamberta R, Weiss JH (1995) In vitro kainate injury to large SMI-32+ spinal neurons is Ca^{2+} dependent. Neuroreport 6:945–948
59. Carriedo SG, Yin HZ, Weiss JH (1996) Motor neurons are selectively vulnerable to AMPA/kainate receptor-mediated injury in vitro. J Neurosci 16:4069–4079
60. Eng LF (1985) Glial fibrillary acidic protein (GFAP): the major protein of glial intermediate filaments in differentiated astrocytes. J Neuroimmunol 8:203–214
61. Donato R (2003) Intracellular and extracellular roles of S100 proteins. Microsc Res Tech 60:540–551
62. Schmidt-Kastner R, Szymas J (1990) Immunohistochemistry of glial fibrillary acidic protein, vimentin and S-100 protein for study of astrocytes in hippocampus of rat. J Chem Neuroanat 3:179–192
63. Sommer I, Schachner M (1981) Monoclonal antibodies (O1 to O4) to oligodendrocyte cell surfaces: an immunocytological study in the central nervous system. Dev Biol 83:311–327
64. Nishiyama A, Lin XH, Giese N, Heldin CH, Stallcup WB (1996) Colocalization of NG2 proteoglycan and PDGFa receptor on O2A progenitor cells in the developing rat brain. J Neurosci Res 43:299–314
65. Sasaki Y, Ohsawa K, Kanazawa H, Kohsaka S, Imai Y (2001) Iba1 is an actin-cross-linking protein in macrophages/microglia. Biochem Biophys Res Commun 286:292–297
66. Ohsawa K, Imai Y, Sasaki Y, Kohsaka S (2004) Microglia/macrophage-specific protein Iba1 binds to fimbrin and enhances its actin-bundling activity. J Neurochem 88:844–856
67. Ahmed Z, Shaw G, Sharma VP, Yang C, McGowan E, Dickson DW (2007) Actin-binding proteins coronin-1a and IBA-1 are effective microglial markers for immunohistochemistry. J Histochem Cytochem 55:687–700
68. Friedman B, Hockfield S, Black J, Woodruff K, Waxman S (1989) In situ demonstration of mature oligodendrocytes and their processes: an immunocytochemical study with a new monoclonal antibody, Rip. Glia 2:380–390
69. Berger T, Frotscher M (1994) Distribution and morphological characteristics of oligodendrocytes in the rat hippocampus in situ and in vitro: an immunocytochemical study with the monoclonal Rip antibody. J Neurocytol 23:61–74
70. Watanabe M, Sakurai Y, Ichinose T, Aikawa Y, Kotani M, Itoh K (2006) Monoclonal antibody Rip specifically recognizes 2',3'-cyclic nucleotide 3'-phosphodiesterase in oligodendrocytes. J Neurosci Res 84:525–533
71. Ayoub AE, Salm AK (2003) Increased morphological diversity of microglia in the activated hypothalamic supraoptic nucleus. J Neurosci 23:7759–7766
72. Wake H, Moorhouse AJ, Jinno S, Kohsaka S, Nabekura J (2009) Resting microglia directly monitor the functional state of synapses in vivo and determine the fate of ischemic terminals. J Neurosci 29:3974–3980
73. Kroemer G, Galluzzi L, Vandenabeele P, Abrams J, Alnemri ES, Baehrecke EH et al (2009) Classification of cell death: recommendations of the Nomenclature Committee on Cell Death. Cell Death Differ 16:3–11
74. De Louw AJ, Van De Berg WD, De Vente J, Blanco CE, Gavilanes AW, Steinbusch HP et al (2002) Developmental apoptosis in the spinal cord white matter in neonatal rats. Glia 37:89–91
75. Andrabi SA, Kim NS, Yu SW, Wang H, Koh DW, Sasaki M, Klaus A et al (2006) Poly(ADP-ribose) (PAR) polymer is a death signal. Proc Natl Acad Sci U S A 103:18308–18313
76. David KK, Andrabi SA, Dawson TM, Dawson VL (2009) Parthanatos, a messenger of death. Front Biosci 14:1116–1128
77. Scott GS, Jakeman LB, Stokes BT, Szabo C (1999) Peroxynitrite production and activation

of poly (adenosine diphosphate-ribose) synthetase in spinal cord injury. Ann Neurol 45:120–124
78. Genovese T, Mazzon E, Muia C, Patel NS, Threadgill MD, Bramanti P et al (2005) Inhibitors of poly(ADP-ribose) polymerase modulate signal transduction pathways and secondary damage in experimental spinal cord trauma. J Pharmacol Exp Ther 312:449–457
79. Wu KL, Hsu C, Chan JY (2009) Nitric oxide and superoxide anion differentially activate poly(ADP-ribose) polymerase-1 and Bax to induce nuclear translocation of apoptosis-inducing factor and mitochondrial release of cytochrome c after spinal cord injury. J Neurotrauma 26:965–977
80. Kawamitsu H, Hoshino H, Okada H, Miwa M, Momoi H, Sugimura T (1984) Monoclonal antibodies to poly(adenosine diphosphate ribose) recognize different structures. Biochemistry 23:3771–3777
81. Love S, Barber R, Wilcock GK (1999) Increased poly(ADP-ribosyl)ation of nuclear proteins in Alzheimer's disease. Brain 122:247–253
82. Kiehn O, Dougherty KJ, Hägglund M, Borgius L, Talpalar A, Restrepo CE (2010) Probing spinal circuits controlling walking in mammals. Biochem Biophys Res Commun 396:11–18
83. Grossmann KS, Giraudin A, Britz O, Zhang J, Goulding M (2010) Genetic dissection of rhythmic motor networks in mice. Prog Brain Res 187:19–37

Chapter 4

Spinal Cord Injury: The Lamprey Model

Andrew D. McClellan

Abstract

Following severe spinal cord injury (SCI), brain locomotor command systems are disconnected from spinal motor networks, resulting in paralysis caudal to the lesion. In higher vertebrates, such as birds and mammals, the CNS is a restrictive environment for axonal regeneration. As a result, axons of descending brain neurons, including reticulospinal (RS) neurons, are unable to grow through a spinal injury site, and there is virtually no behavioral recovery. In contrast, the CNS of lampreys and certain other lower vertebrates is permissive for axonal growth. Thus, following SCI, descending brain neurons regenerate their axons and reconnect with spinal targets caudal to a healed spinal lesion site, resulting in recovery of locomotion and other behavioral functions in a few weeks. The lamprey has a number of very powerful advantages that make it ideally suited for investigating the mechanisms that support axonal regeneration and result in behavioral recovery. The present review will focus on the benefits and unique findings from SCI studies using the lamprey, particularly with regard to recovery of locomotor behavior. Potentially, the lamprey can provide a set of cellular conditions that are required for substantial axonal regeneration and recovery of behavioral functions. These conditions, if created in mammals following SCI, presumably would support axonal regeneration and result in behavioral recovery. Finally, even though the lamprey CNS is permissive for axonal outgrowth following SCI, neural regeneration has some limitations in this animal that reveal potential problems which might occur in the future if the mammalian CNS can be modified to be substantially permissive for axonal regeneration, perhaps closer to that in the lamprey. A better understanding of the mechanisms that support, as well as limit, axonal regeneration in the lamprey could provide insights into methods for improving recovery of function in spinal cord-injured higher vertebrates, including perhaps humans.

Key words: Spinal cord regeneration, Locomotion, Reticulospinal neurons, Axon guidance, Synaptic specificity, Growth cones

Abbreviations

ADP	Afterdepolarization
Alexa 488	Alexa 488 dextran amine
ALV	Anterolateral vagal (cell group)
ARRN	Anterior rhombencephalic reticular nucleus
BL	Body length
cAMP	Cyclic AMP
ChABC	Chondroitinase ABC

Håkan Aldskogius (ed.), *Animal Models of Spinal Cord Repair*, Neuromethods, vol. 76,
DOI 10.1007/978-1-62703-197-4_4, © Springer Science+Business Media, LLC 2013

CNS	Central nervous system
CPG	Central pattern generator
CSPG	Chondroitin sulfate proteoglycan
DCC	Discontinuous current clamp
Di	Diencephalic (cell group)
DiI	1,1′-Dioctadecyl-3,3,3′,3′-tetramethylindocarbocyanine perchlorate
DLV	Dorsolateral vagal (cell group)
DRG	Dorsal root ganglion
EMG	Electromyographic
EPSP	Excitatory postsynaptic potential
fAHP	Fast afterhyperpolarization
FDA	Fluorescein dextran amine
FG	Fluoro-gold
GI	Giant interneuron
HRP	Horseradish peroxidase
HT	Hemi-transection
HVA	High-voltage activated (calcium channel)
LI	Lateral interneuron
MAG	Myelin-associated glycoprotein
MRN	Mesencephalic reticular nucleus
mRNA	Messenger RNA
MRRN	Middle rhombencephalic reticular nucleus
MT	Microtubule
NF	Neurofilament
NOGO	Neurite outgrowth inhibitor
Omgp	Oligodendrocyte myelin glycoprotein
PLV	Posterolateral vagal (cell group)
PRRN	Posterior rhombencephalic reticular nucleus
PS	Propriospinal
PT	Post-transection
RDA	Rhodamine dextran amine
RT-PCR	Real-time polymerase chain reaction
RS	Reticulospinal
sAHP	Slow afterhyperpolarization
SCI	Spinal cord injury
SKKCa	Small conductance calcium-activated potassium
T	Transection
TRDA	Texas red dextran amine

1. Overview

This chapter will review the use of the lamprey, a lower vertebrate, for spinal cord injury (SCI) studies, particularly with regard to recovery of locomotor behavior. Locomotor movements are used by many animals during a variety of behaviors and are critical for survival. Because SCI can severely disrupt locomotion, this motor

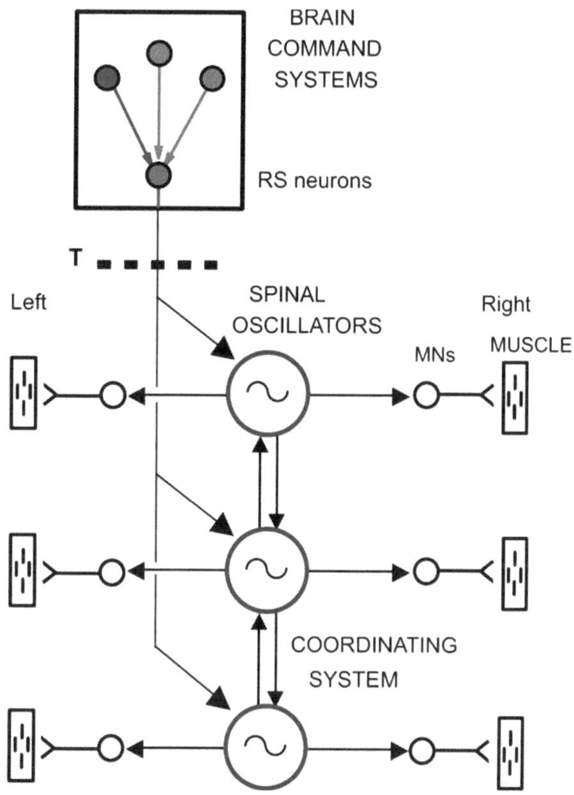

Fig. 1. Model of the locomotor system in the lamprey. Neuronal oscillators are distributed along the spinal cord and are coupled by a coordinating system to form central pattern generator (CPG) networks. The spinal CPG networks produce the basic motor pattern for locomotion such that motoneurons (MNs) activate left and right axial muscles to contract in the proper temporal sequence. Reticulospinal (RS) neurons in a brain command system activate the spinal CPGs and initiate locomotion. Following a rostral spinal transection (T), RS neurons are disconnected from spinal CPGs, resulting in paralysis below the lesion. In the lamprey, the axons of RS neurons regenerate through the healed spinal lesion site and make synapses with spinal neurons, resulting in recovery of locomotor behavior in a few weeks (modified from (12)).

behavior often is used as one of the key outcome measures for testing potential therapies to alleviate the deficits resulting from these injuries.

In the present review, the neural networks that control locomotor function will be divided into three main components (Fig. 1): (a) Neuronal oscillators are distributed along the spinal cord and are coupled by a coordinating system to form central pattern generators (CPGs). These spinal CPGs produce the basic electrical pattern of rhythmic locomotor burst activity that causes muscles to contract in the proper temporal sequence (reviewed in (1–3)). (b) Brain "command" systems initiate locomotor behavior (reviewed in (4)), and reticulospinal (RS) neurons are the output neural elements of the command system that directly activate spinal CPGs. In particular, following severe rostral SCI, disruption of descending axons of RS neurons largely accounts for the loss of voluntary locomotion. (c) Spinal ascending systems transmit locomotor feedback and other important information to the brain to regulate the activity of several descending systems (not shown in Fig. 1; reviewed in (5)).

Severe SCI, such as a substantial spinal contusion or complete spinal transection, disrupts axons from descending brain neurons, including RS neurons, and results in paralysis caudal to the lesion. Also, transmission of ascending sensory information and ANS functions are compromised, but these will not be the focus of the present review. Following a complete rostral SCI, mature "higher" vertebrates, such as birds and mammals, do not behaviorally recover because of little or no axonal regeneration within the CNS (6–11), and for humans, this debilitating condition is a major health problem (6, 9). For mature higher vertebrates after neural injury, the CNS is a restrictive environment for axonal regeneration, mainly because of several proteins associated with myelin (e.g., NOGO, MAG, OMgp) and factors in the glial scar (e.g., CSPGs) that inhibit neural outgrowth (6, 9–11). In addition, neuronal cell death due to secondary injury effects (e.g., hemorrhagic necrosis, inflammation, swelling, release of neurotoxic agents) and delayed demyelination can compromise CNS functions (9), sometimes to a greater extent than the primary effects of the injury.

In contrast, following spinal cord transections in "lower" vertebrates, such as lamprey, fish, and certain amphibians, there is robust axonal regeneration through the healed spinal lesion site and, remarkably, virtually complete behavioral recovery in a few weeks (reviewed in (12–14)). For example, after SCI in these lower vertebrates, injured axons regenerate through the lesion site, sometimes over relatively long distances (15, 16), usually in the correct directions and often in the correct spinal tracts ((17–19); however see (20)), and form synaptic connections with appropriate spinal targets on the opposite side of the lesion site (21, 22).

2. Why Does the Lamprey Recover Locomotion Following SCI?

Following a complete spinal transection in the lamprey, ependymal cells proliferate and form a bridge between the cut ends of the spinal cord (reviewed in (12, 13)). Transected ascending and descending axons form growth cones and regenerate along the ependymal bridge across the injury site. Regenerating axons form synapses on the other side of a healed spinal lesion site, and behavioral functions, such as locomotion, begin to recover. With increasing recovery times, greater numbers of axons extend across a healed transection site and for increasing distances, and within a few weeks there is virtually complete recovery of behavioral functions. However, the capacity for functional axonal regeneration in the CNS of lampreys and other lower vertebrates probably has not been selected for evolutionarily but appears to be a byproduct of other factors or conditions (see below).

There are a number of possible reasons why functional axonal regeneration occurs in the lamprey and certain other lower vertebrates (reviewed in (12, 13)). First, in lower vertebrates, new neurons are added to the CNS, albeit slowly, well after the main phase of neural development (23, 24) as a result of delayed maturation of neuronal precursors (25) and/or neurogenesis (26) (see (27)). Persistent neurodevelopmental mechanisms allow newly added neurons to send out axons and make synapses with appropriate targets. Thus, the permissive conditions and guidance cues that regulate these functions for newly added neurons probably also support axonal regeneration and synaptic reconnection for injured neurons following SCI (13, 22). However, for the lamprey (28, 29) and several other lower vertebrates ((30); also see (13)), axonal regeneration, and not the addition of new neurons, is the main mechanism for restoration of brain–spinal cord projections and recovery of locomotor function after SCI. Second, in lower vertebrates, myelin either is absent ((31), lamprey) or, if present, appears to lack sufficient quantities of the inhibitory proteins that suppress axonal regeneration in higher vertebrates (32). Third, in lower vertebrates, the spinal cord glial scar is not an impenetrable barrier to axonal regeneration (17), and "secondary injury" effects appear to be minimal or nonexistent. Fourth, lower vertebrates are cold-blooded, and their relatively low metabolism and low oxygen needs make them partially resistant to mild anoxia that might result from disruption of blood vessels following SCI. In mammals, including humans, modest hypothermia has a neuroprotective effect following SCI (33, 34) due, in part, to lowering the metabolism of neural tissue in the vicinity of the injury site.

3. Lamprey Species Used for SCI Research

Lampreys are jawless fish in the class Agnatha (sometimes called cyclostomes), which are thought to have appeared ~390–470 million years ago (35), and are considered a key primordial species in the evolution of vertebrates. All of the ~40 living species of lamprey have an initial larval phase (ammocoete), which lasts several years and is characterized by burrowing in the mud of streams or rivers and filter feeding on microorganisms (35). Subsequently, larval animals metamorphose into adults, and for about half of the lamprey species the adults migrate downstream into lakes or the ocean to become parasitic feeders on fish (35). After a few years and substantial growth, the parasitic adults migrate upstream, spawn, and die.

Larval sea lampreys (*Petromyzon marinus*) have been used for almost all of the lamprey studies on SCI and neural regeneration (13, 14). Relatively large numbers of these animals can be collected easily and then maintained in the laboratory for moderately long recovery times following SCI. Neural regeneration and behavioral recovery also occur following SCI in young adult sea lampreys (36). However, because animals at this life stage are parasitic feeders, maintenance of paralyzed or partially paralyzed spinal cord-injured adult animals in the laboratory for long recovery times can be problematic.

4. Conceptual and Biological Framework for Using the Lamprey

Following severe SCI in mature higher vertebrates, axonal regeneration normally is very limited, and there is little or no behavioral recovery. Thus, at present in these animals it is not possible to study the full complement of mechanisms that might be necessary for substantial axonal outgrowth and restoration of locomotor functions. In contrast, the lamprey and certain other lower vertebrates offer the unique opportunity to identify the sets of conditions that support unaided neural repair following SCI and virtually complete recovery of locomotor behavior. Thus, these lower vertebrates can provide a set of target goals for the types of cellular conditions that are required to promote substantial axonal regeneration and recovery of behavioral functions following SCI. These conditions, if created in mammals following SCI, presumably would support axonal regeneration and result in behavioral recovery. Finally, even though the lamprey CNS is permissive for axonal outgrowth, in this animal neural regeneration following SCI has some limitations (see Sect. 6.2.3). These limitations provide insights into potential problems that might occur in the future if the mammalian CNS can be modified to be substantially permissive for axonal regeneration, perhaps similar to the conditions in the lamprey nervous system.

5. Rationale for Using the Lamprey for SCI Research

5.1. Advantages

The lamprey has a number of unique and powerful advantages that make it ideally suited for determining the mechanisms responsible for neural regeneration following injuries of the spinal cord (reviewed in (12–14)): (a) The lamprey CNS can repair itself after

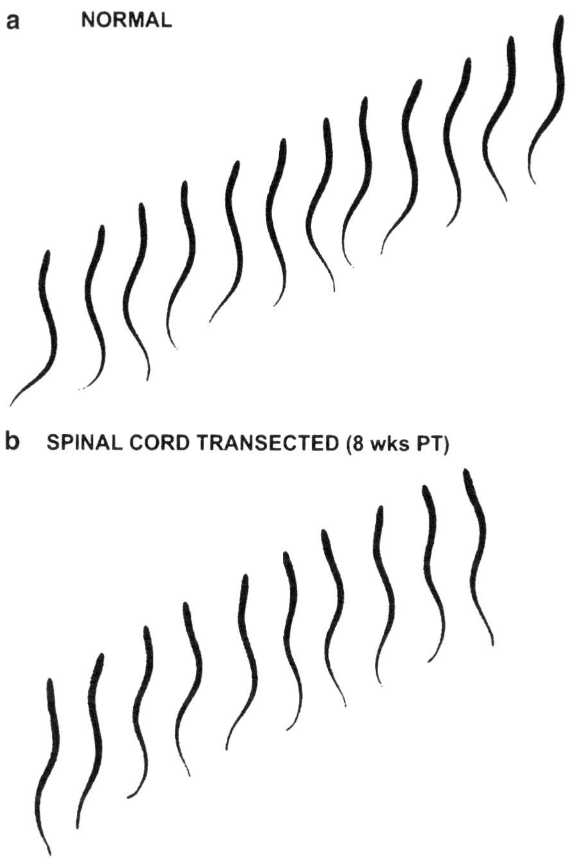

Fig. 2. (**a**) Sequential video images, *left-to-right*, during swimming movements in a normal larval sea lamprey. Swimming movements are characterized by left–right bending at each level of the body, and S-shaped body undulations that propagate toward the tail and propel the animal forward. (**b**) Swimming movements in an animal at 8 weeks following a complete spinal transection at 10% BL. Interframe interval = 133 ms for both sequences.

SCI, and this results in virtually complete behavioral recovery in a few weeks (Fig. 2) ((13); also see (15, 28, 37, 38)). (b) The movements and motor activity during locomotion (swimming) are relatively simple and easy to analyze which allows reliable assessment of locomotor function at various recovery times after SCI (Figs. 2 and 3a). Furthermore, locomotor networks can be studied using in vitro CNS preparations (Fig. 3b) (38–40), which allow control of the ionic and pharmacological makeup of the bath, elimination of mechanosensory inputs, and mechanically stable conditions for intracellular recordings. (c) The lamprey CNS has many basic features in common with the nervous systems of higher vertebrates (Fig. 4a) (41, 42), but because of its primitive nature, it is comparatively simple. As such, in spinal cord-transected lamprey, cellular, synaptic, and integrative properties of brain and spinal cord locomotor networks are more easily analyzed than those in higher vertebrates. Of particular importance, the neural circuits in the

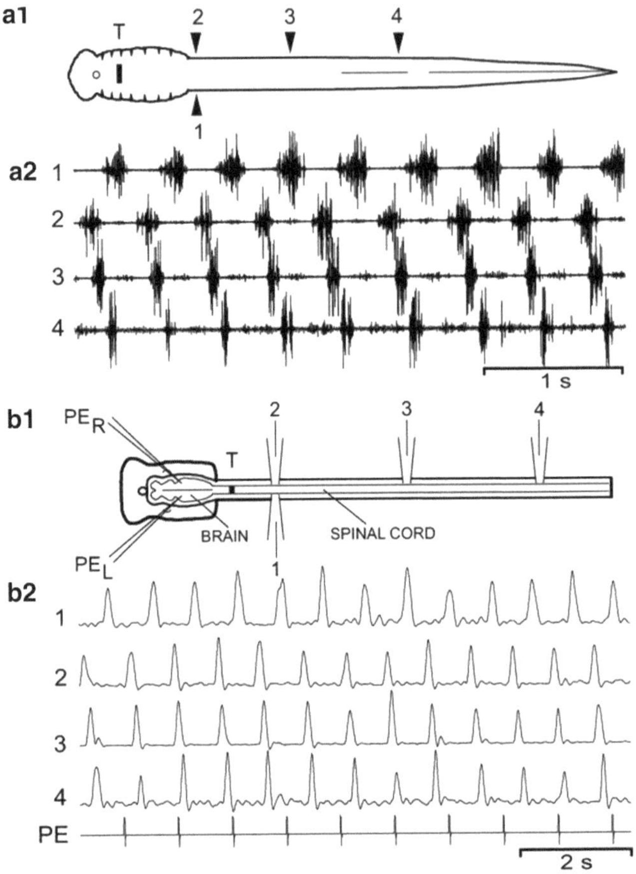

Fig. 3. (**a**) Swimming muscle burst activity in whole animals. (**a1**) Diagram of a larval lamprey that recovered for 8 weeks following a spinal transection (T) at 10% BL. Muscle recording electrodes are located at 20% BL (electrodes 1,2), 40% BL (3), and 60% BL (4). (**a2**) Muscle burst activity during swimming consisting of left–right alternation (1 ↔ 2) and rostrocaudal phase lags (2 → 3 and 3 → 4). (**b**) Ventral root bursts during in vitro swimming activity. (**b1**) Diagram of an in vitro brain–spinal cord preparation from a larval animal that recovered for 32 weeks following a spinal transection (T) at 10% BL. Swimming activity is initiated by pharmacological microstimulation (PE_R, PE_L) in brain locomotor areas and recorded from spinal ventral roots at 20% BL (electrodes 1,2), 40% BL (3), and 60% BL (4). (**b2**) Focal pressure ejection (PE) of excitatory agents in brain locomotor areas initiates locomotor ventral root burst activity (1–4; full-wave rectified and integrated, $t = 50$ ms) consisting of left–right alternation (1 ↔ 2) and rostrocaudal phase lags (2 → 3 and 3 → 4) (modified from (37, 38)).

brain and spinal cord that control locomotor behavior in normal, uninjured lampreys have been intensely studied (43, 44), and there is a wealth of information that can be translated to the study of SCI using this animal. (d) Several classes of large, identifiable neurons in the lamprey brain (45) and spinal cord (44, 46) can be investigated from one animal to the next and are very useful for certain types of studies. For example, the lamprey brain contains ~28 large, *uniquely identified* RS neurons called Müller and Mauthner cells

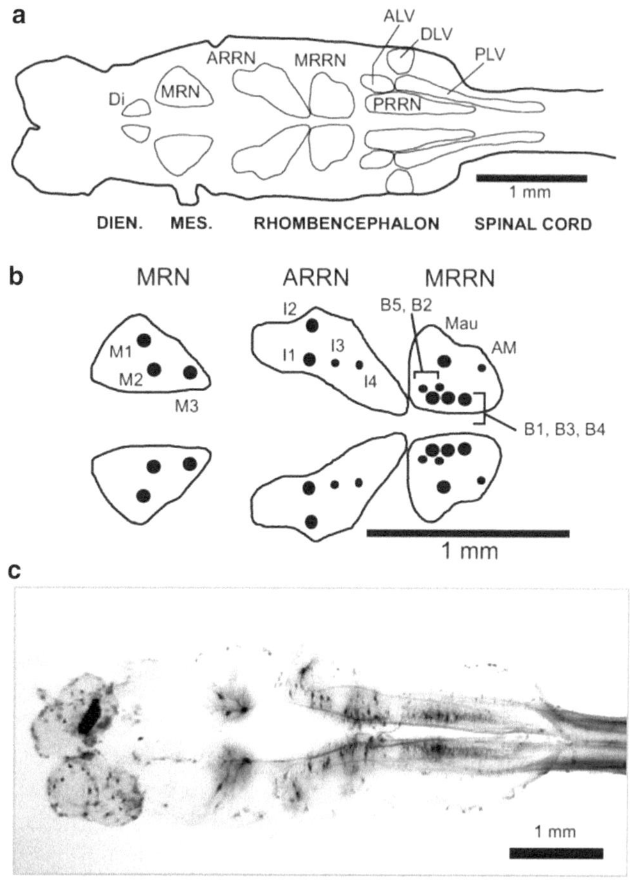

Fig. 4. (**a**) Diagram of a larval lamprey brain (*left*) and rostral spinal cord (*right*) showing contours around the main groups of descending brain neurons (DIEN. = diencephalon, MES. = mesencephalon). Reticular cell groups: MRN—mesencephalic reticular nucleus; ARRN (anterior), MRRN (middle), and PRRN (posterior) rhombencephalic reticular nuclei. Non-reticular cell groups: Di—diencephalic; ALV—anterolateral vagal; DLV—dorsolateral vagal; and PLV—posterolateral vagal. (**b**) Enlargement of right (*top*) and left (*bottom*) reticular nuclei that contain large, uniquely identified RS neurons (Müller cells): M cells (M1–M3) in the MRN, I cells (I1–I4) in the ARRN; and B cells (B1–B5) in the MRRN. Mauthner (Mau) and auxiliary Mauthner (AM) cells are additional identified RS neurons located in the MRRN. Note that large, identified RS neurons in the MRN, ARRN, and MRRN are intermingled with smaller, unidentified RS neurons, which are omitted for simplicity. (**c**) Photomicrograph of a larval lamprey brain showing descending brain neurons in reticular and non-reticular nuclei that were retrogradely labeled by application of HRP to the spinal cord at 20% BL (modified from (15, 16)).

(Fig. 4b) (45). In addition, the brains of larval lampreys also contain over 1,000 smaller, unidentified RS neurons that can be investigated with neurophysiological and anatomical techniques (47, 48). (e) In larval lamprey, the thin CNS (~100–400 μm thick) permits imaging in live or fixed whole-mount tissue (Fig. 4c), which preserves the 3D aspects of the nervous system. (f) The lamprey genome has recently been characterized (http://pre.ensembl.org/

Petromyzon_marinus/index.html), and the lamprey has the potential to become a new model system for molecular neurobiology. (g) Following SCI, axonal regeneration of RS neurons is critical for restoring locomotor function. At present, the lamprey is the *only* animal in which it is feasible to study *all* of the following features of these important neurons: electrophysiological properties (47, 48); synaptic physiology and network function (43, 44, 49); synaptic structure (50, 51); immunohistochemistry (49); gene expression (48, 52); neuroanatomical projections (15, 16); neuron morphology (29); in vivo imaging (53); intracellular calcium dynamics (54); neurite outgrowth in cell culture (55); and neural regeneration following SCI (reviewed in (12, 13)).

5.2. Disadvantages

The larval phase and often the adult phase of lampreys last several years, it is difficult to raise lampreys in the laboratory for long periods of time, and these animals have not been bred in captivity. Therefore, the types of genetic techniques that can routinely be performed with Drosophila, zebrafish, and mice are not available for SCI experiments using lampreys.

Although the permissive conditions for neurite outgrowth in the lamprey CNS are an obvious advantage for certain types of studies (see Sect. 5.1 above), this situation also precludes testing possible manipulations for overcoming the specific mechanisms that inhibit axonal regeneration within the CNS of higher vertebrates. Thus, for SCI studies, the lamprey is mainly useful for determining the sets of conditions that permit functional axonal regeneration and result in behavioral recovery, rather than the mechanisms that hinder neuronal outgrowth in higher vertebrates.

6. Recovery of Locomotor Initiation: Descending Projections

6.1. Lesion Types and Sites, and Survival Times

For lamprey studies, SCIs typically involve complete spinal cord transections to ensure that all descending and ascending axons are severed at the injury site, and that any residual behavioral functions are not due to spared projections. To investigate the restoration of descending brain–spinal cord projections and recovery of initiation of locomotor behavior following SCI, complete transections usually are performed at very rostral spinal levels to disrupt all descending drive from brain locomotor command systems to spinal CPG networks (T in Fig. 1). For example, in studies from the author's laboratory, SCIs in larval sea lamprey (~100 mm typical length) often are performed at 10% body length (BL, normalized distance from anterior tip of head; brain/spinal cord junction ~8% BL), which is located ~2–3 mm below the brain (15, 28, 37, 38). For all surgical procedures, animals are anesthetized in ~100–200 mg/l tricaine methanesulfonate (MS222; Crescent Research Chemicals; Phoenix, AZ USA), which is absorbed via the gills and is a standard anesthetic for fish. The rostral spinal cord is exposed through a ~5–10 mm dorsal incision, and small custom retractor hooks fashioned from suture needles hold the incision open to allow good visibility and free use of the hands. With fine iridectomy scissors, the exposed spinal cord is completely transected, and the completeness of the lesion is confirmed by gently lifting the meninges on one side of the transection to briefly expose the cut, cross-sectional surface of the cord. The incision is manually closed, and animals are returned to their home aquaria to recover for 2–32 weeks at ~23°C. Initially, animals with very rostral spinal transections are paralyzed and unable to swim, and subsequently locomotor behavior gradually recovers as descending axons from brain neurons regenerate through the healed lesion site and reconnect with spinal CPGs (see Sects. 6.2.1, 6.2.2, and 6.2.3). For the lamprey, almost all of the SCI studies described in this review have been carried out at approximately room temperature (~23°C), and at colder temperatures (e.g., 12°C), axonal regeneration is depressed and behavioral recovery is substantially delayed (56).

6.2. Outcome Measures and Key Findings

6.2.1. Kinematic Analysis and Locomotor Muscle Burst Activity in Whole Animals

Following rostral spinal cord transections in larval lampreys, behavioral recovery has been assessed using two main techniques. First, swimming movements, from their earliest appearance following SCI until they approach those of normal animals, have been quantified using kinematic analysis. For example, swimming in normal larval lampreys is characterized by two main features (Fig. 2a) (37): (a) left-right bending of the body at each segmental level; and (b) S-shaped body undulations that propagate toward the tail with increasing lateral amplitude to push against the water and propel the animal forward. Behavioral responses are elicited by brief electrical stimulation of the head or tail (1–10 mA, 2 ms pulses at 100 Hz for 50 ms), and swimming movements are captured with a high-resolution recording system (e.g., 30 frames per second, ~4–8 ms shutter speed). Individual frames are displayed on a computer screen, at least 11 x,y points are digitized along the midline of the body, and these coordinates are used to calculate several features of swimming behavior: cycle times (T) or frequencies ($f = 1/T$) of rhythmic swimming movements; amplitudes of lateral displacement (i.e., left-right bending) versus distance along the body; and mechanical wavelength (λ), which is the length of an S-wave along the body divided by total body length. These parameters for swimming in control lampreys have been compared statistically to those in experimental animals at various recovery times following rostral spinal transections.

Second, for partially recovered spinal cord-injured lampreys, active muscle contractions in the rostral body generate undulatory swimming movements that can passively propagate to the caudal body and produce swimming-like movements, even though muscles in the lower body are inactive (37, 57). Thus, at various recovery times following spinal cord transections it is important to record axial muscle activity (EMGs) during swimming to determine the actual degree of behavioral recovery. For larval lampreys, muscle recording electrodes are fashioned by using a 30 Ga. needle to separately insert two fine wires (56 μm dia.), insulated except at the tip, into musculature at various points along the body (electrodes 1–4; Fig. 3a1). All of the EMG wires are grouped in a bundle and sutured to the body just behind the head, and the wire bundle is led to a light-weight "boom arm" so animals can swim freely in a tank (22 cm × 40 cm with ~2–6 cm of water). Muscle activity during swimming is characterized by two features (Fig. 3a2) (37): (a) left–right alternation of burst activity at each level of the body (electrodes: $1 \leftrightarrow 2$); and (b) a rostrocaudal delay of ipsilateral burst activity ($2 \rightarrow 3$, $3 \rightarrow 4$). During neurophysiology experiments, raw muscle burst activity is amplified (×1,000), filtered (100 Hz to 5 kHz), and stored on tape (Neuro-Data DR890, 11 kHz sampling rate per channel; Cygnus Technologies, Delaware Water Gap, PA USA), and subsequently the data are acquired using a custom eight-channel data acquisition-analysis system (DT-3016 data acquisition board; Data Translation, Marlboro, MA USA). The onset and offset times of locomotor muscle bursts are marked with the data acquisition system, and these values are imported into a spreadsheet to calculate various parameters of swimming muscle burst activity (see (37, 57–59) for definitions): cycle times (T); burst proportions (BP); intersegmental phase lags (Φ_{INT}); and right–left phase values ($\Phi_{RT\text{-}LT}$). These parameters for swimming in control lampreys have been compared to those in experimental animals at different recovery times following rostral spinal cord transections.

Immediately after transection of the rostral spinal cord, lampreys are paralyzed, but at about 2 weeks post-transection (PT), weak undulatory movements and low-amplitude muscle burst activity begin to appear just caudal to the healed spinal lesion site (37). Between 2 and 8 weeks PT, left–right alternating locomotor movements gradually increase in amplitude (see Fig. 2 in (12)) and, remarkably, by 8 weeks PT are not significantly different from those of normal animals (Fig. 2b) (37). At 10–12 weeks PT, mechanical wavelengths (λ) during swimming are very similar to those for normal animals (51). In addition, with increasing recovery times, muscle burst activity occurs at progressively greater distances caudal to the spinal transection site ((37); see Fig. 3 in (12)), so that the general pattern of locomotor activity, with very minor exceptions, returns to normal by 8 weeks PT (Fig. 3a, see "muscle" in Fig. 5d).

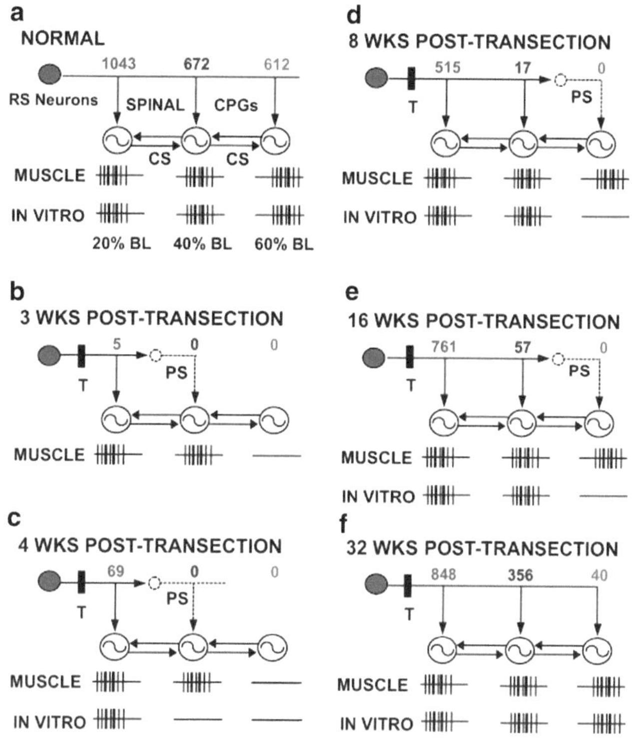

Fig. 5. Summary of locomotor systems in normal lamprey and spinal cord-transected animals. Reticulospinal (RS) neurons in the brain activate spinal oscillators (*circles* with oscillator symbols), which are coupled by a spinal coordinating system (CS). Values above each oscillator are the numbers of RS neurons that project to 20, 40, and 60% BL. "MUSCLE" is locomotor muscle burst activity from whole animals (see Fig. 3a), and "IN VITRO" is locomotor ventral root burst activity from in vitro brain–spinal cord preparations (see Fig. 3b). Healed spinal cord transection sites (T) are at 10% BL. (**a**) In normal animals, RS neurons directly activate the spinal locomotor networks at 20, 40, and 60% BL. (**b**–**f**). Between 3 and 32 weeks PT, locomotor muscle burst activity and ventral root burst activity occur for progressively greater distances along the body below the transection site (T), but with different time courses (see text). Concurrently, RS neurons regenerate their axons for increasingly greater distances below the healed transection site. At early and intermediate recovery times, descending propriospinal (PS) neurons relay descending drive from RS neurons to lower levels of the cord that have not yet received direct input from the brain. At long recovery times, descending PS neurons are in parallel with direct descending pathways from the brain and probably supplement descending drive (not shown; see Fig. 7) (modified from (12)).

6.2.2. Locomotor Activity from In Vitro Brain–Spinal Cord Preparations

In whole animals with rostral SCIs at 10% BL, CNS neural networks as well as peripheral mechanosensory feedback can contribute to the initiation and generation of swimming muscle activity. To determine the contributions of central mechanisms alone, brain-initiated locomotor burst activity has been recorded from in vitro brain–spinal cord preparations (Fig. 3b1), in which sensory feedback is blocked. After removing the skin, musculature, and internal organs along the body, and exposing the dorsal surface of the brain and spinal cord, the preparation is transferred to a recording chamber containing lamprey Ringer's solution (38) maintained at 6–8°C. Also, 15 mg/l D-tubocurarine chloride (Sigma Chemical; St. Louis, MO, USA) is added to the bath to block contractions of remaining body musculature and eliminate movement-related sensory feedback. The chamber is cooled with thermoelectric modules, but refrigerated circulators also are commonly used. With a 0.1 mm dia. stainless steel pin, ventral roots are separated from connective tissue along the lateral edges of the spinal cord. The tips of glass suction electrodes are placed in contact with exposed ventral roots to record locomotor burst activity, which is amplified (×1,000), filtered (10 Hz to 2 kHz), and full-wave rectified/integrated (τ = 50 ms) to better reveal the onsets and offsets of burst activity.

For in vitro brain–spinal cord preparations, spinal locomotor activity is initiated by pharmacological microstimulation in brain locomotor areas (49, 60, 61). Two micropipettes are filled with 5 mM D-glutamate/5 mM D-aspartate in Ringer's solution (pH 7.4), and Fast green is added to visualize the ejection bolus (38, 58, 62). The micropipette tips are inserted into the bath, and while applying pressure ejection pulses (5–10 ms at 1 Hz; 20 psi), the tips are broken off to ~2–5 μm with fine forceps. Typically, each pressure ejection pulse releases a bolus with a diameter of ~25–50 μm, which is equivalent to ~8–65 pl (61). The tips of the micropipettes are then inserted bilaterally ~25–50 μm below the dorsal surface of the brain in locomotor areas, and excitatory agents are pressure ejected into these areas to activate nearby neurons. This stimulation directly or indirectly activates brain locomotor command systems, which project to and excite spinal CPGs to initiate locomotor burst activity (49, 60). At the end of a stimulation sequence (usually ≤1 min), the diameter of the ejection area within the tissue stained with Fast Green is less than ~100–150 μm (width of brain ~1.0–1.5 mm). Following each stimulation sequence a "rest" period of at least 3 min is allowed to reduce fatigue, during which the preparation should be inactive, before stimulation is performed again in the same locomotor area (62). To mark stimulation sites for later determination of their precise locations, the tip of a separate micropipette containing 1–3% Alcian Blue in Ringer's solution is positioned at a stimulation site, and a small amount of the dye is pressure ejected (60, 61). Alcian Blue does not appear to compromise the viability or responsiveness of in vitro preparations during subsequent stimulation trials and remains visible after histological processing.

In vitro brain–spinal cord preparations have been set up from control larval lamprey as well as from experimental animals at 4–32-week recovery times following rostral SCI (38). For control animals, stimulation in brain locomotor areas initiates ventral root burst activity at 20, 40, and 60% BL that is characteristic of locomotor behavior (see "in vitro" in Fig. 5a). For experimental animals (4–32 weeks PT), locomotor burst activity occurs at progressively greater distances caudal to a healed spinal transection site with increasing recovery times ((38); see Fig. 4 in (12)) and gradually returns to normal (Figs. 3b and 5b–f). However, the rate at which this in vitro ventral root activity recovers is much slower than that for EMG activity from spinal cord-transected whole animals. Specifically, for experimental whole animals the pattern of locomotor muscle activity returns to normal by ~8 weeks PT (Figs. 3a and 5d), but for in vitro preparations a normal pattern of locomotor ventral root activity is not present until 24–32 weeks PT (Figs. 3b and 5f). Much of this difference probably is due to the absence of sensory feedback and lower CNS excitability for in vitro brain–spinal cord preparations compared to whole animals. Thus, although the degree of behavioral recovery is dependent to a large extent on axonal regeneration, other mechanisms, such as mechanosensory feedback (see Sect. 9.2.1) and various compensatory mechanisms (see Sect. 6.2.4), also contribute significantly (12, 13, 38, 63). Finally, at 32 weeks PT, regenerated axons from RS neurons can directly activate the spinal CPG networks in the caudal cord under conditions in which conduction via descending propriospinal relay pathways is locally blocked in the rostral and middle spinal cord with a low-calcium Ringer's solution (Fig. 5f; see Fig. 6 in (38))

6.2.3. Axonal Regeneration of Descending Brain Neurons

At different recovery times (3–32 weeks) following spinal transections at 10% BL, the time course of restoration of brain–spinal cord projections has been determined by retrograde labeling of descending brain neurons with horseradish peroxidase (HRP) applied to the spinal cord at 20, 40, or 60% BL. These data have been compared to those for control animals in which tracer is applied to the same levels of the spinal cord. For larval sea lamprey, descending brain neurons are located in eight main cell groups in the brain (Fig. 4a, c), and ~80% of these cells are moderately small, unidentified RS neurons (15). In addition, moderately small RS neurons are both necessary (49) and sufficient (61) for initiating spinal locomotor activity, and these particular neurons are located in four reticular nuclei (Fig. 4a) (41, 42, 45): mesencephalic reticular nucleus (MRN); and the anterior (ARRN), middle (MRRN), and posterior (PRRN) rhombencephalic reticular nuclei.

Through a dorsal incision centered at 20, 40, or 60% BL, the spinal cord is exposed and transected with fine iridectomy scissors. A small piece of Gelfoam (~1 mm^3; Upjohn; Kalamazoo, MI, USA) soaked in a solution of 40% HRP (type VI, Sigma Chemical, St. Louis, MO, USA) and 1% dimethyl sulfoxide (DMSO) in water is applied between the cut ends of the cord. The incision is closed and sealed with a thin line of cyanoacrylate adhesive (Super Glue Gel; Loctite, Rocky Hill, CT, USA), and subsequently animals are returned to their home aquaria. After a ~14-day HRP transport time, brains are removed, and whole-mount tissue is histologically processed in a modified Hanker-Yates solution consisting of 0.10% catechol, 0.05% phenylenediamine, and 0.0375% H_2O_2 in 0.1 M Tris buffer (pH 7.4) for 5–25 min at ~0°C (15, 16, 29). Subsequently, the tissue is dehydrated in an ethanol series, cleared in methyl salicylate, and mounted on slides with Permount (Fisher Scientific, St. Louis, MO, USA). Data analysis is performed with a custom computer-microscope anatomical workstation, which includes stepper motors attached to the *x* and *y* axes of a microscope stage, stepper motor controller board, and joystick, all of which are interfaced with a custom computer program. The joystick controls the movements of the microscope stage, and a crosshair in one of the eyepieces is first moved along the outline of the brain/spinal cord (trace mode) and then aligned with individual labeled neurons (mark mode). The *x,y* coordinates of the tissue outlines and labeled neurons are captured by the computer, which then tabulates the numbers of labeled neurons in different cell groups and also generates a diagram of the brain and neuronal labeling pattern, as if all structures are in the same focal plane. The numbers of labeled descending brain neurons that have extended their axons beyond the lesion site to rostral, middle, or caudal regions of the spinal cord have been compared statistically to the numbers of neurons that project to these same spinal levels for normal animals.

For normal larval lampreys, moderately large numbers of brain neurons project to the spinal cord at 20% BL (~1,250 total descending brain neurons, ~1,043 RS neurons), 40% BL (~900 total neurons, ~672 RS neurons), or 60% BL (~825 total neurons, ~612 RS neurons) (Fig. 5a) (15). Thus, for normal animals, ~60% of both descending brain neurons and RS neurons project for relatively *long distances* to caudal levels of the spinal cord (Fig. 6a). For spinal-transected animals, axons from descending brain neurons, including RS neurons, project for progressively greater distances caudal to a healed spinal transection site with increasing recovery times (Figs. 5b–f and 6b, c) ((15); also see (12)). However, at most recovery times, the distances that RS neurons have regenerated their axons caudal to a spinal lesion site are much less than the projection distances in normal animals. For example, at 8-week recovery times, virtually all of the regenerated axons of RS neurons project just caudal to the lesion site (Fig. 6b). At 32-week recovery times, the numbers of RS neurons that project to the rostral cord are not significantly different from those for normal animals, but projections to the middle and caudal cord are ~50% and ~6%, respectively, of those for normal animals (15). Thus, at this relatively long recovery time, about 60% of descending brain neurons, including RS neurons, that have regenerated their axons do so for relatively *short distances* to rostral levels of the spinal cord (Fig. 6c).

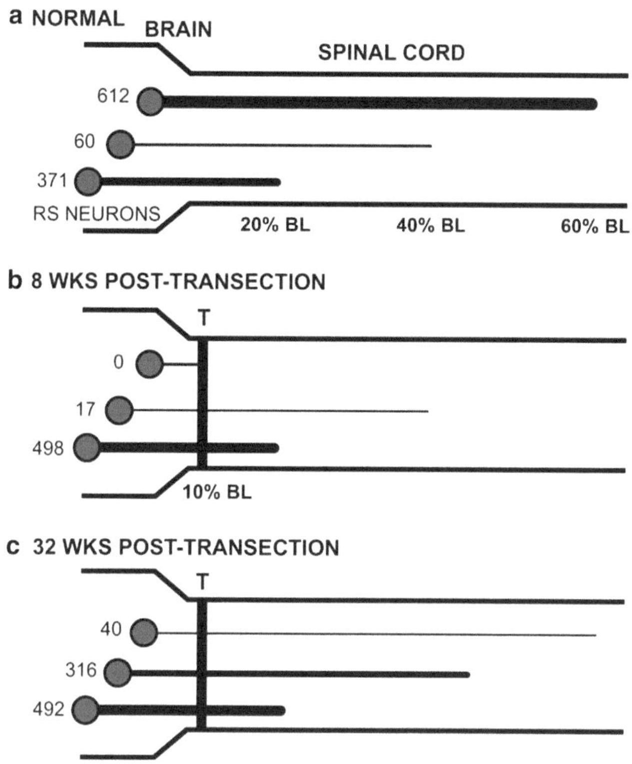

Fig. 6. Summary of projections from RS neuron in normal and spinal cord-transected larval lamprey. The thickness of *horizontal lines* from RS neurons indicates the relative numbers of descending projections. (**a**) In normal animals, about 60% of RS neurons project for relatively long distances to at least the caudal spinal cord (60% BL). (**b**) At 8-week recovery times following spinal transections (T) at 10% BL, ~97% of RS neurons that have regenerated their axons do so for only short distances into the rostral spinal cord (~20% BL). However, at this recovery time in whole animals, locomotor muscle burst activity has returned to normal along the body (see Figs. 3a and 5d). (**c**) At 32-week recovery times, ~60% of RS neurons that regenerate their axons do so for relatively short distances, only as far as the rostral spinal cord (modified from (15, 16)).

Accordingly, even at relatively long recovery times, descending brain neurons display *incomplete axonal regeneration* such that the normal brain–spinal cord projection patterns are not restored (15).

The mechanisms that *limit* neural outgrowth and that are responsible for incomplete axonal regeneration within a permissive CNS very likely are substantially different from those that *inhibit* axonal regeneration in mammals. Specifically, for the lamprey and other lower vertebrates, many regenerating axons of descending brain neurons make synapses a short distance caudal to a healed spinal lesion site, and these synaptic connections are thought to suppress or limit additional axonal extension (12, 13, 64), a phenomenon sometimes referred to as "synaptic capture" (30). For example, in the lamprey, axonal regeneration has been compared following rostral (10% BL) and caudal (50% BL) spinal cord transections (64). Rostral spinal transections, which eliminate virtually all synapses that descending brain neurons make in the spinal cord, are a much stronger stimulus for axonal regeneration than caudal transections, which preserve many of these synapses above the lesion site. We hypothesize that the synapses descending brain neurons make, either above or below a healed spinal lesion site, regulate the degree to which these neurons are stimulated to regenerate (64).

Although the projections of descending brain neurons and their connection patterns with spinal neurons below the transection site are not restored to normal even at 32 weeks after SCI, there still is virtually complete recovery of locomotor behavior in about 8 weeks (reviewed in (10, 11)). Thus, minimal or moderate axonal regeneration can result in substantial locomotor recovery, provided that certain *compensatory mechanisms* are available (see Sects. 6.2.4, 9.2.1, and 11). This is potentially promising information regarding human SCI because it suggests that induction of robust and/or long-distance axonal regeneration might not be necessary for mediating considerable behavioral recovery. A much more modest degree of axonal outgrowth could be sufficient, and this would place less demands on potential therapeutic interventions.

6.2.4. Compensatory Mechanisms: Descending Propriospinal Relay Neurons

Propriospinal (PS) neurons serve a number of functions within locomotor systems (reviewed in (65): (a) ascending and descending PS neurons couple spinal oscillators to coordinate locomotor activity at different levels of the body (Fig. 1; see Sect. 9.2) (3, 66); (b) ascending PS neurons transmit timing and other important feedback information to the brain to modulate the activity of neurons in several descending systems (see Sect. 10) (5); and (c) descending PS neurons relay activity from descending brain neurons to more caudal levels of the spinal cord (Fig. 7a) (63, 67). In normal larval lamprey, the distributions of descending PS relay neurons in the rostral and middle spinal cord with short-to-moderate length axons have been determined by application of HRP to the spinal cord at 40% BL or 60% BL (63). For experimental animals with rostral spinal transections at 10% BL, HRP has been applied to the same spinal levels following recovery times of 4–32 weeks.

In normal lampreys, descending PS neurons are distributed along much of the spinal cord (63), and the majority of these neurons have relatively short axons (<10–15 mm), although a few neurons have moderate-to-long axons (~25–50 mm). Following rostral SCI, it is very likely that some of these neurons relay descending drive from RS neurons to lower levels of the spinal cord (Figs. 5 and 7b). For example, at early recovery times, PS neurons can *substitute* for the absence of direct projections from RS neurons to middle and caudal levels of the spinal cord (Fig. 5b–e). At longer recovery times, PS neurons are in parallel with and probably *supplement* the effects of the relatively few direct descending

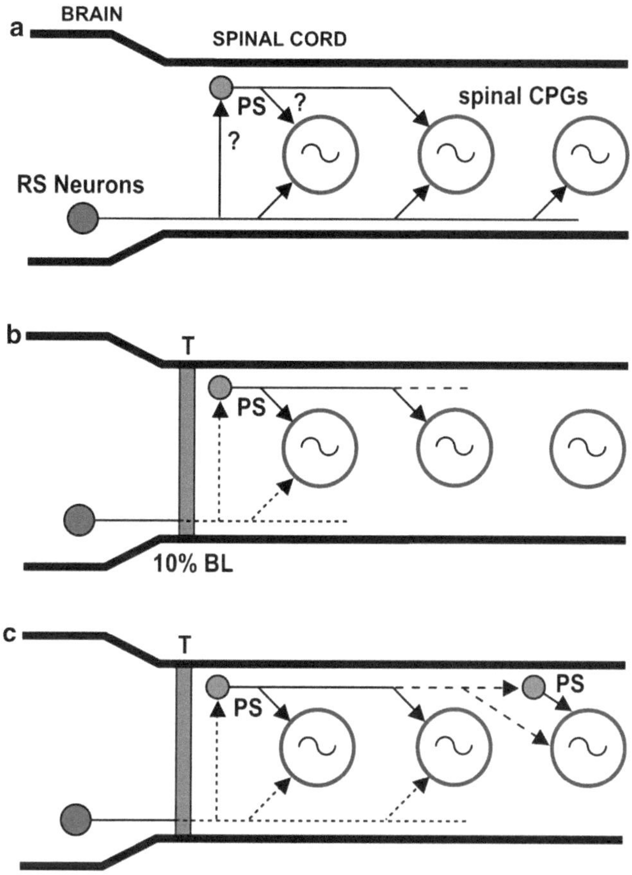

Fig. 7. Model of possible roles of descending propriospinal (PS) relay neurons. (**a**) In normal larval lamprey, RS neurons directly activate spinal CPG networks (*circles* with oscillator symbols) all along the spinal cord. Note that spinal coordinating systems (see Fig. 1) have been omitted for simplicity. Descending PS neurons relay descending drive from RS neurons and contribute to the activation of spinal CPGs. (**b**) At early recovery times (3–4 weeks PT) following spinal transections at 10% BL (T), RS neurons have regenerated their axons for relatively short distances and can only directly activate spinal CPGs in the rostral spinal cord. Some PS neurons relay descending drive from brain command systems to CPG networks in the middle spinal cord (see Fig. 5b, c). (**c**) At intermediate-to-long recovery times (8–16 weeks PT), a few RS neurons directly activate CPGs in the rostral and middle spinal cord (see Fig. 5d, e). Descending PS neurons have sprouted their axons and can directly and/or indirectly relay descending drive to CPGs in the middle and caudal spinal cord (modified from (63)).

projections from RS neurons to these spinal levels (Fig. 7b) (see Fig. 7 in (38); reviewed in (12, 13)). Although neuronal outgrowth of RS neurons contributes substantially to behavioral recovery following SCI, descending PS neurons represent a very important compensatory mechanism that can make up for incomplete axonal regeneration of these descending brain neurons. Interestingly, studies with neonatal rats indicate that when long, direct brain–spinal pathways are disrupted, polysynaptic pathways via descending PS neurons can relay descending drive from the brain to caudal CPG networks and initiate locomotor activity (68). Thus, descending PS neurons represent a potentially important target for therapeutic interventions following SCI (65, 67).

For larval lamprey following spinal transection at 10% BL, significantly greater numbers of *uninjured* descending PS neurons *below* the lesion site project to more caudal spinal levels than in control animals (63). This *plasticity* very likely is due to descending PS neurons that have sprouted and extended their axons to more caudal levels of the spinal cord (Fig. 7b), possibly because of being disconnected from brain centers and/or factors released from other spinal targets below lesion sites (see (69)). Plasticity of descending PS neurons below a spinal injury site would be expected to enhance the compensatory functions of these neurons.

6.2.5. Axonal Regeneration Versus Neural Developmental Mechanisms

In principle, following SCI the restoration of descending projections of RS neurons might be due to *axonal regeneration*, which requires that the axons of these neurons (a) are injured and (b) subsequently extend across the healed spinal lesion site. Alternatively, some axonal growth across a spinal lesion site might be due to *developmental mechanisms* involving uninjured RS neurons, either ones with relatively short axons that were not lesioned by SCI or new neurons that have been added to the nervous system. At present, axonal regeneration appears to be one of the most attractive potential mechanisms for repair following mammalian SCI, and therefore, if the lamprey is to serve as a model system, it is important to demonstrate that axonal elongation is largely due to this mechanism and not development.

In the lamprey brain, Müller and Mauthner cells are large, *uniquely identifiable* RS neurons that project for virtually the entire length of the spinal cord (42, 45). These large RS neurons, which are intermingled with smaller, unidentified RS neurons, are located in three reticular nuclei (Fig. 4b): M cells (M1–M3) in the MRN; I cells (I1–I4) in the ARRN; B cells (B1–B5) in the MRRN; and Mauthner (Mau) and auxiliary Mauthner (AM) cells in the MRRN. In the lamprey, the functions of Müller and Mauthner cells are unclear (reviewed in (70)). Müller cells are rhythmically active during locomotor behavior and can influence burst activity in spinal CPGs, but they are neither necessary nor sufficient for swimming behavior (70, 71), unlike smaller RS neurons (49, 61) (see Sect. 6.2.3). However, Müller cells might contribute to certain aspects of locomotion, such as rapid speed changes, fast directional maneuvers, or rapid postural adjustments (70). In lampreys, stimulation of Mauthner cells elicits weak flexure movements of the body but not the typical C-start escape responses that are initiated by these neurons in fish (reviewed in (45)).

Following spinal cord transections at 10% BL in larval lampreys, HRP retrograde labeling indicates that descending axons of

Müller and Mauthner cells project for progressively greater distances caudal to a healed spinal transection site with increasing recovery times (3–32 weeks) (16). Some of these identified RS neurons can extend their axons for as much as 57 mm below a spinal lesion site (16). Large, identified RS neurons normally have long axons that are injured by rostral spinal transections and subsequently extend across the lesion site, thus satisfying the two criteria for axonal regeneration.

The lamprey brain also contains many moderately small, *unidentified* descending brain neurons, ~80% of which are RS neurons (15). As stated earlier, small, unidentified RS neurons are both necessary (49) and sufficient (61) for initiating spinal locomotor activity. For these neurons, double labeling with fluorescent tracers has been used to distinguish between axonal regeneration and developmental mechanisms following SCI. The use of rhodamine dextran amine (RDA, 10 kDA; Invitrogen, Carlsbad, CA USA) and fluorescein dextran amine (FDA, 3 kDa; Invitrogen) can sometimes result in "bleed through" labeling of descending brain neurons (72), in which one tracer is visible with the optical filter for the second tracer. Obviously, "bleed through" severely compromises the accuracy of double-labeling results. Texas red dextran amine (TRDA, 10 kDa; Invitrogen) and Fluoro-Gold (FG; Fluorochrome, Englewood, CO, USA) do not display "bleed through" labeling (72), although a special application procedure is required to prevent FG from entering the circulation and labeling non-neuronal cells on the dorsal surface of the brain that obscure some of the RS neurons of interest.

For control animals, sequential application of 4% FG in Ringer's solution with 2% DMSO to the spinal cord at 40% BL, followed 2 weeks later by application of 6.25% TRDA in 0.1 M Tris buffer to the cord at 20% BL double labels 91–100% of descending brain neurons, including unidentified RS neurons (28). For experimental animals, FG is applied to the spinal cord at 40% BL, and after allowing 2 weeks for retrograde transport, the spinal cord is completely transected at 10% BL (T) (Fig. 8a). Following recovery times of 8 or 16 weeks, TRDA is applied to the spinal cord at 20% BL, and the resultant percent double labeling of descending brain neurons is ~50% or ~70%, respectively (Fig. 8b). Similar results are obtained for unidentified RS neurons, which account for most of the descending brain neurons (15). Because of technical limitations of double-labeling protocols, the above percentages are underestimates of the descending brain neurons that regenerate their axons (28). For larval lamprey, these results demonstrate that many descending brain neurons, including unidentified RS neurons that initiate locomotion (49, 61), regenerate their axons after SCI (28). Thus, axonal regeneration accounts for the majority of outgrowth of RS neurons following SCI, and very likely contributes substantially to recovery of locomotor function.

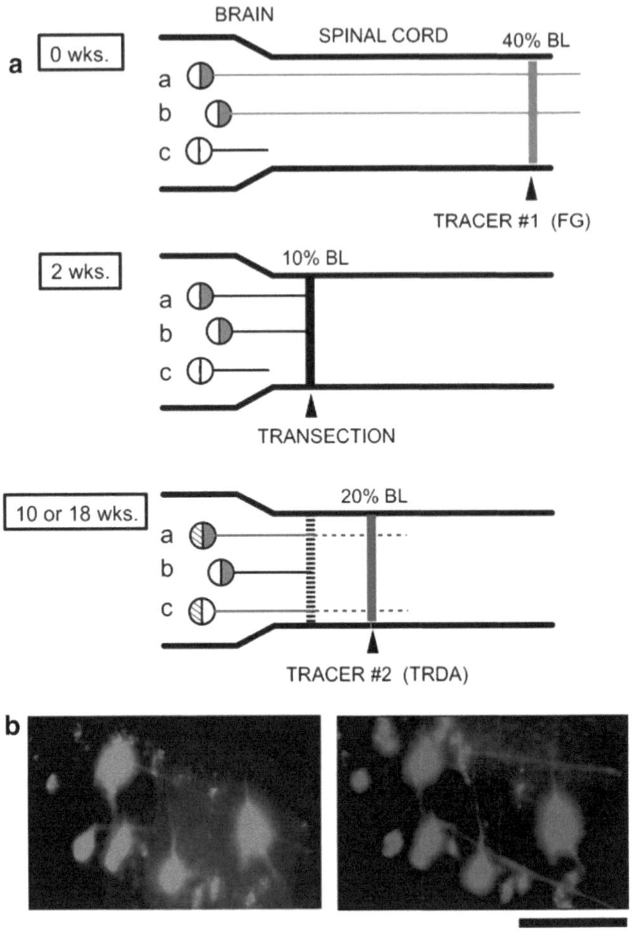

Fig. 8. Protocol for double labeling of lamprey descending brain neurons, including RS neurons. (**a**) Fluorgold (FG, blue-green) is applied to the spinal cord at 40% BL to retrogradely label descending brain neurons. Two weeks later, the spinal cord is transected (T) at 10% BL to lesion the axons of neurons that are labeled with FG. At 8- or 16-week recovery times following spinal transections, Texas red dextran amine (TRDA, red) is applied to the spinal cord at 20% BL. Thus, neurons labeled with both FG (*solid*) and TRDA (*striped*) are those that are injured by rostral spinal transections and subsequently extend their axons across a healed transection site, thus satisfying the two criteria for axonal regeneration (see text). (**b**) Representative examples showing double labeling of RS neurons (left = FG; right = TRDA) in the right ARRN (see Fig. 4a) at a recovery time of 16 weeks (rostral is left in both panels). Scale bar = 100 μm (modified from (28)).

In a different set of experiments, retrograde HRP labeling of descending brain neurons has been performed with "young" (~80 mm average length) and "old" (~125 mm) larval lampreys that differ in age by at least 1 year (29). For "old" larval lampreys, significantly more descending brain neurons, including RS neurons, project to 10 or 20% BL of the spinal cord than for "young" animals, suggesting that new brain–spinal cord projections are added with age. In addition, following recovery from spinal cord transections at 15 or 25% BL, the numbers of descending brain neurons projecting to 10 or 20% BL, respectively (i.e., rostral to the transection sites), are not significantly different than those for normal, unlesioned animals. Thus, the rate of addition of new descending projections is not enhanced significantly by spinal cord transections. In the lamprey, the addition of new projections does not appear to be due to neurogenesis (27, 73), as is the case in some other lower vertebrates (26). The major conclusion from this study is that for larval lampreys, new brain–spinal cord projections are added with age, probably because of the addition of new descending brain neurons that extend their axons into the spinal cord. Importantly, the addition of new brain–spinal cord projections with age occurs at a rate that is at least 7–15 times *slower* than the rate injured descending brain neurons regenerate their axons across a healed lesion site following SCI. Together, these results (29) and those from the double-labeling study described above (28) suggest that for spinal cord-transected larval lamprey, axonal regeneration of RS neurons, and not the slow addition of new brain–spinal cord projections (i.e., developmental mechanisms), accounts for the vast majority of growth of descending axons across a healed spinal transection site. The reconnection of regenerating axons of RS neurons with spinal CPGs contributes substantially to recovery of locomotor function, although other mechanisms are also important (see Sects. 6.2.4, 9.2.1, and 11).

6.2.6. Different Regenerative Capacities of Large, Identified RS Neurons

Although the lamprey CNS is permissive for neural growth, different identified RS neurons regenerate their axons for greatly different distances below a healed spinal lesion site. For example, following rostral SCI certain Müller and Mauthner cells have high capacities for axonal regeneration and extend their axons for moderately long distances (M1, I3, I4, B5, and AM; Fig. 4b), while others have low regenerative capacities (M2, M3, I1, B1, and Mau) (16). Thus, as a general rule, with some exceptions, the largest identified RS neurons tend to have lower regenerative capacities than smaller identified RS neurons and much lower capacities than small, unidentified RS neurons (15, 16). Because these large, identified RS neurons regenerate their axons within a permissive CNS, the differences in regenerative capacity very likely are controlled by intrinsic differences in cell programming. Different regenerative capacities are also observed for different reticular and non-reticular cell groups as a whole that contain unidentified descending brain neurons (15). Thus, a permissive CNS environment does not necessarily lead to successful axonal regeneration, and individual neuron properties also are very important factors.

During the first 4 weeks of recovery following spinal transections at ~15% BL, large, identified RS neurons display a significant reduction of neurofilament (NF) mRNA (52). At ~4–7 weeks, NF expression begins to increase and return toward normal levels for large, identified RS neurons with high capacities for axonal regeneration, but remains depressed for neurons with low capacities. Thus, the restoration of normal NF expression is thought to be part of an intrinsic growth program that is triggered in identified RS neurons with a high capacity for axonal regeneration (52). However, the *purpose* of this intrinsic growth program is unclear because at 4–7 weeks PT, many of these neurons have already extended their axons several millimeters beyond the spinal transection site (16). In addition, even if regeneration is blocked, recovery of NF expression still occurs in identified RS neurons that have high regenerative capabilities.

6.2.7. Experimental Modulation of Regenerative Capacities of RS Neurons

Spinal cord "conditioning lesions" (i.e., transections) have been made at 30 or 50% BL to test the effects on subsequent axonal regeneration of RS neurons (74). After a "lesion delay time" of 2 weeks, a more rostral spinal "test lesion" (i.e., transection) is performed at 10% BL, and then after a "recovery time" of 4 weeks, HRP is applied to the spinal cord at 20% BL to determine if axonal regeneration of descending brain neurons is enhanced.

For conditioning lesions at 30% BL, axonal regeneration of lamprey descending brain neurons is significantly increased compared to spinal-transected animals without conditioning lesions or animals with conditioning lesions at 50% BL (74). Experiments using spinal hemi-transections as conditioning lesions indicate that axotomy rather than diffusible factors released at the injury site is primarily involved in enhancement of axonal regeneration. Results from this study suggest that conditioning lesions "prime" descending brain neurons, including RS neurons, and enhance subsequent axonal regeneration by reducing the initial delay and/or increasing the initial rate of axonal outgrowth. Understanding the cellular and molecular mechanisms for this modulation of axonal regeneration in the lamprey may suggest methods for improving neuronal outgrowth in higher vertebrates in which regeneration is very limited.

7. Properties of Injured Lamprey RS Neurons During Regeneration

7.1. Lesion Types and Sites, and Survival Times

Müller cells in the lamprey brain are large, identified RS neurons with long *ipsilateral* descending axons (42, 45) that are located in three reticular nuclei: MRN; ARRN; and MRRN (Fig. 4b). In Sect. 7.2 below, several aspects of axonal regeneration of injured Müller cells are described: electrical properties; gene expression for ion channels; axon guidance; synaptic specificity; and cell death. For most of these sections, complete spinal transections have been performed. For two of the sections (7.2.1 and 7.2.2), right hemi-transections

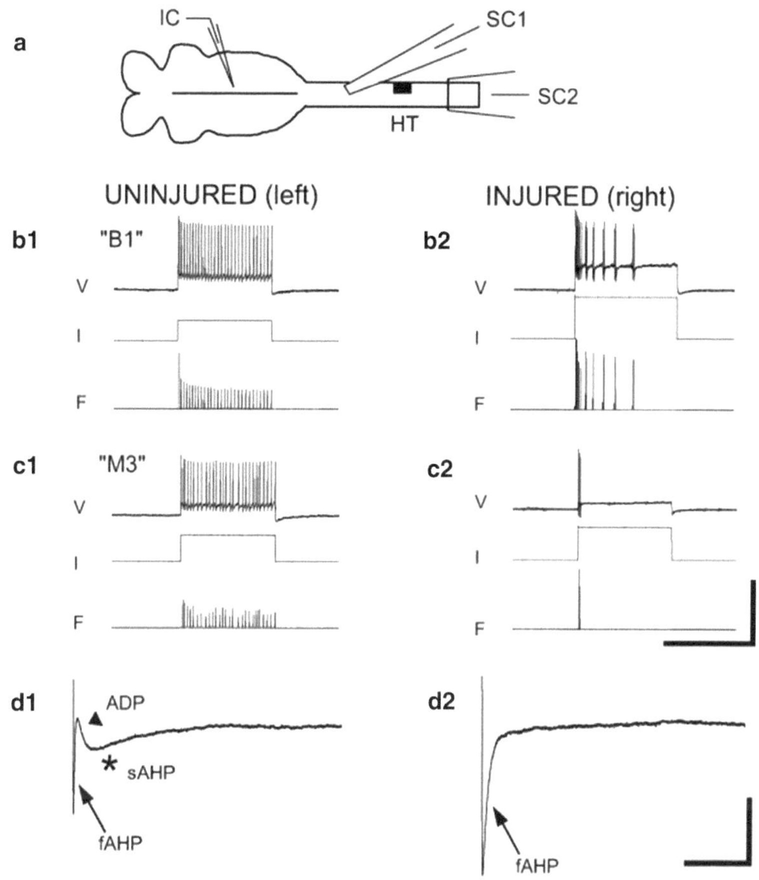

Fig. 9. Spinal cord injury alters the firing patterns and action potential components of injured lamprey RS neurons. (**a**) Isolated brain–spinal cord preparation, showing the brain (*left*) and rostral spinal cord, right spinal cord hemi-transection (HT) at 10% BL (~2–3 weeks post-lesion), intracellular recording micropipette (IC), suction electrode on the right, dorsal surface of the spinal cord above the HT (SC1), and suction electrode around the caudal spinal cord just below the HT (SC2). (**b** and **c**) Membrane potential (*V*), current (*I*), and instantaneous firing frequency (*F*). (b1, c1) Uninjured "B1" and "M3" cells (left side of brain; see Fig. 4b) fire a smooth, continuous train of action potentials in response to depolarizing current pulses, while (b2, c2) the same injured neurons in the same preparation (right side of brain) fire multiple short bursts or a single short burst. (**d**) (d1) Action potentials from an uninjured "M2" cell (Fig. 4b) are followed by afterpotentials with three components: fAHP; ADP; and sAHP (see text). (d2) In an injured "M2" cell (same brain as d1), the ADP and sAHP are absent, and the fAHP is increased, perhaps because of the absence of the ADP. Scale bars: (**b** and **c**) 100 mV, 14 nA, 40 Hz/2 s; (**d**) 4 mV/100 ms (modified from (48)).

(HTs; Fig. 9a) of the spinal cord have been made at 10% BL to injure (axotomize) Müller cells on the right side of the brain, while leaving left cells uninjured (48). Because Müller cells occur in pairs (Fig. 4b), the properties of the same injured and uninjured neurons can be compared in each animal.

7.2. Outcome Measures and Key Findings

7.2.1. Electrical Properties of Uninjured and Injured RS Neurons

Following right spinal HTs, intracellular recordings have been made from pairs of uninjured (left side of brain) and injured (right side) Müller cells (Fig. 4b) (48). At short (2–3 weeks) or long (12–17 weeks) recovery times, animals are anesthetized, and from a ventral approach, the brain and rostral spinal cord are exposed, removed, and transferred to a recording chamber containing lamprey Ringer's solution (Fig. 9a). Suction electrodes are placed rostral to (SC1) and caudal to (SC2) the HTs, and action potentials are elicited in Müller cells to confirm that left cells are uninjured (orthodromic responses in SC2) and right cells are injured (responses in SC1 only). Intracellular experiments are performed on an isolation table to minimize vibrations and to be able to "hold" (record from) neurons for relatively long periods of time. Thin-walled borosilicate glass tubing (1.2 mm O.D.; WPI, Sarasota, FL USA) is pulled on a Brown-Flaming puller (Sutter Instruments; Novato CA USA), and micropipettes are filled with 5 M potassium acetate (KAc; ~60–80 MΩ). The stems of micropipettes are inserted into an electrode holder (MEH1S; WPI), which contains 4 M KCl and which is plugged into an intracellular amplifier probe box (Axoclamp 2A; Axon Instruments, Union City, CA, USA). The probe box is mounted on a sturdy, custom manipulator with a magnetic base that is constructed from optical bench components (Newport; Irvine, CA, USA). A dc motor controls the *z*-axis of the manipulator and is connected to a custom electronic motor drive device that allows rapid advancement or withdrawal of the micropipette (<0.5 μm min. increments). Using conventional current clamp (i.e., "Bridge" mode), short depolarizing current pulses (10 nA, 1–10 ms) are used to elicit single action potentials in RS neurons. With discontinuous current clamp (DCC mode, ~6 kHz sampling rate), longer depolarizing current pulses (0–10 nA, 2 s) are applied to examine firing patterns. The biophysical properties of uninjured RS neurons (left side of brain) have been compared to those of injured neurons (right side) in the same brains following right, rostral spinal HTs.

First, at short recovery times (2–3 weeks), large, uninjured RS neurons (Müller cells) fire a smooth, continuous train of action potentials in response to an applied depolarizing current pulse, while injured neurons fire a single short burst or short repetitive bursts (Fig. 9b, c) (48). Preliminary results indicate that in response to the same amplitude current waveforms (i.e., depolarizing pulses, sine waves), the majority of injured Müller cells fire at *lower* average and peak frequencies (i.e., *reduced electrical excitability*) than uninjured neurons, because these injured neurons typically have higher voltage thresholds, lower input resistances, and display resonance (75, 76). Second, for uninjured RS neurons, the afterpotentials that follow the main depolarizing phase of action potentials have three components (Fig. 9d1) (48): fast afterhyperpolarization (fAHP), which is due to fast potassium channels; afterdepolarization

(ADP), which is not abolished by calcium channel blockers and might be due to persistent sodium channels; and slow AHP (sAHP), which is due to calcium influx via high-voltage-activated (HVA; N- and P/Q-type) calcium channels and calcium-activated potassium channels (SKKCa). For injured RS neurons, the ADP and sAHP are absent or significantly reduced, and the fAHP is significantly larger than for uninjured neurons, perhaps because of a reduced ADP (Fig. 9d2). The substantial reduction of the sAHP and the molecular data described below (see Sect. 7.2.2) suggest that both HVA calcium and SKKCa channels are reduced in injured RS neurons. Third, at relatively long recovery times (12–17 weeks), >85% of injured Müller cells that have regenerated their axons through the spinal lesion site display firing patterns and afterpotentials that are similar to those of uninjured neurons (48). Fourth, small, unidentified injured RS neurons in the MRRN and PRRN (Fig. 4a) display changes in firing patterns and action potential components that are similar to those of injured Müller cells (see Sect. 7.2.2). Fifth, preliminary results suggest that when axonal regeneration and synapse formation with spinal targets are blocked, injured RS neurons usually do not recover normal firing patterns and afterpotentials. Fifth, for *uninjured* RS neurons, blocking calcium channels and the sAHP with 2 mM $CoCl_2$ results in firing patterns that mimic those of injured neurons (48). Thus, for injured lamprey RS neurons, a decrease in calcium channels appears to contribute to the altered firing patterns.

Interestingly, for lamprey RS neurons in culture, manipulations that induce calcium influx inhibit neurite outgrowth or cause growth cone retraction (see Sect. 8.2.2) (55). These cell culture results together with the neurophysiological data above suggest that for injured RS neurons following SCI, decreases in calcium channels and electrical excitability will reduce calcium influx. A reduction in calcium influx would be expected to maintain relatively low intracellular calcium levels and/or limit increases in intracellular calcium in RS neurons, conditions that appear to be important for axonal regeneration (see Sect. 8.2.2) (48, 55).

7.2.2. Gene Expression for Ion Channels of Uninjured and Injured RS Neurons

Using RT-PCR with a combination of both degenerate and specific oligonucleotide primers, partial open reading frames for alpha subunits have been cloned and sequenced for two ion channels from the sea lamprey: HVA calcium channels; and SKKCa channels (48). Spinal hemi-transections (HT) are performed at 10% BL, and animals recover for 1 week or 11–17 weeks. Following recovery, brains are removed, and dorsoventral cuts are made to excise the MRRN and PRRN on the left and right sides of the brains (Fig. 4a, c). It should be noted that ~85–95% of the RS neurons in the MRRN and PRRN have ipsilateral descending axons (70). Thus, most of the RS neurons in the right MRRN and PRRN are injured, while most neurons in left nuclei remain uninjured. Each individual brain region is placed in lysis buffer (RLT-Qiagen, Valencia, CA), total RNA is isolated, reverse transcription is performed, and relative amounts of cDNA are determined using quantitative PCR with validated forward and reverse primers. Fold differences in mRNAs for HVA calcium or SKKCa channels between right and left nuclei of each animal are calculated and averaged.

At 1 week following right spinal cord HTs, mRNA levels for HVA calcium and SKKCa channels in injured RS neurons in the right MRRN and PRRN are significantly reduced by about 50% relative to uninjured RS neurons in left nuclei (48). At 11–17-week recovery times, mRNA levels for these two channels are not significantly different in right and left reticular nuclei. Thus, at early recovery times, the temporary decreases in HVA calcium and SKKCa channel mRNAs for injured RS neurons appear to account for the significant reduction of the sAHP observed for these neurons (Fig. 9d; see Sect. 7.2.1).

7.2.3. Axon Guidance and Tract Selection

Following rostral spinal transections at 20% BL, the orientations and directions of regenerating axons of large, identified RS neurons have been examined anatomically (18). Just rostral to healed spinal lesion sites, Müller or Mauthner axons are impaled with HRP-filled micropipettes (4–8% HRP in 0.2 M KAc; 40–100 MΩ), and HRP is injected iontophoretically (5–25 nA, 200 ms pulses at 3.3 Hz) for 5–25 min. At early recovery times (<30 days) before large, identified RS neurons have regenerated their axons substantially beyond the transection site, the majority of these processes are oriented in the proper direction, similar to their trajectories at longer recovery times (>45 days) (18). Thus, these regenerating axons are directed from the outset, and the final projection patterns apparently are not due to initial sprouting and subsequent retraction of incorrectly oriented neurites. The data suggest that the direction of growth of regenerating descending axons of RS neurons is regulated by guidance cues that persist long after the main phase of neural development.

At 8 weeks after spinal cord transections at 15% BL, when lampreys have recovered substantial locomotor function, partial spinal cord lesions have been performed with iridectomy scissors and fine forceps to spare the lateral spinal tracts above and below the healed spinal transection site. Under these conditions, brain command systems can still activate spinal CPGs and initiate locomotion (19), similar to results from normal animals with similar partial spinal lesions (70, 71). Furthermore, preliminary results from retrograde HRP labeling indicate that the majority of descending brain neurons, particularly small, unidentified RS neurons in the ARRN, MRRN, and PRRN, regenerate their axons in the lateral spinal tracts (19), comparable to their projection patterns in normal animals (70). In contrast, partial lesions that spare the medial spinal tracts above and below healed spinal transection sites usually abolish brain-initiated locomotion (19), like the results from normal animals with similar partial spinal lesions (70, 71). Thus, axonal regeneration of Müller cells, which project in medial spinal tracts (45), does not contribute substantially to behavioral recovery. In addition, preliminary results indicate that relatively few brain neurons regenerate their axons in the medial tracts (19), similar to the projection patterns observed for normal animals (70). Therefore, after spinal cord transections, most descending brain neurons, including small, unidentified RS neurons, regenerate their axons in the same general spinal tracts in which they project in normal larval lamprey (19, 70).

7.2.4. Synaptic Specificity and Structure

Following recovery from spinal transections at ~20% BL, paired intracellular recordings have been performed to test the restoration and specificity of synapses made by identified RS neurons (Müller and Mauthner cells) and spinal lateral interneurons (LIs) (21), which are descending propriospinal neurons whose cell bodies are located below the spinal lesion site (42, 46). Action potentials are elicited in these large, identified RS neurons, and short latency evoked EPSPs recorded from LIs that persist in high-calcium Ringer's solution are used as indicators of monosynaptic connections.

In normal larval lampreys, action potentials in certain large, identified RS neurons (B2, B3, and B4; Fig. 4b) elicit monosynaptic EPSPs in spinal LIs for 58% of the pairs tested (Fig. 10a) (21, 42). At 7–40-week recovery times following spinal transections, LIs below the healed lesion site are monosynaptically activated by the same B cells for only 14% of the pairs examined (Fig. 10b). Thus, these RS neurons do reconnect specifically with appropriate synaptic targets below spinal lesion sites, but with a lower occurrence rate than for normal animals. Other identified RS neurons (e.g., M cells and Mau cells; Fig. 4b) do not connect with spinal LIs in normal animals, and these particular connections are not found in spinal cord-injured animals (Fig. 10a, b).

At 10–12-week recovery times following spinal transections at ~15% BL, EM analysis and Alexa Fluor 488-phalloidin labeling indicate that restored synapses between large, identified RS neurons and spinal neurons display the same basic structure and presynaptic organization of synaptic connections observed for normal animals, but with some deficits ((51); also see (50)). Specifically, regenerating axons of these large RS neurons make relatively few synapses with spinal targets below healed spinal lesion sites, and presynaptic sites within these axons are less complex, have relatively few vesicles, and have smaller active zones compared to those for normal animals. However, large, identified RS neurons are neither necessary nor sufficient for initiating locomotor behavior in either normal lampreys (see Sects. 6.2.3 and 6.2.5) or in recovered spinal-cord injured animals (see Sect. 7.2.3) (19, 70, 71).

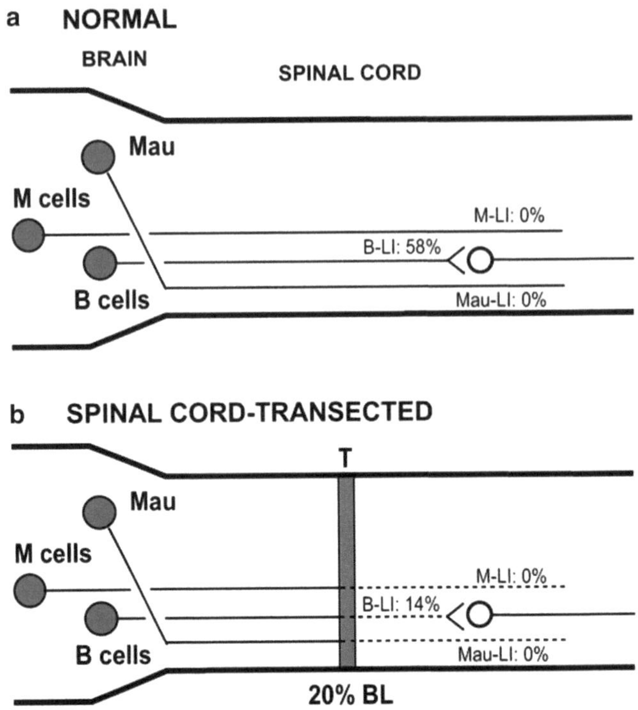

Fig. 10. Synaptic specificity between large, identified RS neurons and spinal cord neurons. (**a**) In normal lampreys, certain Müller cells (B2–B4; see Fig. 4b) evoke monosynaptic responses in spinal LIs in 58% of pairs tested. In contrast, M cells and Mauthner cells (Mau) do not connect with LIs. (**b**) Following spinal transections (T) at ~20% BL and after behavioral recovery (see text), the above B cells monosynaptically excite spinal LIs in 14% of the pairs tested. Inappropriate connections between M cells or Mauthner cells and LIs in the spinal cord are not observed (modified from (21)).

In addition, it is not known if synapses made by these large RS neurons are representative of those of smaller, unidentified RS neurons, which are critical for the initiation of locomotion (49, 61).

7.2.5. Cell Death of Injured RS Neurons

At 32-week recovery times following rostral spinal transections at 10% BL, ~80–85% of descending brain neurons, including small, unidentified RS neurons, extend their axons through the healed lesion site to at least 20% BL (15). However, under similar conditions, only ~50% of Müller and Mauthner cells regenerate their axons to 20% BL (16, 52), suggesting that some of these neurons exhibit poor axonal regeneration and/or undergo apoptosis (cell death). To examine the possibility of cell death, the spinal cords of larval lamprey have been transected at ~15% BL, after which animals recover for 52 weeks (77). Nissl staining or neurofilament-immunostaining in brain whole mounts is used to determine if individual Müller or Mauthner cells are missing. For a different group of recovered, spinal-transected animals, RDA has been applied to the spinal cord at ~20% BL, and 7 days later Alexa 488 dextran amine is applied to the cord at ~9% BL. This procedure double labels identified RS neurons that exhibit long-term survival following spinal transections at ~15% BL (Alexa labeled) and regenerate their axons at least to 20% BL, ~5 mm caudal to the transection site (RDA labeled).

Following rostral spinal transections, several Müller and Mauthner cells (M1, I2, I3, I4, B2, B4, B5, AM) have survival rates >85%, while other cells (M3, I1, B3, Mau) have survival rates <30% (77). Thus, as a general rule, with some exceptions, the largest identified RS neurons tend to have lower survival rates than smaller identified RS neurons. As one might suspect, there is a correlation between the percent survival rate for different Müller and Mauthner cells and the capacity for axonal regeneration of these different neurons (see Sect. 6.2.6).

8. Growth Cones of Regenerating Lamprey Neurons

8.1. Lesion Types and Sites, and Survival Times

In larval lampreys following spinal transections at ~10% BL, the growth cones of regenerating axons of large, identified RS neurons (Müller and Mauthner cells; Fig. 4b) have been examined in histologically processed tissue (78). After recovery times of 3, 4, or 11 weeks, brains and rostral spinal cords are removed and placed in a recording chamber. Micropipettes are filled with 6% HRP in 0.2 M KAc (40 MΩ), and HRP is injected iontophoretically (15 nA, 200 ms pulses at 3.3 Hz for 20 min) into identified RS neurons. The tissue is maintained overnight at 4°C, after which the neural tissue is reacted for HRP using a modified Hanker-Yates protocol, dehydrated, cleared, and mounted on slides. In addition, monoclonal antibodies have been used to detect the presence of NFs and other cytoskeletal elements in growth cones of regenerating RS neurons.

The morphologies of growth cones of lamprey RS neurons in live tissue (53) have been compared to those in histologically processed tissue (see above). Axons of identified RS neurons are retrogradely labeled by application of a small piece of Gelfoam soaked in a solution of 5% RDA (10 kDa; Invitrogen) in 0.1 M Tris buffer to a spinal cord transection at ~15% BL. Following recovery times of 1–3 weeks, the spinal cord at the transection site is exposed, and the tips of regenerating axons of RS neurons are imaged using 2-photon microscopy. For a second group of animals, regenerating axons have been labeled as described above, animals recover for 2–3 weeks, a small piece of Gelfoam soaked in 100 mM dbcAMP, a cAMP analog, in 0.1 M Tris buffer is applied to the transection site, and growth cones are imaged at regular intervals to examine the effects on axonal regeneration. For a third group of animals, 5% RDA and 100 mM 8-OH-cAMP, a cAMP analog, in 0.1 M Tris buffer have

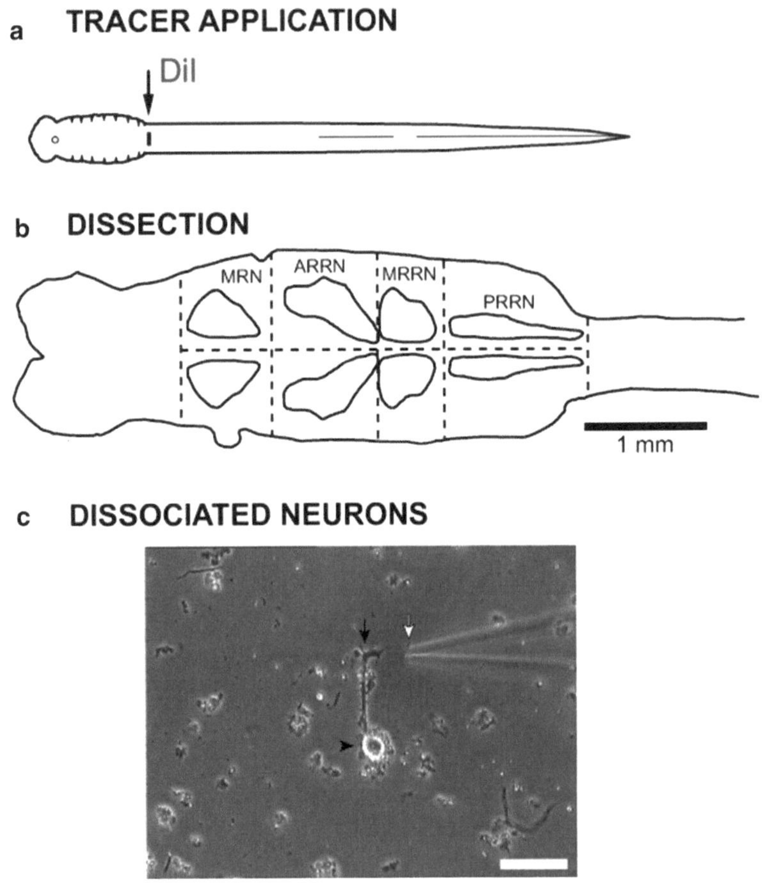

Fig. 11. Isolated lamprey RS neurons in culture. (**a**) Descending brain neurons, ~80% of which are RS neurons, are retrogradely labeled by application of DiI to the spinal cord at 20% BL. (**b**) Approximately 2–4 weeks after tracer application, the brains are removed and cut into tissue blocks (*dotted lines*), and the rostral brain and spinal cord are discarded. The tissue blocks, which contain reticular nuclei (Fig. 4b), are enzymatically treated (see text) and triturated to yield single, DiI-labeled descending brain neurons in culture. (**c**) Photomicrograph showing the cell body (*arrowhead*) of an isolated, DiI-labeled descending brain neuron and a growth cone (*black arrow*) at the end of a neurite. Agents can be applied globally to the bath or focally via micropipettes (*white arrow*) to specific parts of neurons to test the effects on neurite outgrowth (see text). Scale bar = 100 μm (modified from (55)).

both been applied to the transection site, and the effects on axonal die-back have been examined at 1, 2, and 3 weeks PT.

The growth cones of RS neurons also have been studied in cell culture (55), which allows experimental manipulations that are not possible in vivo. Descending brain neurons, ~80% of which are RS neurons (15), are retrogradely labeled by application of a small piece of Gelfoam (~1 mm^3) soaked in a solution of 8% 1,1′-dioctadecyl-3,3,30,30-tetramethylindocarbocyanine perchlorate (DiI; Invitrogen) in dimethyl formamide (DMF) to the spinal cord at 20% BL (Fig. 11a). After a ~2–4-week tracer transport time, the brains are dissected out, cut into small blocks (~200–500 μm; Fig. 11b), and enzymatically digested in 0.125% collagenase 1A (Sigma) for 30 min followed by 0.05% protease (Sigma) for 15 min. The tissue blocks are triturated, and neurons are plated in culture dishes that have a poly-D-lysine substrate (Biocoat; Bectin-Dickinson Labware, Bedford, MA, USA) and that contain a custom, balanced salt L15 media (Gibco; Grand Island, NY, USA) with 10% fetal bovine serum (Gibco). Within a few hours after plating, DiI-labeled brain neurons are present in and attached to the bottom of the culture dishes.

8.2. Outcome Measures and Key Findings

8.2.1. Properties of Growth Cones of Regenerating RS Neurons In Vivo

For histologically processed tissue, the growth cones of regenerating axons of identified lamprey RS neurons are relatively simple, often adopting an oval shape, and lack lamellipodia and long filopodia (78). Furthermore, the growth cones are densely packed with NFs and contain very little F-actin. In live tissue, the growth cones of these RS neurons have many of the same characteristics as those in fixed tissue, including a relatively simple morphology, the presence of NFs, relatively little F-actin, and few microtubules (MTs) (53). These characteristics are quite different from those observed for growth cones of mammalian embryonic neurons in culture, and it has been suggested that the neural developmental mechanisms proposed for pulling axons forward via forces generated by an actin-myosin molecular motor might not be applicable for regenerating axons in vivo (53, 78). In live tissue, activation of cAMP pathways decreases initial axonal die-back. In addition, axonal regeneration of large, identified RS neurons appears to be intermittent, and activating cAMP pathways increases subsequent axonal regeneration by enhancing the rate of elongation during growth periods, but not the fraction of time that axons spend growing (53).

8.2.2. Neurite Outgrowth of RS Neurons in Cell Culture

In cell culture, lamprey descending brain neurons, most of which are RS neurons, have average soma diameters of 28 ± 7 μm (range: 11–55 μm) (55). Thus, the majority of these cultured neurons probably are not Müller or Mauthner cells, which have soma dimensions in vivo of ~40–115 μm (29). Immediately after placement in culture, DiI-labeled neurons usually display processes, which retract during the first 1–2 days. After ~2–4 days in culture, neurons begin to extend processes whose leading edge is characterized by a growth cone (Fig. 11c). Neurites have an average growth rate of ~7.0 μm/h (range: −7 to +34 μm/h), which corresponds favorably to the calculated average axonal regeneration rate in vivo of ~7.4 μm/h during the first 8 weeks following spinal transections at 10% BL (~10 mm/8 weeks; Fig. 5d) (15). The outgrowth rates in culture are characteristic of neurites with a range of growth cone sizes, shapes, and dynamics. However, growth cones of lamprey RS neurons in culture are smaller and simpler in shape compared to those commonly associated with mammalian embryonic neurons in culture.

Focal pressure ejection (~5 ms pulses at 1–2 Hz, 20 psi; each ejection bolus≈25 μm in dia., equivalent to ~8 pl) (Fig. 11c) of 5 mM or 25 mM L-glutamate onto single growth cones of RS neurons in culture inhibits outgrowth of the treated neurite, but other neurites from the same neuron are not inhibited (55). Pressure ejection of medium with high potassium (31 mM) onto growth cones to induce depolarization inhibits neurite outgrowth, an effect that is blocked by bath application of either general calcium channel blockers (2 mM $CoCl_2$ or 100 μM $CdCl_2$) or a specific blocker for N- and P/Q voltage-gated calcium channels (2 μM ω-conotoxin MVIIC). Therefore, depolarization alone induces calcium influx via N- and P/Q type calcium channels that can regulate neurite outgrowth. Application of glutamate to growth cones also inhibits neurite outgrowth under conditions in which N- and P/Q voltage-gated calcium channels are blocked. Thus, calcium influx via ligand-gated ion channels, such as NMDA channels (55), also can regulate neurite outgrowth. Finally, focal pressure ejection of glutamate or high potassium onto the cell bodies of RS neurons inhibits the growth of multiple neurites from the same neuron, demonstrating "global" regulation of neural outgrowth.

For the lamprey, SCI alters the electrical properties of injured RS neurons (48) (see Sect. 7.2.1). Specifically, compared to uninjured RS neurons, injured RS neurons display a reduction of HVA calcium channel expression and decrease in electrical excitability, both of which would be expected to decrease or limit calcium influx. In cell culture, experimental manipulations that induce calcium influx in RS neurons result in inhibition of neurite outgrowth or growth cone retraction. Together, these results suggest that following SCI, the changes in electrical properties of injured RS neurons maintain relatively low intracellular calcium levels and/or limit increases in intracellular calcium, conditions that appear to be important for axonal regeneration (48, 55).

For lamprey RS neurons in culture, some neurites are extending and others are retracting as their growth cones explore the environment. Preliminary results suggest that forskolin, dbcAMP, or IBMX, which results in activation of cAMP pathways, stimulates outgrowth of neurites that are retracting during pre-control periods (79). In contrast, these agents do not significantly enhance outgrowth of neurites that are previously extending during pre-control periods. Thus, for lamprey RS neurons in culture, the growth states of neurites appear to determine their responsiveness to agents that activate cAMP pathways. Finally, stimulation of neurite outgrowth by activation of cAMP pathways appears to be dependent on protein kinase A (PKA) (79).

9. Recovery of Locomotor Coordination: Propriospinal Neurons

9.1. Lesion Types and Sites, and Survival Times

In normal larval lampreys, coordination of spinal locomotor activity mainly is due to three mechanisms (reviewed in (2): (a) bidirectional short distance coupling between nearby spinal oscillators (Fig. 1) that is stronger in the descending direction than the ascending direction (62); (b) bidirectional long distance coupling between more distant oscillators that is relatively weak in both the descending and ascending directions (80); and (c) rostrocaudal gradient in the intrinsic frequencies of burst activity generated by segmental oscillators (81). Some of the ascending and descending PS neurons couple neural oscillators along the spinal cord and ensure that locomotor muscle burst activity is coordinated in different parts of the body (Fig. 1). Mid-body spinal cord transections can disrupt the axons of these PS neurons and abolish coordination of locomotor activity. Axonal regeneration of coordinating neurons has been investigated with neurophysiological and anatomical techniques in larval lampreys that have recovered coordinated locomotor activity following mid-body spinal lesions.

9.2. Outcome Measures and Key Findings

9.2.1. Mechanisms for the Restoration of Locomotor Coordination

For isolated 40 segment sections of spinal cord that include a healed spinal transection site, 15 mg/l D-tubocurarine is present in the bath to abolish possible contractions of remaining musculature along the notochord and block movement-related sensory feedback (82). Application of 500 μM D-glutamate or 50–200 μM NMDA to the bath pharmacologically activates spinal CPGs and elicits locomotor ventral root burst activity above and below the healed spinal lesion site that is coupled 1:1 (e.g., see delay between channels 2→3 in Fig. 3a2, b2,). Thus, spinal coordinating neurons that have regenerated their axons across a spinal transection site can couple rostral and caudal CPG networks (82).

In a related study, two mechanisms appear to be required for restoration of normal coordination of locomotor burst activity following mid-body spinal transections (83). (a) For in vitro brain–spinal cord preparations from animals that have recovered locomotor behavior at 9–25 weeks PT, D-tubocurarine is present in the bath to block movement-related sensory feedback. Stimulation in brain locomotor areas initiates locomotor bursts above and below a spinal lesion site that are coupled 1:1 (e.g., see rostrocaudal delays between channels 2→3 in Fig. 3a2, b2). Thus, restoration of projections of spinal PS neurons alone can coordinate brain-initiated locomotor burst activity on opposite sides of a healed spinal transection site. (b) In whole animals following *acute*, mid-body spinal transections, 50 mg/kg NMDA is injected i.p. to activate spinal CPGs. Under these conditions, locomotor muscle burst activity above and below the transection is coupled 1:1 via mechanosensory feedback alone and in the absence of direct neural coupling between rostral and caudal CPG circuitry (see Box 1 and Fig. 4 in (11)). Importantly, either axonal regeneration of spinal coordinating neurons or mechanosensory feedback alone results in 1:1 coupling of burst activity, but with more variability than in recovered whole animals. Thus, both mechanisms appear to be necessary for recovery of coordination of locomotor burst activity across a healed spinal transection site (83).

9.2.2. Axonal Regeneration of Ascending and Descending Propriospinal Neurons

For normal larval lamprey, the distributions of descending and ascending PS neurons along the spinal cord have been determined by application of HRP to the cord at 60% BL or 40% BL, respectively (84). For experimental animals at 8-week recovery times following spinal cord transections at 50% BL, HRP is applied to one of the above tracer application sites to determine if PS neurons extend their axons across the healed lesion site (84). Finally, following mid-body spinal transections, double labeling has been used to test whether axonal regeneration contributes to the neuronal outgrowth of ascending and descending PS neurons (84), similar to the methods used to examine these issues for RS neurons (see Sect. 6.2.5). Specifically, for double labeling of descending PS neurons in experimental animals, 4% FG in Ringer's solution with 2% DMSO is applied to the spinal cord at 70% BL, 2 weeks later the spinal cord is transected at 45% BL, and after an 8-week recovery period, 6.25% TRDA in 0.1 M Tris buffer is applied to the cord at 55% BL. For double labeling of ascending PS neurons, spinal transections are performed at 45% BL, and the FG and TRDA spinal application sites are 20% BL and 35% BL, respectively. In normal animals, TRDA is applied 2 weeks after FG application, without spinal transections at 45% BL.

For normal animals, descending and ascending PS neurons located in the rostral and caudal spinal cord, respectively, project for at least 30–40 segments (~110 total segments in the spinal cord; (39)). For spinal cord-transected animals at 8-week recovery times, ~20–40% of the normal numbers of descending and ascending PS neurons are retrogradely labeled with HRP (84), suggesting that these neurons exhibit ***incomplete axonal regeneration.*** Finally, double-labeling experiments suggest that following SCI, axonal regeneration contributes significantly to the extension of processes of descending and ascending PS neurons across a healed lesion site, as was the case for descending brain neurons (see Sect. 6.2.5). Axonal regeneration of at least some of these PS neurons very likely contributes to recovery of coordination of locomotor burst activity across a healed spinal lesion site. Thus, following spinal cord transections, complete restoration of projections of descending and ascending PS neurons is not necessary for recovery of coordination of spinal locomotor activity.

10. Restoration of Ascending Spinal Projections to the Brain

10.1. Lesion Types and Sites, and Survival Times

Some spinal neurons with ascending axons transmit locomotor timing and other important feedback information from spinal CPG networks to brain descending systems, including RS neurons (5). This ascending feedback modulates the pattern of activity of descending brain neurons in tune with spinal locomotor burst activity. In mammals, ascending feedback from spinal CPGs modulates brain descending systems via pathways through the cerebellum (5). In contrast, for adult lampreys, lateral projecting axons from some of the ascending spinal neurons make monosynaptic connections with RS neurons (reviewed in (85)).

For normal larval lampreys, the distributions of ascending spinal neurons have been determined by application of HRP to the rostral spinal cord at ~10% BL (brain/spinal cord junction ~8% BL). Similar experiments have been performed with experimental animals at 3–32-week recovery times following spinal transections at 20% BL. These procedures label ascending spinal neurons that project to the very rostral spinal cord and, very likely, to the brain. In addition, restored connections between ascending spinal neurons and RS neurons have been examined.

10.2. Outcome Measures and Key Findings

First, for normal lamprey, ascending spinal neurons are distributed along the spinal cord, although there are many more neurons in the rostral cord with relatively short ascending axons than neurons in the caudal cord with longer axons. In experimental animals with rostral spinal cord transections, preliminary results indicate there is a gradual increase in the numbers of ascending spinal neurons that regenerate their axons across the spinal lesion site with increasing recovery times. Thus, in contrast to fish and amphibians in which there sometimes is little or no regeneration of ascending axons following SCI (reviewed in (13)), in the lamprey there is substantial axonal regeneration of ascending spinal neurons. However, at 32 weeks PT, the average number of labeled ascending spinal neurons is <50% of those for normal animals. Thus, as was the case for RS neurons (see Sect. 6.2.5) and ascending/descending PS neurons (see Sect. 9.2.2), ascending spinal neurons that project to the brain also exhibit *incomplete axonal regeneration.*

Second, synaptic inputs to RS neurons from ascending spinal neurons have been investigated with in vitro brain–spinal cord preparations. Preliminary results indicate that for normal lampreys as well as animals at 32-week recovery times following rostral SCI, focal electrical stimulation of axons in the lateral spinal tracts elicits short latency synaptic responses in RS neurons, similar to those demonstrated in normal adult lampreys (reviewed in (85)). These results suggest that at least some ascending spinal neurons regenerate their axons through a healed spinal lesion site and make synapses with RS neurons.

11. Spinal Cord Injury Alters Neural Networks Below the Lesion

11.1. Lesion Types and Sites, and Survival Times

Although there has been substantial emphasis on regeneration of descending axons following SCI, much less attention has been given to cellular and synaptic plasticity of denervated CPG networks caudal to spinal lesion sites. Plasticity of spinal CPG networks potentially could greatly affect the degree to which regenerated axons of RS neurons as well as descending PS neurons activate these networks and initiate locomotor activity, particularly following incomplete axonal regeneration (see Sects. 6.2.3 and 6.2.4). In the lamprey, spinal cord plasticity has been investigated after animals have recovered for 10 weeks following spinal transections at ~25% BL (86). For this particular type of study, the cut ends of the spinal cord are intentionally separated to limit or block axonal regeneration so that spinal CPGs remain isolated from higher centers.

11.2. Outcome Measures and Key Findings

Behavioral responses are video-taped before performing the spinal lesions, 1 h after the lesions, and thereafter every 2 weeks, for a total recovery time of 10 weeks (86). With in vitro nervous system preparations, activity is recorded from spinal tracts and the cell body layer caudal to spinal lesion sites in response to stimulation rostral to the lesion site. Single and paired intracellular recordings are made from identified and unidentified spinal neurons within 5–10 mm below the spinal lesion.

For the majority of animals, electrical stimulation rostral to spinal lesion sites does not elicit responses caudal to the lesion, suggestive of little or no axonal regeneration, as would be expected for the spinal transection method described above (86). Ironically, in the apparent absence of axonal regeneration, many of these animals appear to recover relatively normal locomotor movements. However, it is unclear how sensitive the method described above is for detecting axonal regeneration, and it is not known if coordinated swimming muscle activity is generated caudal to the spinal lesion sites (see (37, 57)).

Caudal to a spinal transection site in lampreys, neuronal excitability of spinal neurons is increased, as indicated by an increase in discharge of ventral roots in response to cord stimulation, depolarization of resting membrane potentials, increase in neuron input resistances, increase in firing frequencies in response to applied depolarizing current pulses, and the presence of plateau potentials (86). Also, synaptic functions below a spinal lesion are enhanced, including an increase in miniature EPSPs and an increase in the summation of synaptic potentials. Finally, synaptic ultrastructure below a spinal transection site displays several changes, such as a reduction of synaptic clefts and an increase in synaptic vesicle pools. However, it is not known if these various forms of plasticity still occur if axonal regeneration is allowed to proceed normally. Results from a different study indicated that below spinal transection sites at ~75% BL, the chemical component of synaptic responses appears to be increased between giant interneurons (GIs), which are large ascending propriospinal neurons (21). If the general types of cellular and synaptic plasticity described above also occur in lamprey CPG networks below a spinal lesion site (e.g., Fig. 5 in (13)), these changes potentially could provide homeostatic compensation to increase the effectiveness of descending activation from brain command systems.

12. Contributions of the Lamprey to Understanding Mammalian SCI

Following SCI in the lamprey, injured axons regenerate across the lesion site and make synapses with target neurons, resulting in virtually complete recovery of locomotor behavior within a few weeks. Thus, the lamprey provides the opportunity to determine the full complement of mechanisms that lead to unaided axonal regeneration and recovery of locomotion after SCI. Potentially, the lamprey could provide a set of target goals for the types of cellular conditions (e.g., permissive CNS, axon guidance molecules, compensatory mechanisms, little or no secondary injury effects, etc.) that, if present in a mammalian or human spinal cord following SCI, presumably would promote successful neuronal outgrowth and behavioral recovery.

Although the lamprey CNS is permissive for neural outgrowth, not all lamprey neurons display the same degree of axonal regeneration. For example, some large, identified RS neurons have relatively high capacities for regeneration, while others have low capacities (16, 52). Because these differences occur within the same permissive CNS environment, they very likely are due to intrinsic cell-to-cell differences in regenerative capacities. For mammals, similar differences in growth potential for different neuronal cell types recently have been recognized (reviewed in (87)). The lamprey data have important implications for mammalian SCI, because potential manipulations that might make the mammalian CNS permissive for axonal outgrowth may not be sufficient by themselves for promoting behavioral recovery if key neurons of interest have low capacities for regeneration. In such a case, additional manipulations, such as application of neuronal growth factors, might be required to encourage neurons with poor regenerative capacities to extend their processes. These implications stress the need for combinatorial therapeutic approaches to treat mammalian and ultimately human SCI (88–90).

Following SCI, injured lamprey RS neurons display a number of dramatic changes in their properties compared to uninjured neurons, including a decrease in HVA calcium channel expression and calcium channel-mediated components of the action potential (48) as well as a reduction in electrical excitability and altered firing patterns (75, 76). These changes in properties would be expected to limit calcium influx. In cell culture, neurite outgrowth of lamprey RS neurons is inhibited in response to calcium influx (55). Together, these data suggest that for injured lamprey RS neurons following SCI, the changes in properties described above limit calcium influx and maintain intracellular calcium levels in a range that is supportive of axonal regeneration (48, 55). These results from spinal cord-injured lampreys are consistent with studies from mammals. First, for mammalian neurons in culture, contact of growth cones with mature oligodendrocytes, which normally inhibit axonal regeneration following SCI (6–9), induces calcium influx, calcium release from intracellular stores, and growth cone collapse (91), which can be prevented by calcium channel blockers (91, 92). Second, following peripheral "conditioning" spinal nerve lesions in rats, dorsal root ganglion (DRG) neurons display a reduction in electrical activity and L-type calcium channels, both of which are important for regeneration of the central axons of these neurons within the CNS (93). Thus, mechanisms that reduce and/or limit increases in intracellular calcium levels, similar to those described for injured lamprey RS neurons, may be important for axonal regeneration following SCI.

For spinal cord-transected lampreys, descending brain neurons (including RS neurons), PS neurons, and ascending spinal neurons that project to the brain all display *incomplete axonal regeneration*. But, incomplete axonal regeneration can still result in virtually complete behavioral recovery. These data are encouraging regarding mammalian and possibly human SCI because they suggest that experimental manipulations which promote a moderate degree of axonal regeneration could be sufficient for substantial behavioral recovery. Thus, it may not be necessary to induce robust neural outgrowth and restore the original organization of the CNS, and this would significantly lessen the demands on therapeutic manipulations for promoting axonal regeneration in mammals, including humans.

Following SCI in the lamprey, *incomplete axonal regeneration* of descending brain neurons still results in recovery of locomotor function, in part, because of a number of *compensatory mechanisms* that make up for limited neuronal outgrowth (reviewed in (12, 13)). First, in the lamprey some descending PS neurons relay descending drive from brain command systems to lower levels of the spinal cord that have received few or no direct restored projections from the brain (Figs. 5 and 7) ((63); see Fig. 7 in (38)). Interestingly, when direct, long-distance brain–spinal pathways have been disrupted in neonatal rats, CPG networks in the caudal spinal cord can be activated from the brain via descending PS relay neurons (68).

Thus, there appears to be some *redundancy* in the mechanisms by which brain command systems can activate spinal CPGs. Importantly, these descending PS neurons are potential targets for therapeutic interventions following SCI (65, 67). Second, interestingly, for the lamprey, uninjured descending PS neurons *below* a healed spinal injury site appear to sprout and extend their axons, and this *plasticity* probably enhances the effectiveness of these neurons in relaying descending drive from RS neurons to more caudal levels of the cord (63). Third, for larval lamprey, neurons below a spinal transection site exhibit an increase in excitability and synaptic efficacy (21, 86), which if present in spinal CPG neurons following SCI could increase the responsiveness of locomotor networks to descending drive from RS neurons. Fourth, in theory, the few regenerated axons of lamprey RS neurons that project to middle or caudal spinal levels at moderate-to-long recovery times (Fig. 5d–f) might form stronger synapses or more synapses with spinal neurons compared to normal animals (see Fig. 5 in (13)). Following rostral spinal transections, large identified RS neurons (Müller and Mauthner cells) make relatively few synapses with spinal neurons within a few millimeters below the healed lesion site (50, 51), but axonal regeneration of these large RS neurons does not contribute substantially to behavioral recovery (see Sects. 6.2.3 and 6.2.5) (19, 70, 71). However, certain Müller cells (B2–B4) elicit EPSPs in GIs below a healed spinal lesion site that are larger than those in normal animals (Fig. 5 in (21)). In addition, at 32-week recovery times following rostral spinal transections, brain command systems can *directly* activate spinal CPG networks in the caudal spinal cord and initiate locomotor burst activity (see Fig. 6 in (38)), even though only ~6% of the normal numbers of RS neurons project to this level of the cord (Figs. 5f and 6c) (15).

In the future, neuroscientists hopefully will be able to negate many of the inhibitory factors for neuronal outgrowth within the mammalian CNS, minimize secondary injury effects, and minimize demyelination (6–11), thereby making the spinal cord substantially permissive for neuronal outgrowth, perhaps closer to that of the lamprey CNS. However, under these conditions, axonal regeneration within a modified, permissive mammalian CNS very likely will be regulated by similar basic cellular mechanisms as those in the lamprey and, thus, axonal outgrowth probably will be incomplete (see Sect. 6.2.3). Because mammalian locomotion is much more complex than lamprey swimming, the injured mammalian spinal cord very likely would require a greater degree of axonal regeneration for substantial behavioral recovery than the injured lamprey spinal cord. Thus, even if the mammalian CNS were made permissive for neuronal outgrowth, incomplete axonal regeneration probably would significantly limit the degree of recovery. Therefore, it is vitally important to understand the specific cellular and molecular mechanisms that result in incomplete axonal regeneration within a permissive CNS environment and to develop methods for overcoming this limitation. In particular, the mechanisms that *limit* axonal outgrowth in the CNS of lower vertebrates (see Sect. 6.2.3) very likely are substantially different than those that usually *inhibit* axonal regeneration in higher vertebrates (e.g., NOGO, CSPG, etc. (6–11). Thus, the lamprey has the potential to contribute critically important information regarding the mechanisms that are responsible for incomplete axonal regeneration and provide possible manipulations for negating these effects.

Most mammalian spinal cord injury research has focused on acute injuries, but the vast majority of humans with SCIs have had their condition for months or years. When it becomes possible to manipulate the mammalian CNS to make it substantially permissive for axonal regeneration, there are two very important questions that remain unresolved. First, following chronic SCI, when regeneration has been inhibited for long periods of time due to a restrictive CNS environment, if the CNS is then manipulated to become permissive, can descending brain neurons, particularly RS neurons, still mount a successful regenerative response? Second, following SCI and during long-term inhibition of axonal regeneration, are there changes in spinal CPGs (i.e., plasticity) that make them less receptive to reinnervation by regenerating axons of RS neurons? For example, after chronic SCI, vacant synaptic sites on CPG neurons might become occupied, and will regenerating axons of RS neurons later be able to reclaim these synaptic sites? These questions cannot presently be investigated using mammals, but might possibly be examined with the lamprey by blocking axonal regeneration temporarily and then later reversing the block.

13. Summary

The lamprey has a number of key advantages for SCI studies that have stimulated investigations for at least the last 60 years. In particular, much is known about the brain and spinal circuits that control locomotor function in normal lampreys and that can become compromised following SCI. As a result, the lamprey has provided a wide range of important information, from behavioral to molecular, on the mechanisms that support as well as limit neural regeneration within a permissive CNS environment following SCI. These data are interesting in their own right but also have translational implications for the study of mammalian SCI. Many of these unique insights could not have been obtained using other animal models. Because of the many substantial challenges regarding treatment of human SCI, we will need all of the available information from a variety of animal models in order to have the best chance of overcoming this costly and debilitating health issue.

Acknowledgments

The work performed in the author's laboratory was supported by NIH grants NS29043, APA grants MB1-9108 and MA1-9605, SCRF grant NBR 501-5, University of Missouri (MU) Spinal Cord Injury Program grants, MU Research Board grant, and MU Research Council grant. The author would like to thank Lené Holland and Timothee Pale for reading earlier versions of the manuscript.

References

1. Orlovsky GN, Deliagina TG, Grillner S (1999) Neuronal control of locomotion. Oxford, New York
2. McClellan AD (1996) Organization of spinal locomotor networks: contributions from model systems. Comments Theo Biol 4(1):63–91
3. Hill AAV, Masino MA, Calabrese RL (2003) Intersegmental coordination of rhythmic motor patterns. J Neurophysiol 90(2):531–538
4. Jordan LM (1998) Initiation of locomotion in mammals. Ann N Y Acad Sci 860:83–93
5. Arshavsky YI, Gelfand IM, Orlovsky GN (1983) The cerebellum and control of rhythmical movements. Trends Neurosci 6(10):417–422
6. Bradbury EJ, McMahon SB (2006) Spinal cord repair strategies: why do they work? Nat Rev Neurosci 7(8):644–653
7. David S, Lacroix S (2000) Molecular approaches to spinal cord repair. Annu Rev Neurosci 26:411–440
8. Schwab ME (2010) Functions of Nogo proteins and their receptors in the nervous system. Nat Rev Neurosci 11(12):799–811
9. Schwab ME, Bartholdi D (1996) Degeneration and regeneration of axons in the lesioned spinal cord. Physiol Rev 76(2):319–370
10. Silver J, Miller JH (2004) Regeneration beyond the glial scar. Nat Rev Neurosci 5(2):146–156
11. McClellan AD (2011) Special issue on spinal cord injury in Brain Research Bulletin. Brain Res Bull 84(4–5):265–266
12. McClellan AD (1994) Functional regeneration and restoration of locomotor activity following spinal cord transection in the lamprey. Prog Brain Res 103:203–217
13. McClellan AD (1998) Spinal cord injury: lessons from locomotor recovery and regeneration in spinal cord-transected lamprey. Neuroscientist 4(4):250–263
14. Shifman MI, Jin L-Q, Selzer ME (2007) Regeneration in the lamprey spinal cord. In: Becker CG, Becker T (eds) Model organisms in spinal cord regeneration. Wiley, New York, pp 229–262
15. Davis GR, McClellan AD (1994) Extent and time course of restoration of descending brainstem projections in spinal-transected lamprey. J Comp Neurol 344(1):65–82
16. Davis GR, McClellan AD (1994) Long distance axonal regeneration of identified lamprey reticulospinal neurons. Exp Neurol 127(1):94–105
17. Lurie DI, Selzer ME (1991) Preferential regeneration of spinal axons through the scar in hemisected lamprey spinal cord. J Comp Neurol 313(4):669–679
18. Mackler SA, Yin H-S, Selzer ME (1986) Determinants of directional specificity in the regeneration of lamprey spinal axons. J Neurosci 6(6):1814–1821
19. Shaw AC, Holmes T, Johns JL, Thurman S, Jackson AW, Davis GR, McClellan AD (2001) Pathfinding of regenerating descending axons in spinal cord-transected larval lamprey: functional anatomy. Soc Neurosci Abst 27:960
20. Becker CG, Becker T (2001) Regenerating descending axons preferentially reroute to the gray matter in the presence of a general macrophage/microglial reaction caudal to a spinal transection in adult zebrafish. J Comp Neurol 433(1):131–147
21. Mackler SA, Selzer ME (1987) Specificity of synaptic regeneration in the spinal cord of the larval sea lamprey. J Physiol 388:183–198
22. Lee MT (1982) Regeneration and functional reconnection of an identified vertebrate central neuron. J Neurosci 2(12):1793–1811
23. Holder N, Clarke JDW (1988) Is there a correlation between continuous neurogenesis and directed axon regeneration in the vertebrate nervous system? Trends Neurosci 11(3):94–99
24. Taylor JSH, Jack JL, Easter SS (1989) Is the capacity for optic nerve regeneration related to

continued retinal ganglion cell production in the frog? A test of the hypothesis that neurogenesis and axon regeneration are obligatorily linked. Eur J Neurosci 1(6):626–638

25. Farel PB, Wray SE, Meeker ML (1993) Size-related increase in motoneuron number: evidence for late differentiation. Brain Res Dev Brain Res 71(2):169–179
26. Zupanc GKH (2001) Adult neurogenesis and neuronal regeneration in the central nervous system of teleost fish. Brain Behav Evol 58(5):250–275
27. Ferretti P (2011) Is there a relationship between adult neurogenesis and neuron generation following injury across evolution? Eur J Neurosci 34(6):951–962
28. Zhang L, McClellan AD (1999) Axonal regeneration of descending brain neurons in larval lamprey demonstrated by retrograde double labeling. J Comp Neurol 410(4):612–626
29. Zhang L, Palmer R, McClellan AD (2002) Increase in descending brain-spinal cord projections with age in larval lamprey: implications for spinal cord injury. J Comp Neurol 447(2):128–137
30. Duffy MT, Liebich DR, Garner LK, Hawrych A, Simpson SB, Davis BM (1992) Axonal sprouting and frank regeneration in the lizard tail spinal cord: correlation between changes in synaptic circuitry and axonal growth. J Comp Neurol 316(3):363–374
31. Bullock TH, Moore JK, Fields D (1984) Evolution of myelin sheaths: both lamprey and hagfish lack myelin. Neurosci Lett 48(2): 145–148
32. Bastmeyer M, Beckmann M, Schwab ME, Stuermer CAO (1991) Growth of regenerating goldfish axons is inhibited by rat oligodendrocytes and CNS myelin but not by goldfish optic nerve tract oligodendrocyte-like cells and fish CNS myelin. J Neurosci 11(3):626–640
33. Dietrich WD (2009) Therapeutic hypothermia for spinal cord injury. Crit Care Med 37(Suppl 7):S238–S242
34. Levi AD, Casella G, Green BA, Dietrich WD, Vanni S, Jagid J, Wang MY (2010) Clinical outcomes using modest intravascular hypothermia after acute cervical spinal cord injury. Neurosurgery 66(4):670–677
35. Osório J, Retaux S (2008) The lamprey in evolutionary studies. Dev Genes Evol 218(5): 221–235
36. Lurie DI, Selzer ME (1991) Axonal regeneration in the adult lamprey spinal cord. J Comp Neurol 306(3):409–416
37. Davis GR, Troxel MT, Kohler VJ, Grossmann ER, McClellan AD (1993) Time course of locomotor recovery and functional regeneration in spinal-transected lampreys: kinematics and electromyography. Exp Brain Res 97(1):83–95
38. McClellan AD (1994) Time course of locomotor recovery and functional regeneration in spinal-transected lamprey: in vitro brain/spinal cord preparations. J Neurophysiol 72(2):847–860
39. McClellan AD (1988) Functional regeneration of descending command pathways for locomotion in the lamprey demonstrated in the in vitro lamprey CNS. Brain Res 448(2):339–345
40. Davis GR, McClellan AD (1993) Time course of anatomical regeneration of descending brainstem neurons and behavioral recovery in spinal-transected lampreys. Brain Res 602(1):131–137
41. Nieuwenhuys R, Lampreys NC (1998) Petromyzontoidea. In: Nieuwenhuys R, Ten Donkelaar HJ, Nicholson C (eds) The central nervous system of vertebrates. Springer, Heidelberg, pp 397–495
42. Rovainen CM (1979) The neurobiology of lampreys. Physiol Rev 59(4):1007–1077
43. Grillner S (2003) The motor infrastructure: from ion channels to neuronal networks. Nat Rev Neurosci 4(7):573–586
44. Mullins OJ, Hackett JT, Buchanan JT, Friesen WO (2010) Neuronal control of swimming behavior: comparison of vertebrate and invertebrate model systems. Prog Neurobiol 93(2):244–269
45. Rovainen CM (1978) Müller cells, Mauthner cells, and other reticulospinal neurons in the lamprey. In: Faber D, Korn H (eds) Neurobiology of the Mauthner cell. Raven, New York, pp 245–269
46. Buchanan JT (2001) Contributions of identifiable neurons and neuron classes to lamprey vertebrate neurobiology. Prog Neurobiol 63(4):441–466
47. Rouse DT, Quan X, McClellan AD (1998) Biophysical properties of reticulospinal neurons in larval lamprey. Brain Res 779(1–2):301–308
48. McClellan AD, Kovalenko M, Benes JA, Schulz DJ (2008) Spinal cord injury induces changes in electrophysiological properties and ion channel expression of reticulospinal neurons in larval lamprey. J Neurosci 28(3):650–659
49. Paggett K, Jackson AW, McClellan AD (2004) Organization of higher-order brain areas that initiate locomotor activity in larval lamprey. Neuroscience 125(1):25–33
50. Wood MR, Cohen MJ (1981) Synaptic regeneration and glial reactions in the transected spinal cord of the lamprey. J Neurocytol 10(1):57–79

51. Oliphint PA, Alieva N, Foldes AE, Tytell ED, Lau BY, Pariseau JS, Cohen AH, Morgan JR (2010) Regenerated synapses in lamprey spinal cord are sparse and small even after functional recovery from injury. J Comp Neurol 518(14):2854–2872
52. Jacobs AJ, Swain GP, Snedeker JA, Pijak DS, Gladstone LJ, Selzer ME (1997) Recovery of neurofilament expression selectively in regenerating reticulospinal neurons. J Neurosci 17(13):5206–5220
53. Jin LQ, Zhang G, Jamison C, Takano H, Haydon PG, Selzer ME (2009) Axon regeneration in the absence of growth cones: acceleration by cyclic AMP. J Comp Neurol 515(3):295–312
54. McClellan AD, McPherson D, O'Donovan MJ (1994) Combined retrograde labeling and calcium-imaging in spinal cord and brainstem neurons of the lamprey. Brain Res 633(1):61–68
55. Ryan SK, Shotts LR, Hong S-K, Nehra D, Groat CR, Armstrong JR, McClellan AD (2007) Glutamate regulates neurite outgrowth of cultured descending brain neurons from larval lamprey. Dev Neurobiol 67(2):173–188
56. Selzer ME (1978) Mechanisms of functional recovery and regeneration after spinal cord transection in larval sea lampreys. J Physiol 277:395–408
57. McClellan AD (1990) Locomotor recovery in spinal-transected lampreys. Role of functional regeneration of descending axons from brainstem locomotor command neurons. Neuroscience 37(3):781–798
58. McClellan AD, Hagevik A (1997) Descending control of turning locomotor activity in larval lamprey: neurophysiology and computer modeling. J Neurophysiol 78(1):214–228
59. Boyd M, McClellan AD (2002) Variations in locomotor activity parameters versus cycle time in larval lamprey. J Exp Biol 205(23):3707–3716
60. Jackson AW, McClellan AD (2011) Localization, pharmacology, and organization of brain locomotor areas in larval lamprey. Neuroscience 175:235–250
61. Jackson AW, Pino F, Wiebe E, McClellan AD (2007) Locomotor behavior initiated by chemical microstimulation in higher locomotor centers and reticular nuclei in semi-intact preparations of larval lamprey. J Neurophysiol 97(5):3229–3241
62. Hagevik A, McClellan AD (1994) Coupling of spinal locomotor networks in larval lamprey revealed by receptor blockers for inhibitory amino acids: neurophysiology and computer modeling. J Neurophysiol 72(4):1810–1829
63. Rouse DT, McClellan AD (1997) Descending propriospinal neurons in normal and spinal cord-transected lamprey. Exp Neurol 146(1):113–124
64. McClellan AD, Benes JA, Burks F (2005) Factors that regulate axonal regeneration following spinal cord injury in larval lamprey. Soc Neurosci Abst 31:838.14
65. Flynn JR, Graham BA, Galea MP, Callister RJ (2011) The role of propriospinal interneurons in recovery from spinal cord injury. Neuropharmacology 60(5):809–822
66. Juvin L, Simmers J, Morin D (2005) Propriospinal circuitry underlying interlimb coordination in mammalian quadrupedal locomotion. J Neurosci 25(25):6025–6035
67. Jordan LM, Schmidt BJ (2002) Propriospinal neurons involved in the control of locomotion: potential targets for repair strategies? Prog Brain Res 137:125–139
68. Cowley KC, Zaporozhets E, Schmidt BJ (2008) Propriospinal neurons are sufficient for bulbospinal transmission of the locomotor command signal in the neonatal rat spinal cord. J Physiol 586(6):1623–1635
69. Blesch A, Tuszynski MH (2008) Spinal cord injury: plasticity, regeneration and the challenge of translational drug development. Trends Neurosci 32(1):41–47
70. Shaw AC, Jackson AW, Holmes T, Thurman S, Davis GR, McClellan AD (2010) Descending brain neurons in larval lamprey: spinal projection patterns and initiation of locomotion. Exp Neurol 224(2):527–541
71. McClellan AD (1988) Brainstem command systems for locomotion in the lamprey: localization of descending pathways in the spinal cord. Brain Res 457(2):338–349
72. Zhang L, McClellan AD (1998) Fluorescent tracers as potential candidates for double labeling of descending brain neurons in larval lamprey. J Neurosci Methods 85(1):51–62
73. Pizarro IV, Swain GP, Selzer ME (2004) Cell proliferation in the lamprey central nervous system. J Comp Neurol 469(2):298–310
74. Zhang L, Palmer R, McClellan AD (2004) Conditioning lesions enhance axonal regeneration of descending brain neurons in spinal cord-transected larval lamprey. J Comp Neurol 478(4):395–404
75. McClellan AD (2003) Axotomy alters spike frequency adaptation of reticulospinal neurons in larval lamprey. Soc Neurosci Abst 29:42.21
76. McClellan AD (2009) Axotomized reticulospinal neurons in larval lamprey display changes in channel expression and firing activity. Soc Neurosci Abst 34:766.15
77. Shifman MI, Zhang G, Selzer ME (2008) Delayed death of identified reticulospinal

neurons after spinal cord injury in lampreys. J Comp Neurol 510(3):269–282
78. Lurie DI, Pijak DS, Selzer ME (1994) Structure of reticulospinal axon growth cones and their cellular environment during regeneration in the lamprey spinal cord. J Comp Neurol 344(4):559–580
79. Pale T, Frisch EB, McClellan AD (2011) Activation of cAMP stimulates neurite outgrowth of lamprey reticulospinal neurons with little effect on biophysical properties. Soc Neurosci Abst 36:892.27
80. McClellan AD, Hagevik A (1999) Coordination of spinal locomotor activity in the lamprey: long distance coupling of spinal oscillators. Exp Brain Res 126(1):93–108
81. Hagevik A, McClellan AD (1999) Coordination of locomotor activity in the lamprey: role of descending drive to oscillators along the spinal cord. Exp Brain Res 128(4):481–490
82. Cohen AH (1988) Regenerated fibers of the lamprey spinal cord can coordinate fictive swimming in the presence of curare. J Neurobiol 19(3):193–198
83. McClellan AD (1990) Locomotor recovery in spinal-transected lampreys. Regenerated coordinating neurons and mechanosensory inputs couple locomotor activity across a spinal lesion. Neuroscience 35(3):675–685
84. Armstrong J, Zhang L, McClellan AD (2003) Axonal regeneration of descending and ascending spinal projection neurons in larval lamprey. Exp Neurol 180(2):156–166
85. Buchanan JT, Einum JF (2008) The spinobulbar system in lamprey. Brain Res Rev 57(1):37–45
86. Cooke R, Parker D (2009) Locomotor recovery after spinal cord lesions in the lamprey is associated with functional and ultrastructural changes below lesion sites. J Neurotrauma 26(4):597–612
87. Zurn AD, Bandtlow CE (2006) Regeneration failure in the CNS: Cellular and molecular mechanisms. Adv Exp Med Biol 557:54–76
88. Benowitz LI, Yin Y (2007) Combinatorial treatments for promoting axon regeneration in the CNS: strategies for overcoming inhibitory signals and activating neurons' intrinsic growth state. Dev Neurobiol 67(9):1148–1165
89. Fortun J, Hill CE, Bunge MB (2009) Combinatorial strategies with Schwann cell transplantation to improve repair of the injured spinal cord. Neurosci Lett 456(3):124–132
90. Lu P, Yang H, Jones LL, Filbin MT, Tuszynski MH (2004) Combinatorial therapy with neurotrophins and cAMP promotes axonal regeneration beyond sites of spinal cord injury. J Neurosci 24(28):6402–6409
91. Bandtlow CE, Schmidt MF, Hassinger TD, Schwab ME, Kater SB (1993) Role of intracellular calcium in NI-35 evoked collapse of neuronal growth cones. Science 259(5091):80–83
92. Moorman SJ, Hume RI (1993) ω-Conotoxin prevents myelin-evoked growth cone collapse in neonatal rat locus coeruleus neurons in vitro. J Neurosci 13(11):4727–4736
93. Enes J, Langwieser N, Ruschel J, Carballosa-Gonzalez MM, Klug A, Traut MH et al (2010) Electrical activity suppresses axon growth through Cav1.2 channels in adult primary sensory neurons. Curr Biol 20(13):1154–1164

Chapter 5

Dorsal Root Injury for the Study of Spinal Cord Injury Repair

Håkan Aldskogius and Elena N. Kozlova

Abstract

Dorsal root injury provides opportunities for highly reproducible lesions and for detailed anatomical, physiological, and behavioral outcome assessment with high precision and validity. Dorsal root injury models are used to several aspects of relevance to spinal cord injury repair: (1) mechanisms of regeneration failure in the central nervous system and how to overcome it; (2) axon degeneration, as well as myelin degradation and elimination in the central nervous system—their roles and possible manipulations in spinal cord repair; (3) consequences in the spinal cord of mimicking human plexus injuries by dorsal root avulsion, including its effect on neuron survival, inflammatory processes, and vascular dysfunction; and (4) therapeutic strategies which may be translated to the treatment of clinical plexus avulsion injuries. This chapter describes various dorsal root injury models, their relationship to basic and translational aspects of spinal cord injury repair, as well as basic experimental procedures associated with these models in rat and mouse.

Key words: Nerve regeneration, Nerve degeneration, Sensory neuron, Spinal cord, Brachial plexus, Inflammation, Trophic factor, Transplantation

1. Introduction

There are two aspects that make dorsal root injury immediately relevant in the context of spinal cord injury. First, repair of dorsal root injury, wherever it occurs, requires a solution to the failure of injured axons to regenerate in the adult spinal cord, which is also a fundamental challenge after injury to the spinal cord itself. Thus, studies using dorsal root injury models will contribute to understanding how regeneration of injured axons can be achieved in the adult spinal cord. Second, the most devastating form of dorsal root injuries, plexus avulsion injuries entails a substantial damage not only to affected roots, but also to the spinal cord itself. Thus, studies using dorsal root avulsion model will contribute to general aspects which are relevant for spinal cord repair, e.g., the need for

Håkan Aldskogius (ed.), *Animal Models of Spinal Cord Repair*, Neuromethods, vol. 76,
DOI 10.1007/978-1-62703-197-4_5, © Springer Science+Business Media, LLC 2013

neuroprotection and for minimizing the influence of axon growth inhibitory factors. Clinical root avulsion injury typically involves severe damage also to motor neurons and their axons with consequent varying degrees of paralysis. Repairing the motor deficits is however, largely an issue of supporting motor neuron survival and achieving functionally optimal peripheral rather than central neural regeneration (1). This chapter will describe the principal models of dorsal root injuries, their relationship to basic issues and clinical conditions, the specific advantages and disadvantages of the different models as well as key aspects on outcome assessments. The description is largely based on rat and mouse as animal models, the predominating species since long for dorsal root injury studies. Chapter 10 discusses dorsal root injury in non-human primates.

1.1. The Central Issue of Dorsal Root Injury: Failure of Injured Axons to Re-enter the Spinal Cord

For most of their course, dorsal root axons travel in a peripheral nervous system (PNS) environment, but just outside the spinal cord, they enter the dorsal root transitional zone (DRTZ), also referred to as the dorsal root entry zone (DREZ). Following injury, dorsal root axons regenerate steadily, albeit relatively slowly in the PNS compartment (2), but abruptly cease to grow when they reach the DRTZ. The exploration of this distinctive and environmentally dependent behavior by the same axons along their elongation has provided information relevant for spinal cord injury and also for the field of mammalian central nervous system (CNS) regeneration research in general.

1.2. The PNS–CNS Junction of the Mammalian Dorsal Root: Gateway in Development Turns into "Stop Block"

Populations of dorsal root axons enter the spinal cord in waves during embryonic development (3). The entry points are characterized by the presence of a transient population of neural crest-derived cells, boundary cap cells (4–6), which allow entry of axons from the dorsal root into the spinal cord, possibly via a semaphorin dependent mechanism (7, 8). At some time point during late fetal or early postnatal development, a distinct boundary between PNS and CNS is formed at the interface between Schwann cells and astrocytes that appears to be fully mature around 3 weeks after birth in rodents (9–11) but may take several months in, e.g., cats (12). In the neonatal period, regenerating dorsal root axons are able to enter the spinal cord, at least to judge from studies in rodents (13, 14), but whether these findings can be generalized to other mammals is not known. The boundary between the PNS and CNS is easily defined in sections of the dorsal root-spinal cord junction by labeling the PNS compartment with, e.g., anti-laminin antibodies and the CNS compartment with anti-glial fibrillary acidic protein (GFAP; Fig. 1) (15).

After maturation, the prototypic dorsal root ganglion neuron emits one single axon which branches into one process projecting peripherally and one along the dorsal root. At their exit from the dorsal root ganglion, dorsal root axons are collected in a single

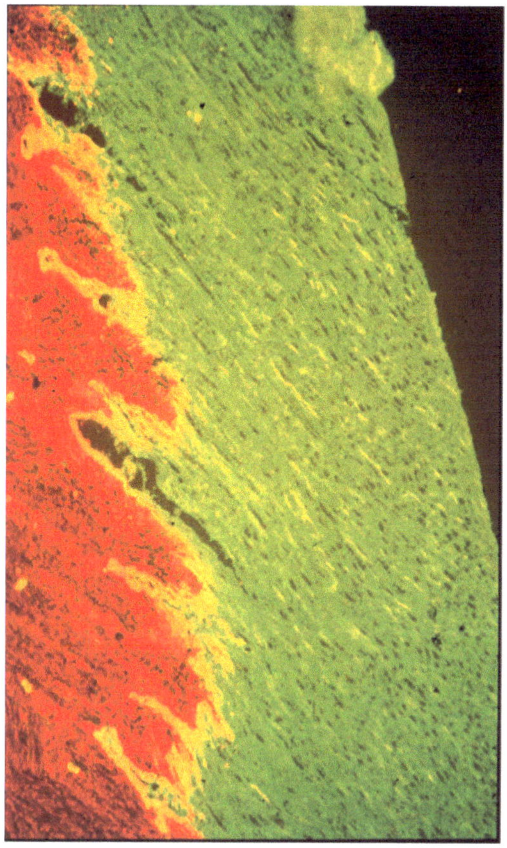

Fig. 1. Intact, lumbar dorsal root of adult rat. Horizontal section through several dorsal rootlets, immunostained with antibodies to laminin (*green*) and glial fibrillary acidic protein (GFAP; *red*). The dorsal root transitional zone is formed by cone-shaped peripheral projections of central nervous tissue (*red*), which interdigitate with the peripheral nervous compartment of the root (*green*) (×200).

well circumscribed bundle, surrounded by Schwann cells, basement membrane and a thin connective tissue sheath. Between the dorsal root ganglion and spinal cord, thin subcomponents, rootlets, emerge and display a divergent course toward the spinal cord, giving the roots a fan-like appearance. The fan-like appearance is most striking at the cervical and upper thoracic levels where distances between dorsal root ganglia and dorsal root entry sites are short and the filaments rapidly separate from each other. Some distance outside the spinal cord surface dorsal root axons encounter a loose network mostly composed of astroglial processes, which project peripherally, intermingle with Schwann cell processes and together with oligodendroglial processes form the central tissue projections or glial dome (9, 10, 16). At the transition from PNS to CNS myelinated axons typically have a node of Ranvier with Schwann cell processes surrounding its peripheral and astroglial processes its central part (9, 10, 17). Since myelination in the CNS is triggered at axonal diameters smaller than in the PNS (18) the larger non-myelinated axons in the PNS become myelinated after their entry in the spinal cord (19).

1.3. Advantages of Dorsal Root Injury for Spinal Cord Repair

The sensory neurons in dorsal root ganglia have been and remain attractive for studies on nerve regeneration from the perspective of the PNS as well as the CNS (20). The location of these neurons outside the spinal cord and the complex CNS neural circuitry has practical advantages, and their peripheral or central axons are easily axotomized. Dorsal root ganglion neurons are also accessible for selective experimental manipulation in vivo, e.g., for electroporation of genetic constructs (21). The physiological properties and functional role of the subset of dorsal root ganglion neurons are known in great detail. Nerve cell body and associated non-neuronal responses can therefore be examined in a relatively simple anatomical and physiological context. The dorsal root injury model can also serve as a useful model in cell replacement therapy, as has been shown with implantation into the dorsal root ganglion cavity of embryonic sensory neurons (22, 23), or stem/progenitor cells (24–26). The popularity of dorsal root injury as a model for the study of spinal cord regeneration aspects has perhaps also a historical explanation, since it was a prime example used by Cajal to demonstrate the failure of injured axons to regenerate in the central CNS in contrast to their unimpeded growth in the PNS (27).

2. Dorsal Root Injury Models: Defining Their Objectives

Three principally different types of dorsal root injury should be distinguished on the basis of the site of injury and their pathophysiological consequences for the spinal cord.

- Injuries that interrupt dorsal root axons at a restricted site along their course in the PNS compartment of the dorsal root.
- Injuries where the dorsal roots are interrupted at their entry into the spinal cord.
- Injuries that pull the dorsal root away from its attachment to the spinal cord.

2.1. Injuries that Interrupt Dorsal Root Axons at a Restricted Site Along Their Course in the PNS Compartment of the Dorsal Root

In terms of models for spinal cord repair a localized dorsal root injury in the PNS compartment of the dorsal root primarily addresses the issue of *axon regeneration* from a permissive (PNS) into a non-permissive (CNS) environment. In this case, injury-induced axon elongation is initiated and the behavior of axons is studied when they reach the interface between the PNS and CNS and attempt to enter the spinal cord. Mechanisms underlying failure of dorsal root axons to re-enter the spinal cord can be explored and approaches designed to attempt to overcome this regeneration failure.

For injury reproducibility, sectioning the dorsal root is the best. However, this approach has shortages for studying regeneration of dorsal root axons—proximal and distal stumps of the dorsal root have to be re-apposed, and fixed to each other, usually by biological material such as fibrin glue. If the repair site is not prepared in an optimal way it may present an obstacle for axons to cross the injury site. Crushing the dorsal root eliminates this constraint, but is more prone to injury variability and raises the issue of how to ensure completeness, in particular as it comes to the non-myelinated axons. Since the connective tissues of dorsal root are more delicate than those of peripheral nerves forceful or repeated crushes may result in a complete separation of proximal and distal root stumps. To establish crush injury of the dorsal root as the model, requires control experiments with careful anatomical examination of the injury site and the portion of the nerve isolated from the nerve cell body. Physiological verification that impulse propagation completely fails across the injury site is an additional control, although a transient conduction block without interruption of axonal continuity may be a potential pitfall (cf. (28)).

2.2. Injuries Where the Dorsal Roots Are Interrupted at Their Entry into the Spinal Cord

Interrupting dorsal root axons at the site of their entry into the spinal cord is a useful injury model for probing interventional strategies which may promote functionally useful growth of dorsal root axons into the spinal cord. The anatomy at the junction between dorsal root and spinal cord limits the type of injury to sectioning the individual rootlets, taking care to minimize damage to the spinal cord itself. The procedure involves exposing target dorsal root(s) via a partial laminectomy, carefully lifting the roots close to their entry site, and thereafter cutting the root filaments. Recent examples using this model are studies in which multiple dorsal roots were sectioned, re-attached, and olfactory ensheathing cells implanted at the injury site to bridge growing dorsal root axons (29, 30). This injury model is in principle a CNS type injury, and more closely related to clinical root avulsion injury than the injury models described above. Still, this injury model does not mimic the abrupt and powerful longitudinal forces subjected to the dorsal roots and their entry sites in clinical root avulsion.

2.3. Injuries that Pulls the Dorsal Root(s) Away from Its Attachment to the Spinal Cord

The models that most closely reflect clinical dorsal root avulsion injury involve pulling dorsal roots away from their attachment with the spinal cord. This type of injury results in damage to the DRTZ and spinal cord which is significantly more extensive than after just cutting or crushing the dorsal roots. This injury can be achieved either by forceful pulling of one or more spinal nerves, e.g., the brachial plexus, outside the vertebral column (31–35), or by exposing dorsal root(s) at their entry into the spinal cord and thereafter pulling them away from the spinal cord (36–38). Pulling spinal nerve(s) breaks both ventral and dorsal roots without visual observation of the actual injury, which may increase its variability. Avulsing only dorsal roots will obviously not reflect the entire pattern of pathophysiological and functional effects of clinical plexus avulsion injury, but allows better reproducibility. Both models have so far been explored mainly for spinal cord pathophysiology and neuropathic pain induced by deafferentation, rather than for the purpose of studying regeneration of injured dorsal root axons.

3. Diversity of Dorsal Root Ganglion Neurons and Their Spinal Cord Projections

The compositions of dorsal root ganglia are quite complex, reflecting the broad range of sensory information conveyed by dorsal root ganglion neurons from the periphery to the spinal cord. These features should be kept in mind in outcome assessments after dorsal root injury. Dorsal root ganglion neurons display distinct differences in structure, content of neurosignaling molecules, receptor expression, functional properties (39–42), axonal course into the spinal cord (43), and central projections of their axon terminals (44–46). A majority of mammalian dorsal root axons are non-myelinated, enter the spinal cord mainly in the lateral compartment of the dorsal root and terminate on neurons in the dorsal horn within the entry segment as well as in several segments rostral and caudal to the level of entry. Myelinated dorsal root axons enter the spinal cord within the medial compartment of the dorsal root and reach the dorsal horn largely by turning laterally from the dorsal funiculus. In general axons of myelinated dorsal root axons terminate more ventrally in the dorsal horn than non-myelinated ones. Importantly, collaterals of sensory axons that mediate information from low-threshold mechanoreceptors in skin, muscle and joints, ascend in the dorsal funiculus to terminate in, e.g., the column of Clarke (proprioceptive information), or the dorsal column nuclei (cutaneous mechanical information and positional information). A smaller fraction of myelinated axons that convey sensory information from muscles, including muscle spindle afferents, reach the ventral horn where they terminate on motor neurons or interneurons. The further routes through the spinal cord of the sensory information from the periphery are described in Chap. 2.

4. The Dorsal Root Injury Model for Exploring Basic Aspects of Spinal Cord Repair

This section will briefly highlight some of the most relevant previous contributions from studies using the dorsal root injury model as well as some future challenges for which this model offers an attractive experimental potential. The failure of injured dorsal root axons to enter the mammalian spinal cord is most likely a consequence of the combination of insufficient regenerative response by axotomized dorsal root ganglion neurons and the emergence of growth inhibitory properties of the DRTZ. Thus, the dorsal root ganglion model is useful for exploring neuronal as well as environmental components underlying mammalian CNS regeneration failure, and how these two processes interrelate. Studies in amphibian show that injured dorsal root axons are able to regenerate into the spinal cord and restore sensorimotor reflex activities (47–51). The mechanisms underlying this regenerative capacity are largely unexplored and deserve more attention.

The preferred model for exploring insufficient neuronal growth responses is to cut or crush dorsal root(s) at a distance from the DRTZ. Lesions at the lumbosacral level offer the best possibility to perform this with minimal direct disturbance of the spinal cord. Typically, one or more of lumbar dorsal roots L4–6 are targeted, since they carry all or almost all of the sensory afferents from the sciatic nerve.

4.1. Exploring Properties of Dorsal Root Ganglion Neurons that Promote Regenerative Growth of Dorsal Root Axons

Axotomized CNS neurons appear to be intrinsically deficient in their capacity to mount and sustain an efficient program for long distance axonal growth. Furthermore, this capacity shows a significant decline with age. Dorsal root ganglion neurons respond in a similar way after dorsal root injury. Axotomy of the central processes of dorsal root ganglion neurons elicits a much weaker nerve cell body growth response than axotomy of their peripheral processes. As a consequence, the rate of regeneration of dorsal root axons is significantly less than that of regenerating peripheral sensory axons, ca. 1 and 3 mm, respectively in the rat (2), regenerating dorsal root axons have limited ability to grow for long distances in a peripheral nerve environment (52, 53), and are practically unable to enter the adult spinal cord. The dorsal root injury model has provided important information on interventions that enhance long distance axonal outgrowth, either by directly reactivating the intrinsic growth potential or by extracellular stimuli. The conditioning lesion paradigm has been paramount in this respect, demonstrating that injury of the peripheral processes prior to an injury of the central processes of the same sensory neurons mobilizes growth-associated molecules in dorsal root ganglion cells and enhances their capacity for elongation of injured dorsal root axons (54–59). This action allows limited number of regenerating axons to enter the spinal cord (60), but these presumably belong to a subpopulation of dorsal root axons, since a conditioning lesion appears to selectively stimulate the growth potential of dorsal root ganglion neurons not expressing calcitonin gene-related peptide or binding isolectin B4 (61). The conditioning lesion paradigm has also provided information on conditions that allow elongation of injured axons in the CNS. Thus, injury to peripheral sensory axons results in significant growth of injured primary sensory axons in the dorsal column (62), an effect that can be induced even a considerable time after the central lesion has occurred (63).

Extracellular delivery of growth factors enhances the potential of dorsal root regeneration into the spinal cord. The most striking effects have been shown with intrathecal delivery of neurotrophin (NT)3, or glial cell line-derived neurotrophic factor (GDNF) (64, 65), or systemic administration of Artemin (66, 67), a member of the GDNF family. These effects may be mediated by intracellular pathways which are at least in part similar as in the conditioning lesion situation (57), but concomitant effects on the properties of non-neuronal cells at the DRTZ may also occur (68). Notably, the time window for neurotrophin-mediated entry of regenerating dorsal root axons is restricted, since delayed treatment failed to induce spinal cord ingrowth (69).

Recently new mechanisms regulating the intrinsic growth state of neurons during postnatal development and after axon injury have been explored. Members of the Krüppel-like family of transcription factors regulate axonal growth development and their re-activation postnatally promotes axon regeneration in vivo (70, 71). PTEN (phosphatase and tensin homolog deleted on chromosome 10), inhibits phosphoinositide 3-kinase (PI3-K)/Akt signaling, a pathway critical for sustained neurite outgrowth. PTEN activation also appears to participate in mediating the growth inhibitory effect of myelin-associated glycoprotein (MAG) (72). Blocking the action of PTEN significantly promotes regenerative outgrowth in vivo from retinal ganglion cells (73, 74), corticospinal neurons (75), and even from peripheral nerve axons (76). Another intracellular signaling system which regulates the capacity for neural regeneration is the Jak-STAT (signal transducer and activator of transcription) pathway. SOCS3 (suppressor of cytokine signaling 3) inhibits this pathway, and relieving this inhibition markedly promotes regeneration of optic nerve axons (77), which is further enhanced by co-inactivation of PTEN (78). Combining PTEN inactivation with cAMP and the macrophage derived factor oncomodulin also enhanced growth of injured optic nerve axons (79), although oncomodulin by itself appears to have little effect on the regenerative capacity of injured dorsal root axons (80).

Another line of exploration has highlighted the fundamental role of microtubule stability for axonal regeneration in the CNS (81). Studies in non-mammalian species demonstrated efficient microtubule-based axon regeneration without the formation of growth cones (82). Treatment with the microtubule stabilizing agent taxol promoted axon regeneration in the mammalian spinal cord (83) and optic nerve (84). Furthermore, inhibition of kinesin-5, a motor protein that counteracts microtubule advancement, promotes sensory neurite extension on a growth inhibitory substrate in vitro (85). The dorsal root injury model is undoubtedly a useful model for exploring the intracellular pathways that are

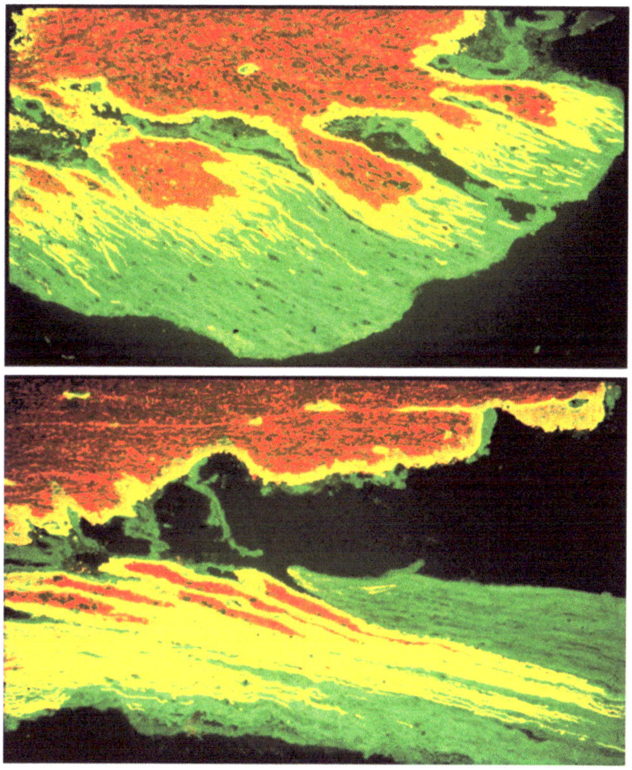

Fig. 2. Lumbar dorsal roots following ipsilateral dorsal root injury. Horizontal section, immunostained with antibodies to laminin (*green*) and GFAP (*red*). Extensive outgrowth of astroglial processes (*red* and *yellow*) into the peripheral compartment is shown at 1 week (*above*) and 2 (below) weeks following injury (×200).

implicated in these effects, how these pathways interact (86) and in the identification of additional, hitherto unknown molecular mechanisms that could be targeted to optimize conditions for regeneration in the spinal cord.

4.2. Exploring the Growth Inhibitory Properties of the DRTZ

Within days after dorsal root injury the DRTZ rapidly presents a virtually full-proof barrier for sensory axons to re-enter the spinal cord. A reasonable explanation for this situation would be the presence of a non-permissive environment for axon growth at the DRTZ, more or less analogous with the inability of adult degenerating CNS tissue in general to support axon growth. After dorsal root injury, DRTZ astrocytes proliferate (87), undergo a marked hypertrophy (88, 89) (Fig. 2), and growth inhibitory proteoglycans as well as myelin-associated growth inhibitors, including Nogo, myelin-associated glycoprotein (MAG) and oligodendrocytes myelin glycoprotein (OMgp) accumulate at this site (90, 91). Blockade of the Nogo-receptor (NgR) using soluble NgR peptide allows regeneration of myelinated axons into the spinal cord, restores connections with postsynaptic neurons and results in some degree of functional recovery (92). However, in other situations when the effect of growth inhibitors are blocked, this needs to be combined with a conditioning lesion of the dorsal root ganglion neurons (93, 94) or combined with potent immunostimulatory agents (95) in order to achieve significant regeneration into the spinal cord. One intriguing feature of the axon-glial interactions at the DRTZ after dorsal root injury is the formation of synapse-like contacts with DRTZ glial cell processes (96–98). The mechanism and significance of this process is still unclear, but its further studies may reveal novel information on the mechanisms preventing regenerating dorsal root axons from entering the spinal cord. Furthermore, exploring the interaction between regenerating dorsal root axons and glial processes my help to reveal novel processes that operate within the spinal cord to prevent long distance extension and/or plasticity of axons that attempt to grow or sprout after injury. In addition, the impact of pericytes and leptomeningeal cells in the emergence of a growth inhibitory environment at the DRTZ warrants exploration in view of their recently described roles in the formation of a scar after spinal cord injury (99, 100).

5. Dorsal Root Injury for Exploring Spinal Cord Injury Pathophysiology

Dorsal root injury is a useful model for exploring the process of axon and myelin breakdown and the associated non-neuronal responses in the spinal cord white matter. The dorsal root injury model is also useful for exploring non-neuronal responses and synaptic plasticity in the partially deafferented spinal cord. Whereas, these aspects are best approached after sectioning one or more dorsal roots, for translation to clinical root avulsion, models of dorsal root avulsion should eventually be the choice.

5.1. Dorsal Root Injury for Examining Mechanisms of Axonal Degeneration and Myelin Breakdown in the Spinal Cord

Section of one or more dorsal roots at the lumbosacral level is usually the best ones for studies on neuronal and non-neuronal processes associated with degeneration of the ascending axons in the dorsal column. An obvious advantage of this model is its absence of direct damage to the spinal cord, implying that its vascular supply and tissue architecture will be essentially untouched.

Axonal degeneration and secondary disintegration of myelin proceed from the injury area toward the spinal cord and along the axonal projections within the spinal cord although terminal parts of the axon appear to undergo degeneration even before their stem axon has disintegrated (101). Astrocytes and microglia undergo marked changes, and interact with infiltrating hematogenous cells in the removal of disintegrating axons and myelin, as well as in the formation of a tissue enriched in filamentous astroglial processes.

Whereas axonal fragments are rapidly removed, myelin debris remains for long periods of time (102–104). In this context, important species variations exist. Remnants of myelin from degenerating heavily myelinated axons remain for decades in humans, and at least for years in, e.g., cats and rabbits. Astroglial processes may even be surrounded by myelin profiles in late stages of Wallerian degeneration in the cat spinal cord (105). One reason for this protracted presence of myelin debris may be an incomplete activation of the molecular mechanisms required for rapid myelin phagocytosis (106–108). The long-term presence of myelin-associated growth inhibitory molecules is likely to be a factor which counteracts growth of axons in the deafferented spinal cord (109). Thus, the dorsal root injury model can help to clarify the impact of the protracted Wallerian degeneration on axon growth in the spinal cord regeneration and possible ways to accelerate the elimination of myelin residues should this be beneficial. Recent studies, using dorsal root injury as experimental system, have also identified mechanisms involved in microglial-mediated clearance of axonal debris, a process which may also influence the conditions for regeneration in the spinal cord (110).

During Wallerian degeneration in the spinal cord white matter of monkeys abnormal profiles termed "myelinoclasts" appear (111). Similar structures were also observed in the cat (112, 113), but not in the rat (103). These profiles were characterized by layers of myelin wrapped around non-neuronal cells, presumably in most cases microglia, which at some later point in time appear themselves to undergo degeneration (113). Thus, the time course of myelin degradation and its associated microglial responses are distinctly different in small (e.g., rat, mouse) compared to large (e.g., cat, monkey, human) mammals, an observation which needs attention in translating findings from experimental models of spinal cord injury repair to humans.

5.2. Dorsal Root Injury for Examining Mechanisms of Neurodegeneration and Repair in the Spinal Cord

Whereas sectioning or crushing dorsal roots has minimal effects on neuron survival in the spinal cord, avulsion of dorsal roots causes a pronounced inflammatory response and extensive neuronal degeneration in the dorsal as well as the ventral horn of the spinal cord (36). Injuries to the spinal cord itself have analogous consequences and exploring ways to counteract these processes in the dorsal root avulsion model will be helpful also for the development of strategies to minimize neuron loss. For clinical translation the dorsal root avulsion model needs eventually to be combined with procedures that induce axon regeneration into the spinal cord and return of sensory function. In this context, re-attaching the avulsed dorsal roots is necessary, such as has been done in recent studies with olfactory ensheathing cell transplants to the DRTZ (114).

6. Experimental Protocols for Rat and Mouse

6.1. General and Anatomical Aspects

The second cervical DRG is the most peripherally located one in relation to the vertebrae and thus more easily exposed than cervical DRGs at other levels. Cervical and thoracic DRGs tend to be located more ventrally with the vertebral column than lumbosacral DRGs. In addition, the rootlets of cervical dorsal roots and the more rostral thoracic ones are highly divergent when coursing from the corresponding DRG toward the spinal cord, making them more difficult to repair after transection compared to lumbosacral dorsal roots.

Dorsal roots are usually exposed via partial laminectomy at the lower cervical or lower lumbar level, i.e., levels corresponding to fore- and hindlimb innervation, respectively. With the short cervical roots the time needed for regenerating dorsal root axons to reach the DRTZ is minimal, thereby reducing the postoperative time needed for outcome analyses. A coincident disadvantage is that a cervical laminectomy will also partially expose the DRTZ and spinal cord. Depending on the issue examined sham operations may therefore be considered to exclude effects from the surgical exposure itself. Sensorimotor forelimb outcome assessments are clearly of particular relevance in non-human primates (see Chapter 10), and may have a greater translational relevance than hindlimb root injuries also in other animal models (30, 114).

Laminectomy to expose lumbar dorsal roots can be made at a considerable distance from the corresponding spinal cord segments, leaving these segments and their surroundings essentially unaffected. Importantly, however, laminectomy may interfere with spinal cord blood flow and thereby affect intraspinal cellular processes (115–117). Sham operations may therefore be necessary, particularly in studies on early and rapid molecular events. With lumbar dorsal root injuries such as root section at the DRTZ or root avulsion, unintended effects on the spinal cord in most instances have to be accounted for by sham operations.

6.2. Preparation and Anesthesia

Animals are anesthetized with isofluran inhalation or by intraperitoneal injection of a mixture of Ketamine, Xylazine and Acepromazine. For mouse: mix 1 ml Ketamine (50 mg/ml), 0.5 ml Xylazine (20 mg/ml), 0.15 ml Acepromazine (10 mg/ml), and 2.35 ml sterile water to make 4 ml solution for use. Recommended dose is 0.08 ml/10 g body weight. For rat: mix 4 ml Ketamine (0.50 mg/ml), 0.5 ml Xylazine (20 mg/ml), 0.3 ml Acepromazine (10 mg/ml), and 5.2 ml sterile water to make 10 ml solution for use. Recommended dose is 0.05 ml/10 g body weight. Solutions have to be kept in darkness. If a longer period of surgery is anticipated, subcutaneous injection of Ringer's lactate solution is recommended prior to surgery. To prolong injection anesthesia, small amounts of Ketamine are added. If inhalation anesthesia is used, the animal will experience pain as soon as anesthesia is discontinued, and in this case analgesic (0.1 mg/kg of buprenorphine i.p.) has to be given prior to or during the early phase of the surgery. Once asleep the animal is placed on a heating pad (caution not to overheat the animal), the skin over the areas to be exposed is shaved and for complete removal of remaining short hairs, a drop of hair removal solution can be used. Efficient cleaning of the shaved skin is achieved by 70 % (v/v) isopropyl alcohol and 10 % (v/v) povidone-iodine solution. Since animals are unable to close their eyes, a drop of ocular lubricant should be placed in each eye to prevent them from drying.

6.3. Surgery

Here, we describe a procedure for exposing dorsal roots and lesion them. Detailed descriptions have been published for producing dorsal root avulsion injuries by pulling spinal nerves (32, 33) or dorsal root themselves (37, 38). The surgical approach is similar in mice and rat, but instruments have to be smaller and tissue manipulations more gentle in mice than rats. In mice, skin incision is made with fine scissors; in rat a scalpel blade can be used. A visual description of the surgical procedure for exposing and crushing lumbar dorsal roots in the mouse has been published (118). Depending on which dorsal roots are targets for the surgery, the following landmarks are helpful:

- The second cervical vertebra has a markedly longer process compared to the first one and those below.
- The second thoracic vertebra has a markedly longer spinous process compared to those above.
- The caudal end of the T13 vertebra corresponds to the L2–3 border.
- The upper part of the iliac crest is at the level of the L5 dorsal root ganglion.

After opening the skin at cervical or high thoracic levels, fat tissues between the shoulder blades are cut and moved to the sides to expose the relevant vertebrae. The dorsal fascia and musculature which insert at the spinous processes of the vertebrae are cut along one side and gently scraped laterally from the dorsal surface of the vertebral laminae which will be partially removed. The intervertebral ligament below the first laminectomy is cut, providing a small opening for a bone cutter. A spinous process (rat) or vertebra (mouse) rostral to the site of laminectomy is kept with clawed forceps to stabilize the spine, and parts of one or more laminae on the right or left side, depending on manual preferences, are gently removed with bone cutter, starting caudally and medially. Bleeding is usually minimal close to the midline, but blood vessels run along the inside of the vertebral canal and are at risk when the laminectomy is extended laterally. For exposure at the lumbosacral level risk of accidental damage to the relatively mobile dorsal roots of cauda equina has to be borne in mind. For the cervical level, attention has to be paid to the vertebral artery which runs within a set of openings in the transverse processes of vertebrae C1–7. For cervical as well as thoracic levels, the distance from the vertebral laminae to the dorsal surface of the spinal cord is close and bone removal therefore requires great caution to avoid accidental spinal cord compression. To secure that dorsal root manipulation prior to their injury will not result in reflex-induced movements a small amount of local anesthetic (lidocaine) is dropped on the root(s).

The injuries are performed as follows:

- Dorsal root(s) cut between the entry of the spinal cord and dorsal root ganglia is done with fine microscissors.
- Dorsal root(s) crush between the spinal cord and dorsal root ganglia is done by squeezing the root(s) for 10 s with fine jeweler's forceps. Because of their delicate tissue too forceful or too long time of squeezing is likely to interrupt the root completely and has to be avoided.
- The entry site of the targeted dorsal root is exposed, the dura carefully opened with microscissors and the root cut just outside and parallel with the spinal cord. This injury will cut through peripheral and well central nervous tissue (central tissue projections).
- To mimic dorsal root avulsion, the entry of the targeted dorsal root is exposed, the dura carefully opened with fine microscissors, and the root kept with fine forceps and pulled caudally parallel with the spinal cord.

At the end of surgery, muscles are closed with a few resorbable sutures (4-0 for rat, 6-0 for mouse), and skin preferably with Michel clips (no. 12040-01, Fine Science Tools). The animals are typically beginning to move around shortly after inhalation anesthesia is terminated or about 1–2 h after Ketamine anesthesia. Until this stage, animals have to be kept warm, e.g., by a heating pad; however, care has to be taken not to overheat.

Animals are observed daily and given analgesics during the first few postoperative days. Multiple dorsal root injuries often result in behavioral signs which may indicate experiences of pain, e.g., mechanical/thermal hypersensitivity, various forms of allodynia (painful experience of normally non-painful stimulus) and/or paw self-mutilation behavior (autotomy). Monitoring these signs is necessary from an animal welfare aspect, and should be an important part of the behavioral outcome assessments after interventions.

7. Notes on Outcome Assessments of Specific Relevance for Dorsal Root Injury Models

Anatomical, physiological, and behavioral outcome assessments following dorsal root injury experiments are largely region-related adaptations of general methodology for analyzing neuron survival, neuronal and non-neuronal responses, neuronal connectivity, and sensorimotor functions in the spinal cord. However, the difficulties to prove anatomically that injured axons have grown (regenerated) into the spinal cord, mediate physiological effects and are responsible for observed functional improvements, are often underestimated. Various forms of interconnections are common between adjacent dorsal roots which provide circumventing conduits from the dorsal root ganglia into the spinal cord (119, 120). As a result axons of some dorsal root ganglion neurons will enter the spinal cord one or more segments above or below the level of the ganglion itself. To account for these interconnections, a minimum of one dorsal root above and one below a repaired root have to be sectioned to allow acceptable anatomical documentation of sensory axon regeneration into the spinal cord. Under these circumstances, regenerating dorsal root axons can be identified using sensory axon markers, neuroanatomical tracers placed on the proximal stump of the dorsal root, injected into the relevant dorsal root ganglion, its spinal nerve, or an appropriate peripheral nerve. Tracer injection into the sciatic nerve is a simple procedure and often the preferred choice for dorsal root regeneration studies at the lower lumbar level. The use of transgenic mice expressing proteins present in regenerating and/or mature axons is attractive. However, expression levels in non-myelinated axons may be too low to be detected in currently used transgenic strains, thus limiting their utility.

When using crush injury, securing the completeness of the injury is obviously crucial. However, to completely exclude the presence of remaining uninterrupted fibers may require detailed anatomical and physiological analyses. A control experiment to prove that return of function is the result of axon regenerating into the spinal cord is to cut the relevant dorsal roots after a steady state of functional improvement has been reached, and verify the immediate loss of previously recovered functions. This control may be particularly important when the crush injury model is used, but is also of obvious relevance in other models of dorsal root injury.

8. Translation of Animal Models to Clinical Plexus Injuries

Dorsal root injury is a useful animal model for several principal aspects relevant to spinal cord injury repair. For dorsal root injuries that occur as a result of plexus avulsion, dorsal root avulsion model in non-human mammals is the model that most closely mimics human plexus injuries. Thus, for translational purposes the use of this injury model and its further refinement in rodents as well as in larger mammals and non-human primates should be encouraged.

Acknowledgments

The author's cited work was supported by the Swedish Research Council, The Biomed II Program of the European Commission, Stiftelsen Olle Engkvist, and Sighnhild Engkvist's Stiftelse.

References

1. Havton LA, Carlstedt T (2009) Repair and rehabilitation of plexus and root avulsions in animal models and patients. Curr Opin Neurol 22(6):570–574
2. Richardson PM, Verge VM (1987) Axonal regeneration in dorsal spinal roots is accelerated by peripheral axonal transection. Brain Res 411(2):406–408
3. Masuda T, Shiga T (2005) Chemorepulsion and cell adhesion molecules in patterning initial trajectories of sensory axons. Neurosci Res 51(4):337–347
4. Niederländer C, Lumsden A (1996) Late emigrating neural crest cells migrate specifically to the exit points of cranial branchiomotor nerves. Development 122(8):2367–2374
5. Golding JP, Cohen J (1997) Border controls at the mammalian spinal cord: late-surviving neural crest boundary cap cells at dorsal root entry sites may regulate sensory afferent ingrowth and entry zone morphogenesis. Mol Cell Neurosci 9(5–6):381–396
6. Golding J, Shewan D, Cohen J (1997) Maturation of the mammalian dorsal root entry zone—from entry to no entry. Trends Neurosci 20(7):303–308
7. Mauti O, Domanitskaya E, Andermatt I, Sadhu R, Stoeckli ET (2007) Semaphorin6A acts as a gate keeper between the central and the peripheral nervous system. Neural Dev 2:28
8. Coulpier F, Decker L, Funalot B, Vallat JM, Garcia-Bragado F, Charnay P, Topilko P (2010) CNS/PNS boundary transgression by central glia in the absence of Schwann cells or Krox20/Egr2 function. J Neurosci 30(17):5958–5967
9. Fraher JP, The SMM (1987) CNS-PNS transitional zone of rat cervical dorsal roots during development and at maturity. A morphological and morphometric study. J Anat 152: 189–203
10. Fraher JP (1992) The CNS-PNS, transitional zone of the rat. Morphometric studies at cranial and spinal levels. Prog Neurobiol 38(3): 261–316
11. Fraher J (2002) Axons and glial interfaces: ultrastructural studies. J Anat 200(4):415–430
12. Berthold CH, Carlstedt T (1977) Observations on the morphology at the transition between the peripheral and the central nervous system in the cat. V. A light microscopical and histochemical study of S1 dorsal rootlets in developing kittens. Acta Physiol Scand Suppl 446:73–85
13. Carlstedt T, Dalsgaard CJ, Molander C (1987) Regrowth of lesioned dorsal root nerve fibers into the spinal cord of neonatal rats. Neurosci Lett 74(1):14–18
14. Carlstedt T (1988) Reinnervation of the mammalian spinal cord after neonatal dorsal root crush. J Neurocytol 17(3):335–350

15. Aldskogius H, Kozlova EN (2002) Strategies for repair of the deafferented spinal cord. Brain Res Brain Res Rev 40(1–3):301–308
16. Berthold CH, Carlstedt T (1977) Observations on the morphology at the transition between the peripheral and the central nervous system in the cat. II. General organization of the transitional region in S1 dorsal rootlets. Acta Physiol Scand Suppl 446:23–42
17. Berthold CH, Carlstedt T (1977) Observations on the morphology at the transition between the peripheral and the central nervous system in the cat. III. Myelinated fibres in S1 dorsal rootlets. Acta Physiol Scand Suppl 446:43–60
18. Hildebrand C, Remahl S, Persson H, Bjartmar C (1993) Myelinated nerve fibres in the CNS. Prog Neurobiol 40(3):319–384
19. Carlstedt T (1977) Observations on the morphology at the transition between the peripheral and the central nervous system in the cat. IV. Unmyelinated fibres in S1 dorsal rootlets. Acta Physiol Scand Suppl 446:61–72
20. Smith GM, Falone AE, Frank E (2012) Sensory axon regeneration: rebuilding functional connections in the spinal cord. Trends Neurosci 35(3):156–163
21. Saijilafu, Hur EM, Zhou FQ (2011) Genetic dissection of axon regeneration via in vivo electroporation of adult mouse sensory neurons. Nat Commun 2:543.
22. Kozlova EN, Seiger Å, Aldskogius H (1997) Human dorsal root ganglion neurons from embryonic donors extend axons into the host rat spinal cord along laminin peripheral surroundings of the dorsal root transitional zone. J Neurocytol 26(12):811–822
23. Levinsson A, Holmberg H, Schouenborg J, Seiger Å, Aldskogius H, Kozlova EN (2000) Functional connections are established in the deafferented rat spinal cord by peripherally transplanted human embryonic sensory neurons. Eur J Neurosci 12(10):3589–3595
24. Åkesson E, Sandelin M, Kanaykina N, Aldskogius H, Kozlova EN (2008) Long-term survival, robust neuronal differentiation, and extensive migration of human forebrain stem/progenitor cells transplanted to the adult rat dorsal root ganglion cavity. Cell Transplant 17(10–11):1111123
25. Aldskogius H, Berens C, Kanaykina N, Liakhovitskaia A, Medvinsky A, Sandelin M, Schreiner S, Wegner M, Hjerling-Leffler J, Kozlova EN (2009) Regulation of boundary cap neural crest stem cell differentiation after transplantation. Stem Cells 27(7):1592–1603
26. König N, Åkesson E, Telorack M, Vasylovska S, Ngamjariyawat A, Sundström E, Oster A, Trolle C, Berens C, Aldskogius H, Seiger Å, Kozlova EN (2011) Forced Runx1 expression in human neural stem/progenitor cells transplanted to the rat dorsal root ganglion cavity results in extensive axonal growth specifically from spinal cord-derived neurospheres. Stem Cells Dev 20(11):1847–1857
27. Ramon y Cajal S (1991) Degeneration and regeneration of the nervous system. In: DeFelipe J, Jones EG (eds) Translated by R. M. May. Oxford University Press
28. Spiegel DA, Seaber AV, Chen LE, Urbaniak JR (1993) Recovery following stretch injury to the sciatic nerve of the rat: an in vivo study. J Reconstr Microsurg 9(1):69–74
29. Ibrahim AG, Raisman G, Li Y (2009) Permanent loss of fore-paw grasping requires complete deprivation of afferent input from a minimum of four dorsal roots of the rat brachial plexus. Exp Neurol 215(1):142–145
30. Ibrahim AG, Kirkwood PA, Raisman G, Li Y (2009) Restoration of hand function in a rat model of repair of brachial plexus injury. Brain 132(Pt 5):1268–1276
31. Spinner RJ, Khoobehi A, Kazmi S, Krumreich JA, Zhao S, Zhang Z et al (2000) Model for avulsion injury in the rat brachial plexus using passive acceleration. Microsurgery 20(2):94–97
32. Cao XC, Ling LJ (2003) Anatomic basis and technical aspects of a new brachial plexus avulsion injury model in the rat. Plast Reconstr Surg 111(7):2488–2490
33. Rodrigues-Filho R, Santos AR, Bertelli JA, Calixto JB (2003) Avulsion injury of the rat brachial plexus triggers hyperalgesia and allodynia in the hindpaws: a new model for the study of neuropathic pain. Brain Res 982(2):186–194
34. Quintão NL, Balz D, Santos AR, Campos MM, Calixto JB (2006) Long-lasting neuropathic pain induced by brachial plexus injury in mice: role triggered by the pro-inflammatory cytokine, tumour necrosis factor alpha. Neuropharmacology 50(5):614–620
35. Paszcuk AF, Dutra RC, da Silva KA, Quintão NL, Campos MM, Calixto JB (2011) Cannabinoid agonists inhibit neuropathic pain induced by brachial plexus avulsion in mice by affecting glial cells and MAP kinases. PLoS One 6(9):e24034
36. Wieseler J, Ellis AL, McFadden A, Brown K, Starnes C, Maier SF, Watkins LR, Falci S (2010) Below level central pain induced by discrete dorsal spinal cord injury. J Neurotrauma 27:1697–1707
37. Chew DJ, Carlstedt T, Shortland PJ (2011) A comparative histological analysis of two models of nerve root avulsion injury in the adult rat. Neuropathol Appl Neurobiol 37(6):613–632

38. Wieseler J, Ellis A, Maier SF, Watkins LR, Falci S (2012) Unilateral T13 and L1 dorsal root avulsion: methods for a novel model of central neuropathic pain. Methods Mol Biol 851:171–183
39. Lawson SN (1995) Neuropeptides in morphologically and functionally identified primary afferent neurons in dorsal root ganglia: substance P, CGRP and somatostatin. Prog Brain Res 104:161–173
40. Lawson SN (1996) Peptides and cutaneous polymodal nociceptor neurones. Prog Brain Res 113:369–385
41. Alvares D, Fitzgerald M. Building blocks of pain: the regulation of key molecules in spinal sensory neurones during development and following peripheral axotomy. Pain 1999; Suppl 6:S71–S85.
42. Lawson SN (2002) Phenotype and function of somatic primary afferent nociceptive neurones with C-, Adelta- or Aalpha/beta-fibres. Exp Physiol 87(2):239–244
43. Light AR, Perl ER (1979) Reexamination of the dorsal root projection to the spinal dorsal horn including observations on the differential termination of coarse and fine fibers. J Comp Neurol 186(2):117–131
44. Hoheisel U, Lehmann-Willenbrock E, Mense S (1989) Termination patterns of identified group II and III afferent fibres from deep tissues in the spinal cord of the cat. Neuroscience 28(2):495–507
45. Jänig W (1996) Neurobiology of visceral afferent neurons: neuroanatomy, functions, organ regulations and sensations. Biol Psychol 42(1–2):29–51
46. Li L, Rutlin M, Abraira VE, Cassidy C, Kus L, Gong S, Jankowski MP, Luo W, Heintz N, Koerber HR, Woodbury CJ, Ginty DD (2011) The functional organization of cutaneous low-threshold mechanosensory neurons. Cell 147(7):1615–1627
47. Katzenstein MB, Bohn RC (1984) Regeneration of transected dorsal root ganglion cell axons into the spinal cord in adult frogs (Xenopus laevis). Brain Res 300(1): 188–191
48. Frank E, Sah DW (1986) Reformation of specific synaptic connections by regenerating sensory axons in the spinal cord of the bullfrog. Neurochem Pathol 5(3):165–185
49. Peng YY, Frank E (1988) Anatomical specificity of regenerated muscle sensory afferents in the spinal cord of the bullfrog. J Neurobiol 19(8):727–742
50. Liuzzi FJ, Lasek RJ (1986) Dorsal root axonal regeneration in the adult frog spinal cord. A model of vertebrate CNS regeneration. Neurochem Pathol 5(3):237–253
51. Mears SC, Frank E (1994) Specific regeneration of cutaneous sensory afferent fibers in the amphibian spinal cord. Exp Neurol 130(1):115–119
52. Richardson PM, Issa VM (1984) Peripheral injury enhances central regeneration of primary sensory neurones. Nature 309(5971):791–793
53. Richardson PM, Verge VM (1986) The induction of a regenerative propensity in sensory neurons following peripheral axonal injury. J Neurocytol 15(5):585–594
54. Neumann S, Bradke F, Tessier-Lavigne M, Basbaum AI (2002) Regeneration of sensory axons within the injured spinal cord induced by intraganglionic cAMP elevation. Neuron 34(6):885–893
55. Qiu J, Cai D, Dai H, McAtee M, Hoffman PN, Bregman BS, Filbin MT (2002) Spinal axon regeneration induced by elevation of cyclic AMP. Neuron 34(6):895–903
56. Wu D, Zhang Y, Bo X, Huang W, Xiao F, Zhang X, Miao T, Magoulas C, Subang MC, Richardson PM (2007) Actions of neuropoietic cytokines and cyclic AMP in regenerative conditioning of rat primary sensory neurons. Exp Neurol 204(1):66–76
57. Hoffman PN (2010) A conditioning lesion induces changes in gene expression and axonal transport that enhance regeneration by increasing the intrinsic growth state of axons. Exp Neurol 223(1):11–18
58. Yang P, Yang Z (2012) Enhancing intrinsic growth capacity promotes adult CNS regeneration. J Neurol Sci 312(1–2):1–6
59. Blesch A, Lu P, Tsukada S, Alto LT, Roet K, Coppola G, Geschwind D, Tuszynski MH (2012) Conditioning lesions before or after spinal cord injury recruit broad genetic mechanisms that sustain axonal regeneration: Superiority to camp-mediated effects. Exp Neurol 235(1):162–173
60. Chong MS, Woolf CJ, Haque NS, Anderson PN (1999) Axonal regeneration from injured dorsal roots into the spinal cord of adult rats. J Comp Neurol 410(1):42–54
61. Kalous A, Keast JR (2010) Conditioning lesions enhance growth state only in sensory neurons lacking calcitonin gene-related peptide and isolectin B4-binding. Neuroscience 166(1):107–121
62. Neumann S, Woolf CJ (1999) Regeneration of dorsal column fibers into and beyond the lesion site following adult spinal cord injury. Neuron 23(1):83–91

63. Ylera B, Ertürk A, Hellal F, Nadrigny F, Hurtado A, Tahirovic S, Oudega M, Kirchhoff F, Bradke F (2009) Chronically CNS-injured adult sensory neurons gain regenerative competence upon a lesion of their peripheral axon. Curr Biol 19(11):930–936
64. Ramer MS, Priestley JV, McMahon SB (2000) Functional regeneration of sensory axons into the adult spinal cord. Nature 403(6767): 312–316
65. Ramer MS, Bishop T, Dockery P, Mobarak MS, O'Leary D, Fraher JP, Priestley JV, McMahon SB (2002) Neurotrophin-3-mediated regeneration and recovery of proprioception following dorsal rhizotomy. Mol Cell Neurosci 19(2):239–249
66. Wang R, King T, Ossipov MH, Rossomando AJ, Vanderah TW, Harvey P, Cariani P, Frank E, Sah DWY, Porreca F (2008) Persistent restoration of sensory function by immediate or delayed systemic artemin after dorsal root injury. Nat Neurosci 11(4):488–496
67. Harvey P, Gong B, Rossomando AJ, Frank E (2010) Topographically specific regeneration of sensory axons in the spinal cord. Proc Natl Acad Sci USA 107(25):11585–11590
68. Hanna-Mitchell AT, O'Leary D, Mobarak MS, Ramer MS, McMahon SB, Priestley JV, Kozlova EN, Aldskogius H, Dockery P, Fraher JP (2008) The impact of neurotrophin-3 on the dorsal root transitional zone following injury. Spinal Cord 46(12):804–810
69. Ramer MS, Duraisingam I, Priestley JV, McMahon SB (2001) Two-tiered inhibition of axon regeneration at the dorsal root entry zone. J Neurosci 21(8):2651–2660
70. Moore DL, Blackmore MG, Hu Y, Kaestner KH, Bixby JL, Lemmon VP et al (2009) KLF family members regulate intrinsic axon regeneration ability. Science 326(5950):298–301
71. Blackmore MG, Wang Z, Lerch JK, Motti D, Zhang YP, Shields CB et al (2012) Kruppel-like Factor 7 engineered for transcriptional activation promotes axon regeneration in the adult corticospinal tract. Proc Natl Acad Sci USA 109(19):7517–7522
72. Perdigoto AL, Chaudhry N, Barnes GN, Filbin MT, Carter BD (2011) A novel role for PTEN in the inhibition of neurite outgrowth by myelin-associated glycoprotein in cortical neurons. Mol Cell Neurosci 46(1):235–244
73. Park KK, Liu K, Hu Y, Smith PD, Wang C, Cai B, Xu B, Connolly L, Kramvis I, Sahin M, He Z (2008) Promoting axon regeneration in the adult CNS by modulation of the PTEN/mTOR pathway. Science 322(5903):963–966
74. Park KK, Liu K, Hu Y, Kanter JL, He Z (2010) PTEN/mTOR and axon regeneration. Exp Neurol 223(1):45–50
75. Liu K, Lu Y, Lee JK, Samara R, Willenberg R, Sears-Kraxberger I, Tedeschi A, Park KK, Jin D, Cai B, Xu B, Connolly L, Steward O, Zheng B, He Z (2010) PTEN deletion enhances the regenerative ability of adult corticospinal neurons. Nat Neurosci 13(9): 1075–1078
76. Christie KJ, Webber CA, Martinez JA, Singh B, Zochodne DW (2010) PTEN inhibition to facilitate intrinsic regenerative outgrowth of adult peripheral axons. J Neurosci 30(27): 9306–9315
77. Smith PD, Sun F, Park KK, Cai B, Wang C, Kuwako K, Martinez-Carrasco I, Connolly L, He Z (2009) SOCS3 deletion promotes optic nerve regeneration in vivo. Neuron 64(5):617–623
78. Sun F, Park KK, Belin S, Wang D, Lu T, Chen G, Zhang K, Yeung C, Feng G, Yankner BA, He Z (2011) Sustained axon regeneration induced by co-deletion of PTEN and SOCS3. Nature 480(7377):372–375
79. Kurimoto T, Yin Y, Omura K, Gilbert HY, Kim D, Cen LP, Moko L, Kügler S, Benowitz LI (2010) Long-distance axon regeneration in the mature optic nerve: contributions of oncomodulin, cAMP, and pten gene deletion. J Neurosci 30(46):15654–15663
80. Harel R, Iannotti CA, Hoh D, Clark M, Silver J, Steinmetz MP (2012) Oncomodulin affords limited regeneration to injured sensory axons in vitro and in vivo. Exp Neurol 233(2): 708–716
81. Hur EM, Saijilafu, Zhou FQ. Growing the growth cone: remodeling the cytoskeleton to promote axon regeneration. Trends Neurosci 2012; 35(3):164–174.
82. Jin LQ, Zhang G, Jamison C Jr, Takano H, Haydon PG, Selzer ME (2009) Axon regeneration in the absence of growth cones: acceleration by cyclic AMP. J Comp Neurol 515(3):295–312
83. Hellal F, Hurtado A, Ruschel J, Flynn KC, Laskowski CJ, Umlauf M, Kapitein LC, Strikis D, Lemmon V, Bixby J, Hoogenraad CC, Bradke F (2011) Microtubule stabilization reduces scarring and causes axon regeneration after spinal cord injury. Science 331(6019): 928–931
84. Sengottuvel V, Leibinger M, Pfreimer M, Andreadaki A, Fischer D (2011) Taxol facilitates axon regeneration in the mature CNS. J Neurosci 31(7):2688–2699

85. Lin S, Liu M, Son YJ, Timothy Himes B, Snow DM, Yu W, Baas PW (2011) Inhibition of Kinesin-5, a microtubule-based motor protein, as a strategy for enhancing regeneration of adult axons. Traffic 12(3):269–286
86. Miao T, Wu D, Wheeler A, Wang P, Zhang Y, Bo X, Yeh JS, Subang MC, Richardson PM (2011) Two cytokine signaling molecules cooperate to promote axonal transport and growth. Exp Neurol 228(2):165–172
87. Liu L, Rudin M, Kozlova EN (2000) Glial cell proliferation in the spinal cord after dorsal rhizotomy or sciatic nerve transection in the adult rat. Exp Brain Res 131(1):64–73
88. Fraher JP (1999) The transitional zone and CNS regeneration. J Anat 194(Pt 2):161–182, Corrected and republished in: J Anat 2000; 196 (Pt 1):137–158
89. Kozlova EN, Lukanidin E (1999) Metastasis-associated Mts1 (S100A4) protein is selectively expressed in white matter astrocytes and upregulated after dorsal root or peripheral nerve injury. Glia 27(3):249–258
90. Giger RJ, Venkatesh K, Chivatakarn O, Raiker SJ, Robak L, Hofer T, Lee H, Rader C (2008) Mechanisms of CNS myelin inhibition: evidence for distinct and neuronal cell type specific receptor systems. Restor Neurol Neurosci 26(2–3):97–115
91. Gardiner NJ (2011) Integrins and the extracellular matrix: key mediators of development and regeneration of the sensory nervous system. Dev Neurobiol 71(11):1054–1072
92. Harvey PA, Lee DH, Qian F, Weinreb PH, Frank E (2009) Blockade of Nogo receptor ligands promotes functional regeneration of sensory axons after dorsal root crush. J Neurosci 29(19):6286295
93. Steinmetz MP, Horn KP, Tom VJ, Miller JH, Busch SA, Nair D, Silver DJ, Silver J (2005) Chronic enhancement of the intrinsic growth capacity of sensory neurons combined with the degradation of inhibitory proteoglycans allows functional regeneration of sensory axons through the dorsal root entry zone in the mammalian spinal cord. J Neurosci 25(35):8066–8076
94. Zhang Y, Zhang X, Wu D, Verhaagen J, Richardson PM, Yeh J, Bo X (2007) Lentiviral-mediated expression of polysialic acid in spinal cord and conditioning lesion promote regeneration of sensory axons into spinal cord. Mol Ther 15(10):1796–1804
95. Busch SA, Horn KP, Silver DJ, Silver J (2009) Overcoming macrophage-mediated axonal dieback following CNS injury. J Neurosci 29(32):9967–9976
96. Carlstedt T (1985) Regenerating axons form nerve terminals at astrocytes. Brain Res 347(1):188–191
97. Liuzzi FJ, Lasek RJ (1987) Astrocytes block axonal regeneration in mammals by activating the physiological stop pathway. Science 237(4815):642–645
98. Di Maio A, Skuba A, Himes BT, Bhagat SL, Hyun JK, Tessler A, Bishop D, Son Y-J (2011) In vivo imaging of dorsal root regeneration: rapid immobilization and presynaptic differentiation at the CNS/PNS border. J Neurosci 31(12):4569–4582
99. Göritz C, Dias DO, Tomilin N, Barbacid M, Shupliakov O, Frisén J (2011) A pericyte origin of spinal cord scar tissue. Science 333(6039):238–242
100. Decimo I, Bifari F, Rodriguez FJ, Malpeli G, Dolci S, Lavarini V, Pretto S, Vasquez S, Sciancalepore M, Montalbano A, Berton V, Krampera M, Fumagalli G (2011) Nestin-and doublecortin-positive cells reside in adult spinal cord meninges and participate in injury-induced parenchymal reaction. Stem Cells 29(12):2062–2076
101. George R, Griffin JW (1994) The proximo-distal spread of axonal degeneration in the dorsal columns of the rat. J Neurocytol 23(11):657–667
102. Bignami A, Dahl D, Nguyen BT, Crosby CJ (1981) The fate of axonal debris in Wallerian degeneration of rat optic and sciatic nerves. Electron microscopy and immunofluorescence studies with neurofilament antisera. J Neuropathol Exp Neurol 40(5):537–550
103. George R, Gfiffin JW (1994) Delayed macrophage responses and myelin clearance during Wallerian degeneration in the central nervous system: the dorsal radiculotomy model. Exp Neurol 129(2):225–236
104. Zhang J, Jones M, DeBoy CA, Reich DS, Farrell JA, Hoffman PN et al (2009) Diffusion tensor magnetic resonance imaging of Wallerian degeneration in rat spinal cord after dorsal root axotomy. J Neurosci 29(10):3160–3171
105. Bignami A, Ralston HJ III (1968) Myelination of fibrillary astroglial processes in long term Wallerian degeneration. The possible relationship to 'status marmoratus'. Brain Res 11(3):710–713
106. Liu L, Persson JK, Svensson M, Aldskogius H (1998) Glial cell responses, complement, and clusterin in the central nervous system following dorsal root transection. Glia 23(3):221–238
107. Rotshenker S (2009) The role of Galectin-3/MAC-2 in the activation of the innate-immune

function of phagocytosis in microglia in injury and disease. J Mol Neurosci 39(1–2):99–103
108. Rotshenker S (2011) Wallerian degeneration: the innate-immune response to traumatic nerve injury. J Neuroinflammation 8:109
109. McPhail LT, Borisoff JF, Tsang B, Hwi LP, Kwiecien JM, Ramer MS (2007) Protracted myelin clearance hinders central primary afferent regeneration following dorsal rhizotomy and delayed neurotrophin-3 treatment. Neurosci Lett 411(3):206–211
110. Hosmane S, Tegenge MA, Rajbhandari L, Uapinyoying P, Kumar NG, Thakor N et al (2012) Toll/interleukin-1 receptor domain-containing adapter inducing interferon-β mediates microglial phagocytosis of degenerating axons. J Neurosci 32(22):7745–7757
111. Daniel PM, Strich SJ (1969) Histological observations on Wallerian degeneration in the spinal cord of the baboon, Papio papio. Acta Neuropathol 12(4):314–328
112. Bignami A, Ralston HJ 3rd (1969) The cellular reaction to Wallerian degeneration in the central nervous system of the cat. Brain Res 13(3):444–461
113. Aldskogius H (1974) Indirect and direct Wallerian degeneration in the intramedullary root fibres of the hypoglossal nerve. An electron microscopical study in the kitten. Adv Anat Embryol Cell Biol 50(1):7–78
114. Kachramanoglou C, Li D, Andrews P, East C, Carlstedt T, Raisman G, Choi D (2011) Novel strategies in brachial plexus repair after traumatic avulsion. Br J Neurosurg 25(1):16–27
115. Anderson DK, Nicolosi GR, Means ED, Hartley LE (1978) Effect of laminectomy on spinal cord blood flow. J Neurosurg 48(2): 232–238
116. Anderson DK, Means ED, Waters TR (1980) Spinal cord energy metabolism in normal and postlaminectomy cats. J Neurosurg 52(3): 387–392
117. Hales JR, Yeo D, Stabback S, Fawcett AA, Kearns R (1981) Effects of anesthesia and laminectomy on regional spinal cord blood flow in conscious sheep. J Neurosurg 54(5):620–626
118. Skuba A, Himes BT, Son YJ (2011) Live imaging of dorsal root axons after rhizotomy. J Vis Exp 55:e3126
119. Jacob M, Weddell G (1975) Neural intersegmental connection in the spinal root and ganglion region of the rat. J Comp Neurol 161(1):115–123
120. Moriishi J, Otani K, Tanaka K, Inoue S (1989) The intersegmental anastomoses between spinal nerve roots. Anat Rec 224(1):110–116

Chapter 6

No-Laminectomy Spinal Cord-Transected Murine Model

Pierre A. Guertin

Abstract

Most animals of contused, compressed, or transected spinal cord injury require that a laminectomy is performed. However, increasing evidence suggests that laminectomies may induce undesirable effects including neuropathic pain, spinal instabilities, longer anesthetic and related consequences, as well as lordosis and other biomechanical and behavioral problems. Here, we present a method for spinal cord transection in adult mice in which laminectomies do not need to be performed. Under isoflurane anesthesia, a partial section of the interspinous ligaments dorsally between the ninth and tenth thoracic vertebrae was performed in order to allow extra fine scissors to be inserted between the corresponding vertebrae and a complete transection to be properly achieved. Once the transection completed (overall procedures take approximately 10–15 min), the surgical area was closed with sutures clips and animals were allowed to recover on heating pads. Histological data from early and late chronic spinal cord-transected mice showed that complete paraplegia with this approach resulted of a 0.25–1 mm scar occurring at the lesion site. Several studies conducted in our laboratory provided evidence that it is a simple, rapid, reliable and reproducible model that may be particularly useful for studies on behaviors, motor and locomotor movements requiring large cohorts of completely paraplegic mice—e.g., for drug screening in vivo and studies aimed at characterizing neuronal and non-neuronal adaptive changes following rehabilitation.

Key words: Spinal cord transection, Mouse, Paraplegia, Complete spinal cord injury

1. Introduction

Chronic spinal cord injury (SCI) is associated with the development of serious medical problems. Generally, SCI patients who survive the first 24 h will, indeed, rapidly develop within a few weeks to a few years, cardiovascular problems, type II diabetes, sarcopenia, osteoporosis, immune deficiencies, and other life-threatening problems (1–10). The cellular mechanisms underlying these so-called secondary complications remain unclear and no drug treatment has been developed to specifically treat these medical concerns more extensively found in severe SCI cases (completely or near completely injured patients corresponding

Håkan Aldskogius (ed.), *Animal Models of Spinal Cord Repair*, Neuromethods, vol. 76,
DOI 10.1007/978-1-62703-197-4_6, © Springer Science+Business Media, LLC 2013

to ASIA-A and ASIA-B based on the *American Spinal Injury Association* classification (see Chap. 1)). This may be associated with the fact that animal models ideally suited to investigate these problems have not been developed (11–15).

Here, we describe an animal model of complete paraplegia in which so-called secondary complications (health problems) and chronic adaptations post-injury could be studied with ease. In order to investigate the cellular, metabolic, and systemic changes associated with severe and chronic SCI, we developed a "no-laminectomy" completely spinal cord-transected mouse model. Given the critical need of patients and clinicians for innovative treatments designed specifically for chronic SCI (16–18), the experimental procedures were adapted to optimize the preparation in several ways—minimizing surgical time, increasing survival, achieving complete lesion, reducing physical and psychological stress for the animals, etc. This model is ideally suited also for drug screening studies on large cohorts of paraplegic animals with minimal variability due to lesion (unlike most incompletely SCI models) which greatly ease comparisons between drug candidates.

Other no-laminectomy approaches have been described elsewhere also with the main objective of developing a simple and reliable animal model of complete SCI (14). However, the present animal model developed in our laboratory is the first in which a complete spinal cord transection intervertebrally is performed in non-laminectomized mice.

2. Animal Model and Surgical Procedures

2.1. Rationale

The model has been reported for the first time a few years ago (19) although a detailed description has only been sent for publication recently (20). Mice were chosen over other species for several reasons. Firstly, mice are associated with a significant number of commercially available genetic tools including knock-out and transgenic murine models (15). Secondly, space required for housing and animal care costs are typically lower than for other species which is also an important factor for researchers working on large cohorts of animals (e.g., groups can be housed together in relatively small cages). Thirdly, mice are probably among the least expensive species to purchase. Altogether, these factors are generally considered as important specifically for studies aimed at characterizing mechanisms and/or conducting drug screening studies on large cohorts of animals (20).

2.2. Pre-op and Surgery

In brief, all experimental procedures are conducted in accordance with the Canadian Council on Animal Care guidelines. Mice are generally housed 4–5 animals per cage in a controlled-temperature

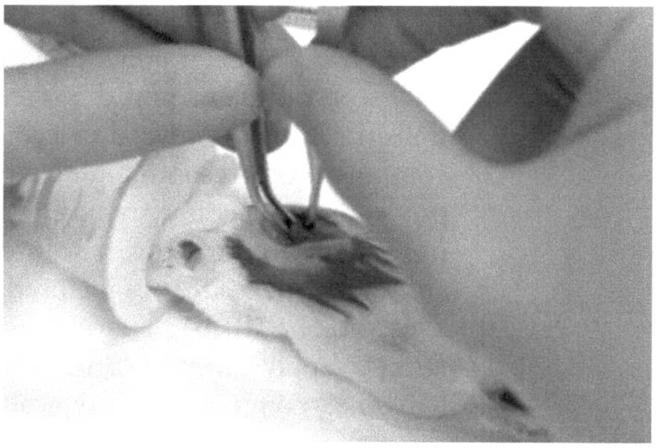

Fig. 1. Spinal cord transection. The vertebral column is gently tilted up with forceps allowing microscissors to cut intervertebral ligaments and to be inserted intervertebrally for a transection of the spinal cord.

environment (22 ± 3°C), maintained under a 12 h light:dark cycle with free access to water and food (Teklab global 18% protein rodent diet no. 2018, Harlan Teklab, Madison, WI). For surgery, pre-operative care is provided 30 min prior to anesthesia including subcutaneous injections of 1.0 ml of lactate-Ringer's solution, 0.1 mg/kg of buprenorphine, and 5 mg/kg of Baytril, an antibiotic (Bayer, Toronto, ON). Complete anesthesia is achieved using 2.5% isoflurane initially in a cage of induction. Anesthetized animals are then shaved dorsally (2 cm) from the mid-dorsal area to the neck. Afterwards, each animal is maintained under complete anesthesia using a specially adapted facial mask delivering directly 2.5% isoflurane to the animal. The shaved area is cleaned with 70% (*v/v*) isopropyl alcohol and, then with 10% (*v/v*) povidone-iodine solution whereas eyes are protected from dryness using ocular lubricant (Lacri-Lube®, Allergan Inc., Markham, ON). The first incision of the skin is made using extra fine scissors (no. 14077-10, Fine Science Tools, North Vancouver, BC) over 2 cm along the midline between the mid-dorsal area and the neck. Fat tissues (interscapularis fat) are cut and removed to expose the high-thoracic segments. The latissimus dorsi fascia is cut bilaterally to expose the vertebral column between the fourth and sixth thoracic vertebrae. Curved forceps are used to hold that area of the vertebral column that has been previously cleaned from fascia and muscles to improve the grip. Forceps are used to gently tilt up that part of the vertebral column—it eases the transection of the intervertebral ligaments between the ninth and the tenth vertebrae (Fig. 1). This last part is critical to offer an open access for very fine microscissors (no. 15000-08, Fine Science Tools) that need to be inserted between the ninth and tenth thoracic vertebrae. The spinal cord is cut and the inner vertebral walls are explored and entirely, but delicately, scraped three or four times with fine scissors tips (Fig. 1, no.

14077-10, Fine Science Tools) in order to sever any small fibers which had not been previously cut. It is important to scrape carefully to avoid damaging the intervertebral ligaments located ventrally (i.e., this would lead to a complete dislocation of the vertebral column and corresponding bleeding). Throughout the transection procedures, bleeding, typically minor, is controlled by applying pressure with cotton tips. The interscapularis fat is carefully replaced and the opened skin area is closed using three or four Michel suture clips (no. 12040-01, Fine Science Tools). Michel suture clips are faster to apply and are normally associated with less infection problems (i.e., compared with typical sutures with treads). This overall surgical procedure is conducted under aseptic conditions using only perfectly cleaned surgical tools (continuously sterilized throughout the surgery using a portable quartz beads-sterilizer) and materials.

2.3. Recovery and Post-op

Anesthesia is interrupted and mice are placed in a large cage with a heating pad put underneath. It is critically important to use only minimal heating intensity (35°C) to avoid rapid dehydration, heat shock, and death during the recovery period. Generally, the animals will recover completely within 15 min although we normally leave them on the heating pad overnight with free access to food and water. The recovery procedure is critical to ensure a high percentage of survival post-surgery (typically around 95% in our laboratory). The following day, the animals are replaced in their initial cage with their initial mates to reduce potential aggressions and fights.

Postoperative care is provided a few hours after surgery as well as every day for the next 4 days. It includes injections of lactate-Ringer's solution (2 × 1 ml/day, s.c.), buprenorphine (2 × 0.1 mg/kg/day, s.c.), and Baytril (5 mg/kg/day). Bladders are also manually emptied twice a day until a spontaneous return of some micturition reflexes. For voiding, the bladder has to be delicately pressed between the thumb (side of the bladder) and two fingers (e.g., the index and one other finger placed on the other side of the bladder). This maneuver requires time and experience. In male mice, it is specifically challenging since, in addition, penises have to be continuously maintained against a paper towel to improve successful voiding (i.e., seems to contribute to the expulsion of urine outside the urinary tract probably by capillary action). The belly and sexual organs are also cleaned daily using paper towels and chlorhexidine gluconate solution (0.05% *v/v*) to prevent urinary tract infection. Normally, with this procedure, mice that survive the first 24 h, will remain relatively healthy for a long period of time (i.e., several months). Finally, Michel suture clips are removed after 10 or 14 days post-surgery. Cages have to be cleaned regularly (if possible changing cages every 3 or 4 days) and mice have to be washed, as mentioned above, on a daily basis to prevent urinary tract infection.

All in all, once anesthetized, this surgical procedure takes no longer than 5 min whereas another 10 or 15 min is typically required for animals to recover from anesthesia. Reasons for using a no-laminectomy approach are multiples. Firstly, since we aim at obtaining a complete transection, a minimally invasive approach such as this one (i.e., inserting microscissor between two vertebrae) can be performed with ease. Obviously, this would be more complicated in studies where an incomplete spinal cord lesion is seeked (e.g., contused or crushed spinal cord). Secondly, avoiding the laminectomy part saves time, reduces periods of anesthesia, and, thus, probably contributes to increase survival rate, recovery, and health post-surgery. Thirdly, a no-laminectomy approach prevents many subsequent problems generally associated with laminectomies (e.g., which are known to reduce vertebral structure integrity and to induce lordosis, epidural fibrosis, altered biomechanics, pain-related behaviors, etc.).

3. Spinal Cord and Systemic Adaptations

Within a few days to a few weeks post-surgery, spinal cord-transected mice undergo a wide variety of adaptations at the spinal cord level (sublesionally) as well as the system level (e.g., muscles, bones, hormones, blood, etc.). Many of these adaptations have been summarized elsewhere (21–24). In brief, immediately following surgery, a complete and irreversible loss of sensory and voluntary motor function is found in these animals. A 1–2-mm-wide scar is found at the site of transaction which is responsible for a complete interruption of descending and ascending signals between the lumbosacral segments and the brain (Fig. 2).

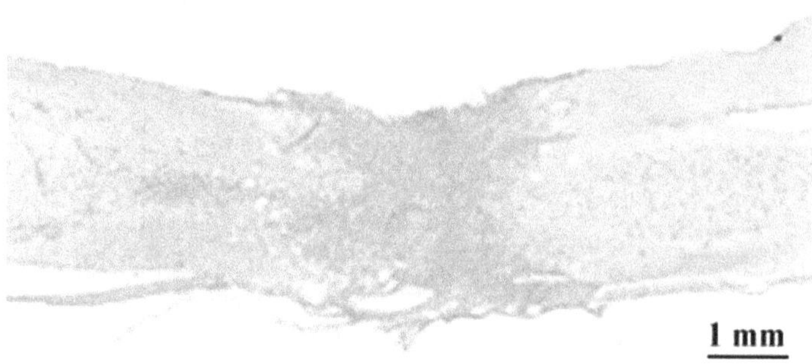

Fig. 2. Spinal cord histology. *Luxol blue* and *cresyl violet* staining of a longitudinal section of the spinal cord from a spinal cord-transected mouse 1 week post-surgery.

3.1. Spinal Cord Adaptations Distally from Epicenter (Sublesionally)

Although, in other animal models (e.g., incompletely injured following contusion or crush), several cellular events have been reported to occur at the site of injury (epicenter) (25–30), comparable processes have not been characterized in spinal cord-transected models probably because it is irrelevant—i.e., regeneration after a complete transection has never been reported in mammals.

In turn, several cellular events have been found to occur distally from the site of injury (27). Recent findings from this animal model and other comparable models have shown that neurons located caudally from the epicenter (so-called sublesional neurons) remain alive and functionally viable (31–35). This said, sublesional neurons were reported also to undergo a number of plasticity and adaptive events within a few days to a few months post-trauma (22, 32, 34, 35). The next sections provide a list of the main sublesional adaptive changes reported in animal models of SCI and, specifically, in this no-laminectomy spinal cord-transected mouse model.

3.1.1. Immediate Early Genes and Other Cytokines

Immediate early genes (IEGs) constitute a large family of genes well-known as early regulators of cell growth, differentiation signals, learning, and memory (28, 30, 36–38). Most recently, some IEGs were found also to be modulated post-trauma in sublesional spinal cord areas (27). For instance, Landry and colleagues have reported in low-thoracic spinal cord-transected mice that *c-fos* and *nor-1* were respectively increased and decreased within a few days in the segments L1–L2, specifically in the dorsal horn and intermediate zone areas (39). In cervical or thoracic spinal cord-transected rats, increased fos-immunoreactive levels were also found as early as at 2.5 h post-trauma in lumbar laminae I–IV, VII, VIII, and X. Changes in the lumbar spinal cord of rostrally transected animals is of special interest since some of these segments (e.g., L1–L2 in mice) were shown to contain critical central pattern generator (CPG) elements (40). Given that IEGs are better known for their role in CNS development and plasticity, spontaneous changes of IEG expression (i.e., specifically *c-fos* and *nor-1*) in L1–L2 segments may be considered as among the first sublesional cellular events associated with altered cellular functions and properties post-SCI. This said, some of these changes may be associated also with other phenomena than plasticity or reorganization of spinal motor and locomotor networks. For instance, *c-fos* and nor-1 were used as markers in experimental models of pain and transient global ischemia suggesting a role in several functions (36, 38). In other studies, c-fos transcripts were found to drastically increase for up to 6 h near the site of lesion and to augment later in some distally located neurons (28). Such a regional- and time-dependent increase in IEG expression, which was found also in this study (39), points toward distinct roles for IEGs in cell death at the injury site versus in reorganization and plasticity at a distance from the epicenter.

Numerous regulatory changes are likely to be subsequently activated after altered IEG expression. These may include *Tumor Necrosis Factor-alpha* (TNF-alpha), *preprodynorphin*, and nitric oxide synthase (NOS), C1qb, Galectin-3 and p22(phox) which expression levels were found to increase in lumbar spinal cord segments for several days soon after a low-thoracic transection or injury in rats (29).

3.1.2. Transmembranal Receptors

Other key elements including transmembranal receptors may be considered as good candidates for plasticity and reorganization of motor and locomotor networks located sublesionally following a spinal cord-transection (and probably to some extent also after partial injuries). For instance, glycine receptors and GAD-67 (enzyme in GABA synthesis) were found displaying increased expression levels in lumbar spinal cord segments 3 months after a low-thoracic transection (32, 41). Increased 5-$HT_{1A/7}$ receptor expression was detected using autoradiography with (^{3}H) 8-OH-DPAT in lumbar spinal cord segments of spinal cord-transected cats (42). In spinal cord-transected mice, we found using *in situ* hybridization increased 5-HT_{1A} mRNA levels in L1–L2 segments in 5-HT_7-deficient mice compared with wild-types (43). This was interpreted as evidence suggesting that even greater changes may occur post-trauma in the absence of functionally closely-related genes. Results in mice revealed also increased 5-HT_{2A} mRNA levels in lumbar segments (laminae VII, VIII, and IX) several days after a low-thoracic transaction (22).

3.1.3. Neurotrophic Factors

Although not specifically examined in spinal cord-transected mice, factors well-recognized for their role in development and plasticity have also been found to be modulated post-transection. Indeed, NGF, BDNF, and NT-3 were found displaying up-regulated levels of expression in sublesional areas of low-thoracic-transected rats (32). In contrast, BDNF and NT-3 as well as synapsin-1 were found to display lower levels of expression in the lumbar cord of low-thoracic-transected rats that were also deafferented which may suggest that these decreasing levels were associated specifically with low levels of muscle reflex activity due to deafferentation (22, 32).

3.2. Systemic Adaptations

A plethora of adaptive changes associated with several systems has been described in the spinal cord-transected mouse model. Interesting, most of these changes correspond with comparable adaptations found specifically in severe SCI cases (see Chap. 1).

3.2.1. Body Weight and Muscular System

We found that the body weight of this animal model significantly decreases soon after transection. A rapid loss typically ranging between 20 and 25% of initial body weight was found after only 1 week post-surgery (21). No significant return to normal levels was reported after 4 weeks although a moderate augmentation of approximately 1 g/week was found. Post-mortem measurements of individual limb masses revealed greater losses in hindlimbs (–28%) than in forelimbs (–21%). These relatively rapid adaptive changes were somewhat comparable to those generally reported soon after trauma in patients. Reasons underlying this initial weight loss are not fully understood but may involve reduced physical activity (or complete immobility), metabolic changes and hormonal deregulation. In turn, after a few years, this initial weight loss is often transformed into a weight gain leading to overweight and obesity problems in patients.

As mentioned above, hindlimbs are specifically affected in spinal cord-transected paraplegic animals. Results from our laboratory showed that part of this weight loss is due to a specific decrease in muscle mass. For instance, measurements from *soleus* muscles (ankle extensor) revealed a 32% decrease in mass at 1 week post-transection in mice (21). This decrease in mass was found to correspond to a proportional decrease in strength (absolute maximal tetanic force). Interestingly, in 2-month-spinal cord-transected mice, slow-twitch fibers (type I) from the *soleus* were found to progressively acquire some of the biochemical and contractile properties of fast-twitch fibers (fiber-type conversion). Again, evidence of comparable changes has been found in chronic SCI patients (2–4, 8).

3.2.2. Skeletal System

SCI is also associated with a rapidly increasing risk of fracture in patients. In fact, nearly all SCI individuals experience a significant loss of bone mineral tissue (up to 30% in the femora) within only a few months (7, 10). Comparable changes are found in spinal cord-transected mice, where histomorphometric data from our laboratory revealed a drastic decrease in trabecular bone volume (–22%), thickness (–11%), and number (–15%) within only 1 month post-surgery (44).

After 2 months, densitometric measurements from mouse femoral bones using dual-energy X-ray absorptiometry reported a significant decrease (approximately 15% lower) of bone mineral density (BMD) and bone mineral content (BMC). There is evidence from mice suggesting that rapid bone loss in such conditions may involve a sharp decrease of osteoblastic activity and a rapid increase of osteoclastic activity (i.e., low osteocalcin and high acid phosphatase levels) (45).

3.2.3. Vascular System

Soon after SCI, there is a well-known risk of developing deep venous thrombosis (DVT) in the lower limbs and, hence, cardiovascular and pulmonary complications in patients (1). However, the specific mechanism underlying DVT formation following SCI remains poorly understood. Our model may be used to study this specific problem. We found using in vivo confocal microscopy that deep vein morphology is largely changed as rapidly as after 1 week post-spinal cord-transection (46). The femoral and saphenous veins were clearly found to largely increase (>1.5-fold) in diameter which may suggest that DVT formation after SCI is associated with venous stasis (reduced blood flow) due to blood vessel enlargement.

3.2.4. Immune, Blood, and Bone Marrow Systems

Immune deficiencies may lead to life-threatening complications after SCI. To examine possible factors that may contribute to this pathological condition, we used an automatic blood analyzer to characterize changes in red and white blood cell counts in spinal cord-transected mice. The results showed that total leukocyte numbers decreased by 35% at 1 week post-surgery and remained low subsequently (47). Specifically, lymphocyte numbers were reduced by as much as 53% at 1 week post-transection whereas monocytes and neutrophils generally remained unchanged. In turn, eosinophil counts were found to gradually decrease by 81% after 4 weeks. Analyses from bone marrow samples revealed similar changes.

Blood lipid profiles examined with clinical chemistry analyzer revealed decreased concentrations of cholesterols (−25%), triglycerides (−45%), low-density lipoproteins (LDL, −55%), high-density lipoproteins (HDL, −14%) but not platelets. Comparable data (e.g., low LDL-triglyceride levels) have been reported in acute SCI patients (47). Erythrocyte, platelet, hemoglobin, and hematocrit levels were either unchanged or moderately decreased (mild anemia) at least for 1 month post-transection.

3.2.5. Anabolic Hormone Systems

Serum levels of testosterone, GH, DHEA, PTH, and insulin were studied using ELISA in control and spinal cord-transected mice. We found early transient changes in testosterone (−40% at 1 week) and GH (threefold increase) levels during the first 2 weeks post-transection (47). Other hormones undergo apparently sustained changes such as DHEA (−75%), PTH (−75%), and insulin levels (−84.5% at 1 week) which were consistently reduced throughout the time period studied (1 month post-surgery). Comparable changes have been found in humans since shortly after trauma, men were found displaying decreased testosterone levels for several weeks whereas increased GH and decreased PTH levels were reported in late chronic SCI patients (6). It remains unclear what role these hormonal changes may play in health degradation post-SCI. However, results in mice suggest that some of these changes may participate to immune deficiencies since high temporal correlations were found between specific anabolic hormones and immune cell count changes.

The many anatomic, systemic, and metabolic changes or adaptations reported here suggest that this mouse model may be useful to study SCI-related secondary complications typically found also in severely SCI patients. These detailed quantitative data measured in mice are critically important to properly assess and compare in subsequent studies the potential effects of innovative drug treatment candidates to prevent or reduce these adaptations and related health problems in chronic SCI individuals.

4. Specific Approaches Developed to Measure Drug Treatment-induced Recovery in Spinal Cord-transected Mice

As mentioned earlier, several adaptations may occur distally from the epicenter and, as reported in Chap. 2, several spinal cord circuits may be affected given their known localization in segments of spinal cord generally spared (lumbar and sacral segments) after SCI (most patients are injured at the cervical or thoracic level). Although, clear evidence exists that neuronal circuits controlling locomotion, micturition, or ejaculation (48) remain functional even several months after a spinal cord transection in mice, behavioral assays specifically developed to study their activity following drug administration remain largely lacking (11, 49, 50).

Most other vital functions remain relatively unaffected (cardiac, respiratory, feeding, voiding functions are not significantly impaired) after a low-thoracic spinal cord transection but sublesional circuits involved in locomotion, micturition, and ejaculation become either quiescent or improperly active (i.e., often reflexively) (48).

4.1. Locomotion

4.1.1. Description of ACOS

In order to examine pharmacological approaches aimed to re-activate properly each of these networks, corresponding quantitative assays had to be developed. For locomotor CPG reactivation, we created a novel quantitative assay called ACOS which stands for *Average Combined Score* (51). In brief, it allows assessing separately locomotor-like movements (LM) and non-locomotor movements (NLM) in the hindlimbs of spinal cord-transected mice. LM and NLM incidence, frequency, and amplitude are normally assessed during 4-min (video-recorded or not) prior to drug administration and after drug injection. A representative single score can be conveniently obtained for each animal by combining arithmetically the assessed values mentioned above as follows: ACOS = (NLM + (2 × LM)) × amplitude). One LM is defined as an entire step-like cycle consisting of an extension phase followed by a flexion phase occurring in both hindlimbs consecutively (i.e., bilaterally alternated or *out-of-phase* relation). Weight-bearing is not assessed and does not need to be expressed for LMs to be considered as such. One NLM is defined as one non-bilaterally coordinated movement (i.e., not followed by a flexion-extension on the other side). They include unilateral movements, jerks, brief sequences of fast-paw shaking (typically lasting 1–2 s/episode and counted as one NLM), twitches, and kicks. *Amplitude* is assessed simply by assigning one of three values; 0—if no movement is observed; 1—if the amplitude of most movements is less than half the range of motion of normal steps; 2—if the amplitude of most movements is at least more than half the range of motion of normal steps. Note that amplitude is assessed for LM and NLM indistinctively. *Incidence* is not measured per se but is obtained when averaging ACOS scores from all mice of a group. It corresponds to the number of mice (out of all mice tested in a group) in which NLMs or LMs were observed. Plantar foot placement and body-weight support may be reported as either present or not although they do not contribute to the equation used to obtain ACOS (i.e., in general, spinal cord-transected mice cannot be made to display weight-bearing stepping without drug or even with most drugs (for details, see Sect. 5.1)). Note that, in the equation, LM is multiplied by a factor of "2" for a very simple reason—even if the experimenter counts each LM as a one movement, an LM is a bilaterally alternating movement which as such is in fact constituted of two movements (one in each hindlimb). Consequently, LM values are multiplied by a factor of "2" whereas NLMs are not (typically unilateral movements) (34, 51).

4.1.2. Validation of ACOS

To validate the approach, we compared scores obtained with ACOS versus with other approaches often used in incompletely SCI animal models (e.g., rats) namely the BBB BMS, AOB, and Farooque scales (12, 31, 49, 50).

In summary, ACOS scores were found to progressive increase in untreated (no drug treatment or training) spinal cord-transected mice during at least 5 weeks whereas the other methods detected only a small increase of hindlimb movements for 2 or 3 weeks without any significant progress at weeks 4 and 5 post-surgery. Given that hindlimb movements occurring spontaneously in untreated animals are rather small and occasional, the results indicate that ACOS is more sensitive than the other assays for detecting even small changes found either in males or females examined in open-field or treadmill conditions (51, 52).

These results further confirmed the existence of a limited but significant phenomenon referred to as "spontaneous recovery" occurring without intervention (no drug, no training) in chronic spinal cord-transected animals (most probably attributable to spinal cord reorganization or plasticity events in this animal model since no regeneration has ever been found in spinal cord-transected mammalian models). Although each method could detect small levels of recovery, only the ACOS method was discriminative enough to detect progressive changes up to 5 weeks post-transection.

In conclusion, the ACOS rating scale is more discriminative than other methods for assessing small and subtle changes in hindlimb movements. This is why it has become an ideally suited assay to compare even subtle drug-induced effects on locomotor CPG activation (see Sect. 5.1). This said, it does not assess weight bearing, equilibrium or plantar foot placement capabilities which is why for higher performance levels it may be useful to use appropriate complementary assays such as the AOB (31).

4.2. Micturition, Ejaculation, and Defecation

Assays specifically designed for measuring a recovery of micturition, ejaculation and defecation in spinal cord-transected animals remain incompletely developed. Attempts were made recently in our laboratory to develop simple and reliable assays for each of these functions. Our work toward this objective is in progress. Preliminary results from our laboratory are expected to be presented shortly in peer-reviewed articles. This said, as with ACOS for locomotion, it is as critically important to develop quantitative and reliable assays that can conveniently be used in vivo for properly quantifying drug-activating or modulating effects on these other networks (33, 52). Recent data from our laboratory using preliminary versions of such assays suggest that we may have found promising spinal network-specific drugs for each of these three other functions. Intellectual property protection is currently being seeked for the last three candidate products which prevents us from providing further details in this publication.

5. Advantages and Disadvantages of the Model for Drug Screening on Sublesional Ciruit Re-activation

5.1. Advantages

There are several advantages associated with the use of this no-laminectomy intervertebrally spinal cord-transected mouse model. As mentioned earlier, it is quick, cost-effective, and minimally invasive (i.e., minimal stress and deformation of the vertebral column). Thus, large cohorts can conveniently be tested rapidly in drug screening studies aimed to identify drug treatment capable of re-activating selectively and potently lumbosacral neuronal networks and related systems.

It is generally accepted that a proper study aimed to compare different drugs require a well-standardized model and experimental conditions. The species, breeding conditions, surgical procedures, day of testing, dose regimen, and assessment methods have to be chosen with great care and maintained throughout testing (i.e., from one drug to another).

With this in mind, we decided to perform our tests specifically in male CD1 mice at 1 week post-surgery. Males were chosen based upon statistics revealing that approximately 80% of all SCI patients are men. We chose CD1 mice (Charles River Laboratories, St. Constant, Quebec) because they are well-known, standardly produced, affordable and easy to maintain animals (moderately aggressive). They are also size-wise among the largest strains of commercially available mice which is a clear advantage for surgical procedures on structures as small as the spinal cord. In contrast with larger mammalian species, mice can generally be housed 4–5 animals per cage which constitutes an advantage when designing and performing large-scale drug-screening studies. In the literature, most experimental animal models of SCI are compression or contusion models.

Although these models were shown to be useful for studying specifically inflammation, cell death, regeneration, or sprouting at or across the lesion, they are generally known for introducing rather large unspecific variability issues (i.e., either associated with the surgeon, the SCI device, the animal species or strains, etc.) (12, 13, 18). Using a completely spinal cord-transected mouse model (Th9/10 level) constitutes a fair approach to reduce that variability. Indeed, this model is reliable and highly reproducible. As described earlier in this chapter, this type of lesion leads to significant secondary complications that are generally comparable to those reported in severe SCI patients. In order to further reduce unspecific variability, we developed surgical procedures where the spinal cord transection is performed without a laminectomy. As mentioned earlier, laminectomies are associated with lordosis and impaired biomechanics, increasing risks of infections, vascularity, and degenerative processes which may obviously affect subsequent experiments (53, 54). Finally, we further reduced variability by using isoflurane for anesthesia rather than non-volatile agents such as ketamine and xylazine (i.e., reduces mortality post-surgery, faster and better recovery, unpublished observations). Using this approach, the surgical procedures are completed in less than 5 or 10 min per animal which is a fair advantage over other methods. This approach leads to a complete lesion shown histologically forming a 1–2 mm thick scar (with no cavity or liquid formation, see Fig. 2) that completely and irreversibly interrupts regeneration, sprouting or all synaptic communications across the lesion.

For experiments aimed at re-activating sublesional networks such as the locomotor CPG, the completeness of the lesion constitutes another key advantage of this animal model (i.e., removal of potential indirectly brain-mediated effects on lumbosacral networks which would make it difficult to interpret the results) (33).

Experiments consistently conducted at 1 week post-surgery allows both enough time for full recovery post-surgery and little-to-no expression of spontaneous hindlimb movements which eases analysis (i.e., movements acutely induced by drugs vs. movements expressed due to time-dependent reorganization and plasticity at the spinal cord network level) (34).

5.2. Disadvantages and Limitations

One of the main disadvantages is associated with the fact that it is a non-clinically relevant type of injury. SCI patients are rarely diagnosed with a sharp and complete laceration of their spinal cord (i.e., although it happens occasionally in the case of injury with weapons—e.g., guns or knifes). Consequently and as mentioned earlier, it is not ideally suited to study repair and regeneration strategies of the injury per se (epicenter). This animal model is better suited to study in a standardized approach, sublesional and systemic adaptations or pharmacological re-activation of spared networks. Along this idea, it is limited by the fact that effects assessed in completely spinal cord-transected models may not reflect what the effects of a drug may be in an incompletely injured subject (with some spared descending and ascending fibers where combined brain-mediated and spinal network-mediated effects are likely to be found). Effects in partial SCI would be best investigated in a corresponding incompletely injured model such as the rat contused or crushed spinal cord model (11, 12).

Limitations are also associated with post-operative cares which require experienced and dedicated individuals. For instance, for unclear reasons, expressing (voiding) the bladder in male mice is particularly difficult (compared with females). For both male and female, manual bladder expression is normally performed by gently squeezing the bladder with two or three fingers. However, in males, successfully achieving this requires also the penis to be simultaneously touching paper towels while squeezing (i.e., probably using the capillary action potential of the paper towel to further force urine expulsion). Individuals with less experience may destroy accidentally the bladder (explosion) with often dramatic consequences (rapid death). Also, urinary tract infection seems to occur more frequently in males than females which is why post-op care (cleaning penises and cages regularly) is critical. Natural aggressiveness (associated with fights and wounds among each group) found in males and rarely in females may also constitute another disadvantage of using male mice.

5.3. Successfully Identified Drugs for CPG Activation

In the 1980s, great therapeutic hopes for ASIA-A and ASIA-B patients have emerged from studies performed in completely paraplegic cats showing that basic hindlimb stepping can be promoted, to some extent, following administration (*i.t.* or *i.p.*) of clonidine, an alpha-2 noradrenergic agonist (55). In patients, clonidine and/or a few monoaminergic drugs have even been shown, in some conditions to promote episodes of overground in previously wheelchair-bound ASIA-C or ASIA-D patients although comparable effects in ASIA-A and ASIA-B patients have never been reported. In fact, some experiments have shown that clonidine facilitates locomotor function recovery through primary afferent-mediated reflex actions by decreasing spinal reflexes and hence spasticity and clonus (56).

Therefore, it remained imperative to pursue research efforts aimed to identify and develop other therapeutic strategies and drugs that could safely and potently restore locomotor function (and, hence, yield health problems) in chronic SCI patients and, more so, in ASIA-A and ASIA-B patients. In 2003, we launched a large research program using this animal model aimed, using a drug screening approach, to identify the first potent locomotor CPG-activating treatment.

In brief, we found that a few classes and subtypes of BBB permeable molecules were capable to elicit partial CPG-activating effects (i.e., no real walking but, instead, locomotor-like movements (LMs) that resemble sweeping—successive flexions and extensions coordinated in both hindlimbs without weight bearing). For instance, we found that 5-HT1A, 5-HT2A, 5HT7, and D1 receptor agonists as well as DA/NA precursor were among the best classes and subclasses of ligands for partial CPG activation in spinal cord-transected mice (33, 57–59).

However, we concluded also that none of the drugs tested were capable of full CPG activating effect and, hence, of weight-bearing stepping movement induction. That is why we began testing drug combinations using our best ligands. With this new sets of experiments, we found that drug combinations including dopaminergic and serotonergic compounds (e.g., L-DOPA combined with a decarboxylase inhibitor such as carbidopa, and a 5-HT1A receptor agonist such as 8-OH-DPAT or buspirone, etc.), could best elicit CPG activation and weight bearing stepping movements in untrained and otherwise non-stimulated animals (60, 61). Interestingly, the idea that drug combinations can elicit superior CPG-activating effects was also supported by comparable findings in in vitro isolated spinal cord preparations (i.e., drug combinations including NMDA, serotonin, and dopamine were found to best activate CPG neurons in vitro (40).

We recently received significant support from a special NIH program called *Rapid Access to Interventional Development* (http://nihroadmap.nih.gov/ raid/ApprovedPr14.aspx) to conduct some of the key preclinical studies (i.e., toxicity and safety pharmacology in rats) with the latest and final version of our combination therapy for CPG activation called *Spinalon* (L-DOPA, carbidopa, and buspirone). NIH-RAID experts in toxicology have acknowledged that this combination is likely to be safe and ideally suited for clinical studies (i.e., each drug is already FDA-approved for diseases other than SCI). Further supporting this, the preclinical data obtained by the NIH through the RAID program have recently shown no evidence of drug–drug interaction or toxicity in vitro and in vivo in rats. Administered orally in spinal cord-transected mice, *Spinalon* was found to potently elicit weight bearing stepping and also, most interestingly, health benefits associated with restored physical activity several times per week (60, 61).

All in all, this novel drug product discovered following drug screening experiments (data assessed with ACOS and AOB scales) in non-laminectomized spinal cord-transected male CD1 mice could become the first treatment capable of temporarily eliciting locomotor activity on a treadmill in severely SCI patients. It would be expected to prevent or reduce physical inactivity-related health problems often found in ASIA-A and ASIA-B patients (e.g., sarcopenia, osteoporosis, immune deficiencies, multiples infections, depression, hormonal deregulation, anemia, etc.)

6. Concluding Remarks

We have developed and fully characterized a well-suited animal model to study spinal cord and systemic adaptations typically found in motor-complete SCI patients (ASIA-A and ASIA-B). It may be used to properly conduct and facilitate drug screening studies in large cohorts of paraplegic animals designed to identify promising drug treatments that can restore specific functions (associated with sublesional neuronal networks) and/or to reduce or prevent health problems associated with chronic SCI.

References

1. Apstein MD, George BC (1998) Serum lipids during the first year following acute spinal cord injury. Metabolism 47:367–370
2. Burnham R, Martin T, Stein R, Bell G, MacLean I, Steadward R (1997) Skeletal muscle fibre type transformation following spinal cord injury. Spinal Cord 35:86–91
3. Gerrits HL, De Haan A, Hopman MT, van Der Woude LH, Jones DA, Sargeant AJ (1999) Contractile properties of the quadriceps muscle in individuals with spinal cord injury. Muscle Nerve 22:1249–1256
4. Grimby G, Broberg C, Krotkiewska I, Krotkiewski M (1976) Muscle fiber composition in patients with traumatic cord lesion. Scand J Rehabil Med 8:37–42
5. Mechanick JI, Pomerantz F, Flanagan S, Stein A, Gordon WA, Ragnarsson KT (1997) Parathyroid hormone suppression in spinal cord injury patients is associated with the degree of neurologic impairment and not the level of injury. Arch Phys Med Rehabil 78: 692–696
6. Naftchi NE, Viau AT, Sell GH, Lowman EW (1980) Pituitary-testicular axis dysfunction in spinal cord injury. Arch Phys Med Rehabil 61:402–405
7. Ragnarsson KT, Sell GH (1981) Lower extremity fractures after spinal cord injury: a retrospective study. Arch Phys Med Rehabil 62: 418–423
8. Scott WB, Lee SC, Johnston TE, Binkley J, Binder-Macleod SA (2006) Contractile properties and the force-frequency relationship of the paralyzed human quadriceps femoris muscle. Phys Ther 86:788–799
9. Bauman WA, Spungen AM, Adkins RH, Kemp BJ (1999) Metabolic and endocrine changes in persons aging with spinal cord injury. Assist Technol 11(2):88–96
10. Zehnder Y, Lüthi M, Michel D, Knecht H, Perrelet R, Neto I et al (2004) Long-term changes in bone metabolism, bone mineral density, quantitative ultrasound parameters, and fracture incidence after spinal cord injury: a cross-sectional observational study in 100 paraplegic men. Osteoporos Int 15:180–189
11. Basso DM, Beattie MS, Bresnahan JC (1996) Graded histological and locomotor outcomes after spinal cord contusion using the NYU weight-drop device versus transection. Exp Neurol 139:244–256
12. Farooque M (2000) Spinal cord compression injury in the mouse: presentation of a model including assessment of motor dysfunction. Acta Neuropathol (Berl) 100:13–22
13. Kwon BK, Oxland TR, Tetzlaff W (2002) Animal models used in spinal cord regeneration research. Spine 27:1504–1510
14. Sheng H, Wang H, Homi HM, Spasojevic I, Batinic-Haberle I, Pearlstein RD et al (2004) A no-laminectomy spinal cord compression injury model in mice. J Neurotrauma 21(5):595–603
15. Steward O, Schauwecker PE, Guth L, Zhang Z, Fujiki M, Inman D et al (1999) Genetic approaches to neurotrauma research: opportunities and potential pitfalls of murine models. Exp Neurol 157:19–42
16. McKerracher L, David S (2004) Easing the brakes on spinal cord repair. Nat Med 10:1052–1053
17. Plunet W, Kwon BK, Tetzlaff W (2002) Promoting axonal regeneration in the central nervous system by enhancing the cell body response to axotomy. J Neurosci Res 68(1):1–6
18. Silver J, Miller JH (2004) Regeneration beyond the glial scar. Nat Rev Neurosci 5(2):146–156
19. Guertin PA (2004) Role of NMDA receptor activation in serotonin agonist-induced air-stepping in paraplegic mice. Spinal Cord 42(3):185–190

20. Guertin PA (2008) A technological platform to optimize combinatorial treatment design and discovery for chronic spinal cord injury. J Neurosci Res 86(14):3039–3051
21. Landry E, Frenette J, Guertin PA (2004) Body weight, limb size, and muscular properties of early paraplegic mice. J Neurotrauma 21(8): 1008–1016
22. Lapointe NP, Ung RV, Guertin PA (2007) Plasticity in sublesionally located neurons following spinal cord injury. J Neurophysiol 98(5):2497–2500
23. Ung RV, Lapointe NP, Guertin PA (2008) Early adaptive changes in chronic paraplegic mice: a model to study rapid health degradation after spinal cord injury. Spinal Cord 46(3): 176–180
24. Ung RV, Rouleau P, Guertin PA (2011) Functional and physiological effects of treadmill training induced by buspirone, carbidopa, and L-DOPA in clenbuterol-treated paraplegic mice. Neurorehabil Neural Repair 26(4): 385–94
25. Byrnes KR, Garay J, Di Giovanni S, De Biase A, Knoblach SM, Hoffman EP et al (2006) Expression of two temporally distinct microglia-related gene clusters after spinal cord injury. Glia 53:420–433
26. Di Giovanni S, Knoblach SM, Brandoli C, Aden SA, Hoffman EP, Faden AI (2003) Gene profiling in spinal cord injury shows role of cell cycle in neuronal death. Ann Neurol 53(4): 454–468
27. Lukacova N, Kolesarova M, Kucharova K, Pavel J, Kolesar D, Radonak J et al (2006) The effect of a spinal cord hemisection on changes in nitric oxide synthase pools in the site of injury and in regions located far away from the injured site. Cell Mol Neurobiol 26:1365–1383
28. Ruggiero DA, Anwar M, Kim J, Sica AL, Gootman N, Gootman PA (1997) Induction of c-fos gene expression by spinal cord transection in the rat. Brain Res 763:21–29
29. Song G, Cechvala C, Resnick DK, Dempsey RJ, Rao VL (2001) GeneChip analysis after acute spinal cord injury in rat. J Neurochem 79:804–815
30. Yakovlev AG, Faden AI (1994) Sequential expression of c-fos protooncogene, TNF-alpha, and dynorphin genes in spinal cord following experimental traumatic injury. Mol Chem Neuropathol 23:179–190
31. Antri M, Orsal D, Barthe JY (2002) Locomotor recovery in the chronic spinal rat: effects of long-term treatment with a 5-HT2 agonist. Eur J Neurosci 16:467–476
32. Gomez-Pinilla F, Ying Z, Roy RR, Hodgson J, Edgerton VR (2004) Afferent input modulates neurotrophins and synaptic plasticity in the spinal cord. J Neurophysiol 92:3423–3432
33. Guertin PA (2009) Recovery of locomotor function with combinatory drug treatments designed to synergistically activate specific neuronal networks. Curr Med Chem 16(11): 1366–1371
34. Ung RV, Lapointe NP, Tremblay C, Larouche A, Guertin PA (2007) Spontaneous recovery of hindlimb movement in completely spinal cord transected mice: a comparison of assessment methods and conditions. Spinal Cord 45(5): 367–379
35. Li XL, Zhang W, Zhou X, Wang XY, Zhang HT, Qin DX et al (2007) Temporal changes in the expression of some neurotrophins in spinal cord transected adult rats. Neuropeptides 41: 135–143
36. Bai ZT, Liu T, Pang XY, Chai ZF, Ji YH (2007) Suppression by intrathecal BmK IT2 on rat spontaneous pain behaviors and spinal c-Fos expression induced by formalin. Brain Res Bull 73:248–253
37. Cruz CD, Ferreira D, McMahon SB, Cruz F (2007) The activation of the ERK pathway contributes to the spinal c-fos expression observed after noxious bladder stimulation. Somatosens Mot Res 24:15–20
38. Kim Y, Hong S, Noh MR, Kim SY, Huh PW, Park SH et al (2006) Induction of neuron-derived orphan receptor-1 in the dentate gyrus of the hippocampal formation following transient global ischemia in the rat. Mol Cells 22:8–12
39. Landry ES, Rouillard C, Levesque D, Guertin PA (2006) Profile of immediate early gene expression in the lumbar spinal cord of low-thoracic paraplegic mice. Behav Neurosci 120(6):1384–1388
40. Nishimaru H, Kudo N (2000) Formation of the central pattern generator for locomotion in the rat and mouse. Brain Res Bull 53:661–669
41. Talmadge RJ, Roy RR, Edgerton VR (1996) Alterations in the glycinergic neurotransmitter system are associated with stepping behavior in neonatal spinal cord transected rats. Society for Neuroscience Abstracts 22:1397.
42. Giroux N, Rossignol S, Reader TA (1999) Autoradiographic study of alpha1- and alpha2-noradrenergic and serotonin1A receptors in the spinal cord of normal and chronically transected cats. J Comp Neurol 406(3):402–414
43. Landry ES, Lapointe NP, Rouillard C, Levesque D, Hedlund PB, Guertin PA (2006) Contribution of spinal 5-HT1A and 5-HT7 receptors to locomotor-like movement induced by 8-OH-DPAT in spinal cord-transected mice. Eur J Neurosci 24(2):535–546

44. Picard S, Lapointe NP, Brown JP, Guertin PA (2008) Histomorphometric and densitometric changes in the femora of spinal cord transected mice. Anat Rec (Hoboken) 291(3):303–307
45. Rantakokko J, Uusitalo H, Jamsa T, Tuukkanen J, Aro HT, Vuorio E (1999) Expression profiles of mrnas for osteoblast and osteoclast proteins as indicators of bone loss in mouse immobilization osteopenia model. J Bone Miner Res 14:1934–1942
46. Rouleau P, Guertin PA (2007) Early changes in deep vein diameter and biochemical markers associated with thrombi formation after spinal cord injury in mice. J Neurotrauma 24(8): 1406–1414
47. Rouleau P, Ung RV, Lapointe NP, Guertin PA (2007) Hormonal and immunological changes in mice after spinal cord injury. J Neurotrauma 24(2):367–378
48. Guertin PA, Steuer I (2009) Key central pattern generators of the spinal cord. J Neurosci Res 87(11):2399–2405, Review
49. Basso DM, Beattie MS, Bresnahan JC (1995) A sensitive and reliable locomotor rating scale for open field testing in rats. J Neurotrauma 12:1–21
50. Ma M, Basso DM, Walters P, Stokes BT, Jakeman LB (2001) Behavioral and histological outcomes following graded spinal cord contusion injury in the C57Bl/6 mouse. Exp Neurol 169:239–254
51. Guertin PA (2005) Semiquantitative assessment of hindlimb movement recovery without intervention in adult paraplegic mice. Spinal Cord 43:162–166
52. Lapointe NP, Ung RV, Bergeron M, Cote M, Guertin PA (2006) Strain-dependent recovery of spontaneous hindlimb movement in spinal cord transected mice (CD1, C57BL/6, BALB/c). Behav Neurosci 120(4):826–834
53. Cabukoglu C, Güven O, Yildirim Y, Kara H, Ramadan SS (2004) Effect of sagittal plane deformity of the lumbar spine on epidural fibrosis formation after laminectomy: an experimental study in the rat. Spine (Phila Pa 1976) 29(20):2242–7
54. Kosta V, Kojundzić SL, Sapunar LC, Sapunar D (2009) The extent of laminectomy affects pain-related behavior in a rat model of neuropathic pain. Eur J Pain 13(3):243–248
55. Rossignol S, Giroux N, Chau C, Marcoux J, Brustein E, Reader TA (2001) Pharmacological aids to locomotor training after spinal injury in the cat. J Physiol 533(Pt 1):65–74
56. Rémy-Néris O, Denys P, Daniel O, Barbeau H, Bussel B (2003) Effect of intrathecal clonidine on group I and group II oligosynaptic excitation in paraplegics. Exp Brain Res 148(4): 509–514
57. Lapointe NP, Guertin PA (2008) Synergistic effects of D1/5 and 5-HT1A/7 receptor agonists on locomotor movement induction in complete spinal cord-transected mice. J Neurophysiol 100(1):160–168
58. Lapointe NP, Rouleau P, Ung RV, Guertin PA (2009) Specific role of dopamine D1 receptors in spinal network activation and rhythmic movement induction in vertebrates. J Physiol 587(7):1499–1511
59. Ung RV, Landry ES, Rouleau P, Lapointe NP, Rouillard C, Guertin PA (2008) Role of spinal 5-HT2 receptor subtypes in quipazine-induced hindlimb movements after a low-thoracic spinal cord transection. Eur J Neurosci 28(11): 2231–2242
60. Guertin PA, Ung RV, Rouleau P (2010) Oral administration of a tri-therapy for central pattern generator activation in paraplegic mice: proof-of-concept of efficacy. Biotechnol J 5(4):421–426
61. Guertin PA, Ung RV, Rouleau P, Steuer I (2011) Effects on locomotion, muscle, bone and blood induced by a combination therapy eliciting weight-bearing stepping in non-assisted spinal cord-transected mice. Neurorehab Neural Repair 25(3):234–242

Chapter 7

Spinal Cord Injury: The Rabbit Model

Nadežda Lukáčová, Jaroslav Pavel, and Ján Gálik

Abstract

This chapter gives reasons for the choice of rabbit in experimental neurobiology and includes descriptions of both the spinal cord injury models and behavioral neuroscience methods. The models provide mechanisms for hypothesis generation that can be investigated at molecular and cellular levels. New insights obtained at both levels will result in a better understanding of various changes in nervous tissue after spinal cord injuries and ultimately will improve prevention and treatment modalities. The functional justification of the techniques presented in this chapter will allow the readers to design their own experiments in the rabbit.

Key words: Anesthesia, Ischemia, Trauma, Neurological outcome

1. Introduction

Rabbits are commonly used in biomedical research. Currently, many strains of rabbit are available. Laboratory rabbits belong to the order Lagomorpha and are collectively referred to as lagomorphs (1). The most popular strain for research purposes is New Zealand White, a medium-sized (weighing between 3 and 5 kg) outbred rabbit. The second in importance as a laboratory strain is American Dutch, a small strain, weighing between 0.9 and 1.8 kg. The Flemish Giant strain weighs between 0.5 and 9 kg is used for research only occasionally. There are many other strains such as American Chinchilla, Satin, Rex, Silver Martin, American Checkered Giant, Polish, and Lop.

1.1. General Advantages of Rabbits for Spinal Cord Injury Research

A number of advantages make rabbits a widely used animal in biomedical research. Their size, ease of handling, and the relative ease of blood collection due to their large ear vessels make them suitable for many types of experiments (2). The rabbits are suitable for long-term experiments, because most of them live for 5–8 years, but some individuals live to the age of 10 years or more. Historically, there has

Håkan Aldskogius (ed.), *Animal Models of Spinal Cord Repair*, Neuromethods, vol. 76,
DOI 10.1007/978-1-62703-197-4_7, © Springer Science+Business Media, LLC 2013

been a strong tendency to use albino varieties. This probably reflects the aim of early research workers towards the "standardization" of their animals as, superficially at least, albinism does indicate a degree of uniformity (2).

1.2. Anesthesia of Rabbits

Successful anesthesia of rabbits requires careful attention to all aspects of perioperative care. There are a few specific issues that are of particular importance, for example, the relative difficulty of endotracheal intubation and the apparent increased susceptibility to respiratory arrest (3). When *tracheotomy* is needed, it is usually performed at the level of the second and third tracheal cartilages. The animals are intubated with an endotracheal cannula and ventilated on a volume percentage of 25% O_2 and 75% N_2O with a volume-cycled ventilator. *Intermittent intravenous administration of the anesthetic agents* through an ear vein catheter can be used to maintain an adequate level of anesthesia without the need for endotracheal intubation and mechanical ventilation (4). Here, we should bear in mind the fact that rabbit veins are very thin-walled and are prone to hematoma formation. For *long-term* experiment, the rabbit should be premedicated with ketamine (40 mg/kg) and xylazine (8 mg/kg, i.m.) and anesthetized with 2–3% isoflurane in room air, delivered *via* face mask during operation.

1.3. Advantages/ Possibilities for Exploring Specific Aspects of Spinal Cord Injury

An understanding of the functional organization of the nervous system is necessary for the diagnosis, prognosis, and treatment of neurological problems. One reason why the rabbit is used in neurological research is that this animal has unique segmental arterial blood supply to the spinal cord from the infrarenal aorta, and neurological signs of rabbits do not differ significantly from a neurological work-up in other companion animal species and should follow a similar pattern (5). An additional reason for using this animal species is that the rabbit's spinal cord extends along the entire length of the spine, unlike that in, e.g., dogs and many other mammals used for experimental neurobiology (6). This means that a low spinal injury, particularly a low lumbar spinal injury, will affect not only lower motor neurons and the primary sensory afferents, but also descending and ascending spinal cord pathways (5). Depending on the site and severity of the injury, there may be interference with bladder control, anal function, hindlimb reflexes, and sensation. At neurological examination, the absence of response to deep pain stimulation indicates a poor prognosis though pain tests are not always reliable in the rabbit, since signs of pain may be concealed and the behavioral response therefore may be difficult to assess. A withdrawal reaction restricted to the hindlimbs may be a purely local reflex, not necessarily reflecting a functionally intact spinal cord nociceptive system. Rabbits are very nervous and sensitive animals with a low pain treshold, and a light sensitivity that is eight times greater than that of humans.

2. Lesion Types

The vertebral formula of the rabbit is C7T12L7S4C16; although in some animals, there are 13 thoracic vertebrae. Greenaway et al. (7) used 64 rabbits in his study.

Following humane euthanasia, the animals were radiographed and dissected. The number of thoracic (T) and lumbar (L) vertebrae was 12T/7L in 44%, 13T/6L in 32%, and 13T/7L in 23%. The 13th ribs were rudimentary, asymmetric, or both in 15 animals. The spinal cord terminated at the level of the second sacral vertebra (S2) in 79%, within the first sacral vertebra (S1) in 19%, and within the third sacral vertebra (S3) in 1.7% of the sample. Surgical procedures that are occasionally performed on the spinal cord of rabbits include transient ischemia and traumatic lesions performed at different segmental levels. The models are discussed in the relevant sections below.

2.1. Spinal Cord Ischemia—General Remarks

The study of focal spinal cord ischemia has been impeded for many years by the lack of animal models that are both reproducible and simulate human ischemic strokes. To circumvent these problems, Zivin and DeGirolami (8) have developed a rabbit spinal cord infarction model. The infarction of the caudal lumbar cord was produced by temporary occlusion of the abdominal aorta below the renal arteries. The changes in spinal cord infarction were qualitatively similar to infarction of most areas of the central nervous system. The rabbit has unique segmental arterial blood supply to the spinal cord from the infrarenal aorta, because it possesses a homosegmental aorta in which individual spinal segments are supplied by abdominal aorta at each segmental level, with minimal to no intraspinal collateral system (8). Man, dog, and cat do not often suffer damage to the lower cord following abdominal aortic compromise because one or more major arterial branches from the aorta above the left renal artery usually supply the lumbar and sacral cord (in man, the artery of Adamkiewicz). In the rabbit, quinea pig, and rat, small segmental branches arise from the entire length of the abdominal aorta to supply lumbosacral segments of the cord. There is apparently sufficient collateral blood supply in the quinea pig and rat to prevent consistent spinal cord infarction when the abdominal aorta is occluded (9). In order to achieve significant spinal cord ischemia in lower thoracic and lumbosacral segments of animals with typical heterosegmental organization of the spinal cord supply (monkey, dog, sheep, pig, rat), it is necessary to occlude the descending thoracic aorta (10). However, to achieve comparable spinal cord ischemia in lumbosacral segments of rabbit, it is sufficient to occlude only the abdominal aorta (10). In this model, consistent paraplegia is achieved after ischemic intervals as short as 10–15 min (8, 11–21). This model has numerous features which makes it particularly useful for neurochemical, neuropathological, and neurophysiological studies (8). The above mentioned anatomical differences in spinal vasculature can potentially be important in determining the outcome after transient aortic occlusion.

Before embarking on complicated animal experiments, it may even be wise to consider employing in vitro models of ischemia. Only few studies have been published utilizing the spinal cord slices (22, 23) maintaining ischemia in vitro, which may help to study the pathological changes appearing under in vivo conditions. The spinal cord slices in vitro can be used for the study of the mechanisms which play a crucial role during aglycemia and ischemia in vivo. It is known that in vitro models provide many technical advantages. For example, the efficacy of neuroprotectants to tissue slices can be tested in a living system devoid of the blood–brain barier. Under such circumstances, electrophysiological, chemical, and metabolic measurements can be performed. The control of extracellular fluid composition and ionic substitution can be made allowing to study the different mechanisms under various experimental conditions. The temperature of the slice preparations can be precisely controlled. Moreover, the changes in the blood flow and the influence of anesthesia during in vivo experiments are eliminated using the tissue slices (24). In vitro ischemia of the tissue slices can be produced superfusing medium without glucose and oxygen. Within minutes of oxygen and/or glucose deprivation, cellular ATP levels are decreased (25) and excitatory amino acids, glutamate and aspartate, are released. The continuous glutamate release influencing *N*-methyl-d-aspartate (NMDA) receptor leads to Ca^{2+}-influx (26), and the activation of the Ca^{2+}-dependent enzyme cascades including neuronal and endothelial isoforms of nitric oxide synthases (nNOS and eNOS), which generate nitric oxide (NO) from l-arginine (27) in neurons and endothelial cells. On the other hand, NO is produced by inducible isoform of NOS (iNOS), which is independent of calcium and is expressed after exposure to cytokines and/or lipopolysaccharide in glial and macrophages cells (23, 28). Kolesárová et al. (23) compared the changes of Ca^{2+}-dependent NOS activity in the lumbosacral spinal cord after ischemia in vivo and in vitro, both followed by reoxygenation periods in vitro. The results have shown that oxygen–glucose deprivation followed by reoxygenation in vitro is valid and sensitive means to monitor ischemia/reperfusion changes in vivo. The measurement of constitutive NOS activity in this study implies that 15 min ischemia in vivo and 45 min ischemia in vitro causes reversible changes.

2.2. Preparation of Spinal Cord Slices and Performing Ischemia In Vitro

After the excision of the rabbit's backbone (segments L4–S2), the spinal cord should be quickly extruded into an ice-cold (2–4°C) artificial cerebrospinal fluid (ACSF), pH = 7.4, that has been pre-oxygenated with 95% O_2 and 5% CO_2. The composition of the

ACSF used (in mM) is: NaCl, 120; KCl, 2; $CaCl_2$, 2; $NaHCO_3$, 26; KH_2PO_4, 1.18; $MgSO_4$, 1.19; and glucose, 11 (29). The spinal cord is quickly cut approximately to 400 μm thick transverse slices by razor blade, and the slices are placed into plastic flasks (25 ml capacity) containing 20 ml of ACSF completely submerged and superfused at a rate of 2–3 ml/min with ACSF, that is saturated with 95% O_2 and 5% CO_2. The flasks are held in a shaking water bath at 37°C for 30 min. This is widely used preparations of the spinal cord that is suitable for maintaining spinal cord ischemia. After preincubation in ACSF, ischemia can be induced by oxygen/glucose deprivation, i.e., the slices are moved into flasks containing glucose-free ACSF that is continuously gased with 95% N_2 and 5% CO_2 for 45 or 60 min. Thereafter, the in vitro reoxygenation (usually 15, 30, 60, or 90 min) can be induced by the incubation of spinal cord slices in the ACSF containing glucose and continuously saturated with 95% O_2 and 5% CO_2. The control spinal cord slices are incubated in ACSF with glucose, 95% O_2, and 5% CO_2 during the whole procedure. The spinal cord slices can then be used for neurochemical, histochemical, and immunohistochemical studies.

2.3. Spinal Cord Ischemia In Vivo

A widely used method for induction of spinal cord ischemia without the opening of abdomen is intraaortic balloon occlusion performed via a catheter inserted via femoral arteriotomy. This approach offers an easy and minimally invasive procedure for producing transient spinal cord ischemia.

In a rabbit weighing 2.5–3.0 kg, a femoral arteriotomy is performed under sterile conditions, 3–4 cm distal to the inguinal ligament. To monitor distal blood pressure (DBP), the left femoral artery is cannulated with an 18-gauge polytetrafluoroethylene (PTFE) catheter. Before catheter insertion, heparin 500 IU is administered intravenously. Aliquots of arterial bood samples (1 ml) taken shortly before ischemia, during ischemia, and 5 min after aortic occlusion are drawn for measurement of blood gasses and pH. Afterwards, a 5-Fogarty arterial embeloctomy catheter (120805 F Baxter Healthcare Corporation, USA) is inserted 15 cm into the right femoral artery (30). This results in a balloon location 0.5–1.5 cm distal to the left renal artery. The extra-artieral end of the catheter is connected with a syringe filled with 1.5 ml of air, and the balloon at the intra-arterial end of catheter is quickly inflated with the syringe with 1.5 ml of air during the time of ischemia. The efficiency of the inflation is indicated by the loss of distal aortic pressure. During ischemia, the body temperature of the rabbits should be monitored with a rectal thermometer, and maintained at 37°C with a heating pad. After operation, the animals are given analgesics and antibiotics. Before starting the experiment, it is very useful to make some pilot studies to determine the position of the balloon in the aorta after autopsy, since the exact baloon position depends on variations in vasculature anatomy, strain of rabbits and differences in their body weight.

2.4. The Rostro-Caudal Gradient of Spinal Cord Blood Flow

The rostro-caudal gradient of spinal cord blood flow (SCBF) as well as differences in dorsal versus ventral perfusion are likely to reflect different metabolic demands of neuronal pools in the spinal cord. Such differences can be of importance particularly under conditions of partial ischemia, because the regions displaying the highest baseline blood flow (central gray matter) will presumably be most sensitive to transient ischemic insult (10). Such a selective neuronal vulnerability can be expected to correspond with the clinical deficits resulting from the specific loss of function associated with these neurons. Selective degeneration or loss of inhibitory interneurons in the intermediate zone (laminae V–VII) of lumbosacral segments causes spastic paraplegia and hyperreflexia, while the loss of alfa-motoneurons causes flaccid paraplegia (i.e., lower motor neuron lesion) (10).

2.5. Spinal Cord Trauma

Traumatic injury of the spinal cord represents a serious clinical problem associated with lifelong disability and a broad range of secondary complications, the processes that stem from the initial mechanical insult and affect cells far beyond the original lesion in both antero- and retrograde directions (31). Although spinal cord trauma causes loss of neuronal cell bodies as well as myelinated axons, the dysfunction of the white matter tracts is the factor that determines most of the clinical symptoms. It is known that each white matter column represents axonal pools that are heterogenously structured, containing various ratios of long ascending and descending axons including a significant component represented by reciprocal propriospinal pathways. Generally, after hemisection the remaining contralateral part of the spinal cord is able to transmit signals from the hindlimbs to unilateral as well as bilateral supraspinal loci (32). For example, the contralateral caudally affected area following anterolateral spinothalamic tractotomy can still be subserved by the uninterrupted spinocervicothalamic tract and bilaterally projecting spinoreticular tract, since both mediate nociception. In addition, there are short propriospinal pathways which are able to relay information to supraspinal centers from one side of the spinal cord to the other, since animals given thoracic overhemisections on opposite sides of the cord demonstrate supraspinal responses to noxious stimuli to the hindlimb. An example of a neuroanatomic substrate for the unexpected forelimb behavior is the numerous individual nociceptive spinothalamic cells clustered in the first few cervical spinal segments which respond to peripheral noxious stimuli from the whole body. In addition, laminae I neurons, which are activated by nociceptive and thermoreceptive activity, are known to project ipsilaterally and bilaterally to regions in the brainstem that control behavioral states. Thus, there are several intact pathways that provide neuroanatomic substrates for the transmission of hindlimb, as well as forelimb, somatosensation to supraspinal centers following spinal hemisection (32, 33).

For the assessment of motor functions after spinal cord injury, it is relevant to realize that although the pyramidal tract of the rabbit is located in the lateral funiculus, it does not descend caudal to the C3 spinal cord segment (34, 35). The rubrospinal tract of the rabbit on the other hand appears to show a similar organization as the human pathway by being located in the central part of the lateral funiculus and terminating in the cervical spinal cord (35). The site of location of the rabbit rubrospinal tract coincides with a discrete, relatively small area containing a small number of NADPH diaphorase exhibiting neurons (36). The control of trunk and hindlimb movements are thus dependent on other pathways from the brainstem and their cortical afferents.

3. Outcome Measures—Functional Aspects of Recovery

The protocols presented here are intended to serve as a guide to specific methods for the established behavioral neuroscientist. The degree of neurological impairment after spinal cord injury can be graded on a three-point scale according to Zivin et al. (37), on a five-point grading scale according to the method of Tarlov (38) or Drummond and Moore (39), and on a six-point grading scale according to Johnson et al. (40). All protocols are well-validated, quantitative, easily replicated, rapid, and simple. Two individuals without knowledge of the treatment should grade neurological function independently.

A three-point scale of neurological score according to Zivin et al. (37):

Grade 0: No neurological impairment; the animals hop normally and evade visual threats.

Grade 1: Partial neurological impairment; the animals drag their hindlimbs and show subnormal response to visual threats and painful stimuli. Bowel and bladder functions are variable. Animals showing any degree of impairment (from barely detectable to quite severe) are included in this group.

Grade 2: Complete neurological impairment; the animal's hindlimbs are completely paralysed, unresponsive to painful stimuli, and the animal is unable to pass urine and feces.

A five-point grading scale according Tarlov's criteria (38):

Grade 0: Paraplegia with no lower-extremity motor function; no movement of the hindlimbs.

Grade 1: Poor lower-extremity motor function; perceptible movement of the joints of hindlimbs.

Grade 2: Good movement of the joints but inability to stand.

Grade 3: Ability to stand and walk.

Grade 4: Complete recovery; ability to hop.

A five-point scale of neurological score according to Drummond and Moore (39):

Grade 4: Normal motor function.

Grade 3: Ability to draw legs under the body and hop but not normally.

Grade 2: Some lower-extremity function with good antigravity strength but inability to draw legs under the body.

Grade 1: Poor lower-extremity motor function and weak antigravity movement only.

Grade 0: Paraplegic with no lower-extremity function.

A six-point grading scale according to the method of Johnson et al. (40):

Grade 0: Hindlimb paralysis.

Grade 1: Severe paraparesis.

Grade 2: Functional movement, no ability to hop.

Grade 3: Ataxia, disconjugate hop.

Grade 4: Minimal ataxia.

Grade 5: Normal function.

4. Conclusion

The arterial blood supply of lumbar and sacral segments of the spinal cord in the rabbit is organized differently compared to, e.g., rat, dog, or humans. It is well-documented that the above-mentioned portions of the spinal cord in rat, dog, and human are supplied by arterial blood coming not only from circular arteries around spinal cord but a considerable amount of blood is supplied by longitudinally arranged arteries and, between them, Adamkiewicz artery plays a significant role. Therefore, any surgical intervention aimed to diminish or completely block the blood supply to lower spinal cord segments in rat, dog, or human must be arranged in a such way that the ligation of abdominal aorta should be performed above the exit of Adamkiewicz artery, thus achieving complete ischemia of lumbar and sacral spinal cord segments. In this context, it should be mentioned that the blood supply of lower spinal cord in the rabbit is organized on the principle of radicular arteries with one radicular artery supplying one spinal cord segment. Therefore, each segment of the lumbar and sacral spinal cord in the rabbit possess one supplying segmental artery, and due to this organization, the final result of abdominal aorta ligation just below the left renal artery causes a complete or almost complete ischemia of lumbar and sacral spinal cord segments. Based on these circumstances, the rabbit model has distinct advantages over many other animal models for exploring mechanisms underlying ischemia induced spinal cord injury.

References

1. Hrapkiewicz K, Medina L (2007) Rabbits. In: Clinical laboratory animal medicine: an introduction, 3rd edn. Wiley-Blackwell, John Wiley & Sons, Inc., Hoboken, NJ
2. Donnelly H (1995) Quality in laboratory animals. In: Tuffery AA (ed) Laboratory animals—An introduction for experimenters, 2nd edn. Wiley, Chichester, pp 181–203
3. Malley D (2000) Handling, restraint and clinical techniques. In: Flecknell P (ed) Manual of rabbit medicine and surgery. British Small Animal Veterinary Association, UK, pp 3–9
4. Moore WM Jr, Hollier LH (1991) The influence of severity of spinal cord ischemia in the etiology of delayed-onset paraplegia. Ann Surg 213:427–431
5. Boydell P (2000) Nervous system and disorders. In: Flecknell P (ed) Manual of rabbit medicine and surgery. BSAVA, Gloucester, pp 57–61
6. Orendacova J, Cízková D, Kafka J, Lukácová N, Marsala M, Sulla I et al (2001) Cauda equina syndrome. Prog Neurobiol 64:613–637
7. Greenaway JB, Partlow GD, Gonsholt NL, Fisher KR (2001) Anatomy of the lumbosacral spinal cord in rabbits. J Am Anim Hosp Assoc 37:27–34
8. Zivin JA, DeGirolami U (1980) Spinal cord infarction: a highly reproducible stroke model. Stroke 11:200–202
9. DeGirolami U, Zivin JA (1982) Neuropathology of experimental spinal cord ischemia in the rabbit. J Neuropathol Exp Neurol 41:129–149
10. Maršala M (1998) Spinal cord blood flow and metabolism in transient spinal ischemia. In: Stålberg E, Sharma HS, Olsson Y (eds) Spinal cord monitoring. Springer, Wien, New York, pp 5–25
11. Cheng MK, Robertson C, Grossman RG, Foltz R, Williams V (1984) Neurological outcome correlated with spinal evoked potentials in a spinal cord ischemia model. J Neurosurg 60:786–795
12. Vacanti FX, Ames A 3rd (1984) Mild hypothermia and Mg++ protect against irreversible damage during CNS ischemia. Stroke 15: 695–698
13. Marsala M, Vanický I, Radoňák J, Kliesenbauerova E, Maršala J (1992) Postischemic hyperoxia enhances vulnerability in the rabbit spinal cord ischemia model. Restor Neurol Neurosci 6:283–291
14. Jacobs TP, Kempski O, McKinley D, Dutka AJ, Hallenbeck JM, Feuerstein G (1992) Blood flow and vascular permeability during motor dysfunction in a rabbit model of spinal cord ischemia. Stroke 23:367–373
15. Schreiberová A, Kisucká A, Hricová L, Kucharíková A, Pavel J, Lukáčová N (2012) The vulnerability of nitrergic neurons to transient spinal cord ischemia: a quantitative immunohistochemical and histochemical study. J Mol Histol 43(2):203–213.
16. Schreiberová A, Lacková M, Kolesár D, Lukácová N, Marsala J (2006) Neuronal nitric oxide synthase immunopositivity in motoneurons of the rabbit's spinal cord after transient ischemia/reperfusion injury. Cell Mol Neurobiol 26:1483–1494
17. Lukácová N, Jalc P, Marsala J (1998) Phospholipid composition in spinal cord regions after ischemia/reperfusion. Neurochem Res 23:1069–1077
18. Lukácová N, Halát G, Chavko M, Marsala J (1996) Ischemia-reperfusion injury in the spinal cord of rabbits strongly enhances lipid peroxidation and modifies phospholipid profiles. Neurochem Res 21:869–873
19. Marsala M, Danielisová V, Chavko M, Hornáková A, Marsala J (1989) Improvement of energy state and basic modifications of neuropathological damage in rabbits as a result of graded postischemic spinal cord reoxygenation. Exp Neurol 105:93–103
20. Martiniak J, Saganová K, Chavko M (1991) Free and peptide-bound amino acids as indicators of ischemic damage of the rabbit spinal cord. J Neuropathol Exp Neurol 50:73–81
21. Marsala J, Kluchova D, Marsala M (1997) Spinal cord gray matter layers rich in NADPH diaphorase-positive neurons are refractory to ischemia-reperfusion-induced injury: a histochemical and silver impregnation study in rabbit. Exp Neurol 145:165–179
22. Deshpande SB, Jha A (2004) Aglycemia and ischemia depress monosynaptic and polysynaptic reflexes in neonatal rat spinal cord in vitro by involving different types of 5-hydroxytryptamine receptors. Neurosci Lett 372:167–172
23. Kolesárová M, Pavel J, Lukáčová N, Kolesár D, Maršala J (2006) Effect of ischemia in vivo and oxygen-glucose deprivation in vitro on NOS pools in the spinal cord: comparative study. Cell Mol Neurobiol 26:1281–1294
24. Croning MDR, Haddad GG (1998) Comparison of brain slice chamber designs for investigations of oxygen deprivation in vitro. J Neurosci Methods 81:103–111
25. Dobrota D, Matejovicova M, Kurella EG, Boldyrev AA (1999) Na/K-ATPase under

oxidative stress: molecular mechanisms of injury. Cell Mol Neurobiol 19:141–149
26. Lehotsky J, Kaplan P, Matejovicova M, Murin R, Racay P, Raeymaekers L (2002) Ion transport systems as targets of free radicals during ischemia reperfusion injury. Gen Physiol Biophys 21:31–37
27. Dawson TM, Dawson VL, Snyder SH (1992) A novel neuronal Messenger molecule in brain: the free radical, nitric oxide. Ann Neurol 32:297–311
28. Knowles RG, Moncada S (1994) Nitric oxide synthases in mammals. J Biochem 298: 249–258
29. Garthwaite G, Brown G, Batchelor AM, Goodwin DA, Garthwaite J (1999) Mechanisms of ischemic damage to central white matter axons: a quantitative histological analysis using rat optic nerve. Neuroscience 94:1219–1230
30. de Haan P, Kalkman CJ, Vanicky I, Jacobs MJ, Drummond JC (1998) Effect of mildhypothermia and the 21-aminosteroid U-74389 G on neurologic and histopathologic outcome after transient spinal cord ischemia in the rabbit. J Neurosurg Anesthesiol 10:86–93
31. Edgerton VR, Roy RR (2002) Paralysis recovery in humans and model systems. Curr Opin Neurobiol 12:658–667
32. Christensen MD, Everhart AW, Pickelman JT, Hulsebosch CE (1996) Mechanical and thermal allodynia in chronic central pain following spinal cord injury. Pain 68:97–107
33. Lee JH, Price RH, Williams FG, Mayer B, Beitz AJ (1993) Nitric oxide synthase is found in some spinothalamic neurons and in neuronal processes that appose spinal neurons that express Fos induced by noxious stimulation. Brain Res 608:324–333
34. Armand J (1984) The pyramidal tract. Recent anatomic and physiologic findings. Rev Neurol (Paris) 140(5):309–329
35. Hobbelen JF, Gramsbergen A, van Hof MW (1992) Descending pathways and the hopping response in the rabbit. Behav Brain Res 15:217–221
36. Marsala J, Marsala M, Lukacova N, Ishikawa T, Cizkova D (2003) Localization and distribution patterns of nicotinamide adenine dinucleotide phosphate diaphorase exhibiting axons in the white matter of the spinal cord of the rabbit. Cell Mol Neurobiol 23:57–92
37. Zivin JA, DeGirolami U, Hurwitz EL (1982) Spectrum of neurological deficits in experimental CNS ischemia. A quantitative study. Arch Neurol 39:408–412
38. Tarlov IM (1957) Spinal cord compression: mechanism of paralysis and treatment. Charles C Thomas, Springfield, IL, p 147
39. Drummond JC, Moore SS (1989) The influence of dextrose administration on neurologic outcome after temporary spinal cord ischemia in the rabbit. Anesthesiology 70: 64–70
40. Johnson SH, Kraime JM, Graeber GM (1993) Effects of flunarizine on neurological recovery and spinal cord blood flow in experimental spinal cord ischemia in rabbits. Stroke 24: 1547–1553

Chapter 8

The Cat Model of Spinal Cord Injury

Alain Frigon

Abstract

The cat can recover a robust pattern of hindlimb locomotion following partial or complete spinal cord injuries. As a result, it has been instrumental in our understanding of spinal networks controlling and regulating locomotor activity. Thanks to our feline friend, spinal cord-injured humans are now trained on a treadmill to promote the recovery of walking. This chapter highlights some of the landmark studies using the cat that have led to a better understanding of the adaptive changes that take place after spinal cord injury, as well as the underlying mechanisms governing locomotor control and recovery.

Key words: Gait, Central pattern generator, Sensorimotor, GABA, Glycine, Noradrenaline, Serotonin

Terminology

Central pattern generator (CPG)	Neuronal network that produces rhythmic activity without extraneous inputs
Pattern generator (PG)	Neuronal network that produces rhythmic activity with tonic and/or phasic sensory inputs
Spinal cat	A cat with a complete spinal cord injury
Spinalization	A transection or complete spinal cord injury
Spinal locomotion	Hindlimb walking after complete spinal cord injury

Abbreviations

AP-5	2-Amino-5-phosphonovaleric acid
CPG	Central pattern generator
EMG	Electromyography
ENG	Electroneurography
5-HT	Serotonin
GABA	Gamma-aminobutyric acid
L	Lumbar
LG	Lateral gastrocnemius
NA	Noradrenaline
NMDA	*N*-methyl-D-aspartate

Håkan Aldskogius (ed.), *Animal Models of Spinal Cord Repair*, Neuromethods, vol. 76,
DOI 10.1007/978-1-62703-197-4_8, © Springer Science+Business Media, LLC 2013

PG	Pattern generator
PICs	Persistent inward currents
SCI	Spinal cord injury
Srt	Sartorius
St	Semitendinosus
T	Thoracic
TA	Tibialis anterior
VL	Vastus lateralis

1. Introduction

In terms of contemporary animal models of spinal cord injury (SCI), the cat is by far the oldest and most documented, particularly, as it relates to the neurophysiology of the spinal cord and to the control and adaptation of locomotion before and after various types of spinal lesions. Since the late nineteenth century, several laboratories around the world have focused on the spinal neurophysiology and neural control of locomotion in the cat, which has profoundly impacted the field of locomotor recovery following SCI. The goals of this chapter are (1) to present the biological and conceptual framework for using the cat, (2) to describe some of the experimental methodology, and (3) to summarize key findings derived from work in the cat and their importance for SCI research. The intent is not to provide an exhaustive historical account of all studies that have used the cat but to highlight the main contributions of the cat to the field of SCI research.

2. Biological and Conceptual Framework

The cat (*Felis catus*) is a small carnivorous mammal, but much larger than frequently used animal models of SCI, such as the rat and mouse. Its relatively large size provides the opportunity to perform detailed biomechanical analyses in locomotor and postural studies. The cat, like humans, also has a high center of gravity during locomotion, as opposed to rodents, which can lower their bellies to the ground to improve stability. Its larger size also facilitates the recording and stimulation of multiple structures simultaneously. Also, because of its larger size, many experimental techniques were first developed in the cat before being transposed to smaller animal models.

The cat was domesticated thousands of years ago. An obvious advantage of domestication is that cats can easily be handled and trained by humans to perform a variety of motor tasks, such as walking on a treadmill or over ground at consistent speed. Treadmill walking in rodents is generally performed with electroshock therapy or by constraining the animal in a device. The ability to study an animal walking freely, without undue stress, is a clear advantage for locomotor studies. Cats can be trained to walk at various speeds and inclines on a treadmill, and can also perform more difficult locomotor tasks, such as obstacle avoidance (1–3), walking across a horizontal ladder (4, 5), or on interspersed elevated pegs (6, 7).

Anatomically, the spinal cord of cats is organized much like that of humans. Most notably, the main part of the corticospinal tract is located in the dorsolateral spinal cord. In rodents, such as mice and rats, the main part of the corticospinal tract is located below the dorsal columns. The location of specific ascending and descending pathways becomes important when comparing deficits associated with injury to specific quadrants of the spinal cord across different species.

Cost can be seen as a disadvantage in using the cat as an animal model of SCI. However, the higher cost per cat and per diem for care is largely offset by the lower number of animals required to answer specific questions, some of which can only be investigated in a mammal with a high center of gravity. A relatively large mammalian model, such as the cat, is important for translating results to humans, particularly for locomotor and postural studies.

3. Experimental Methodology

Different types of experimental preparations in the cat have benefited SCI research. The following sections summarize the conceptual framework and methodology of three different cat preparations used to study functional recovery after SCI and/or spinal cord neurophysiology.

3.1. The Chronically Implanted Cat

Implanting electrodes in animals for long-term investigations was pioneered and perfected in the cat. Such chronic studies are essential in determining changes that occur within the central nervous system (CNS) over time after SCI. Figure 1 summarizes the methodology for the chronically implanted cat, which can vary slightly depending on the laboratory.

The cat is first trained to walk on a treadmill at various speeds for several consecutive minutes. Once the animal can accomplish this task consistently, it is implanted with electrodes to record muscle activity (EMG, electromyography). All surgeries are done under general anesthesia with 1–3 % isoflurane. Electrodes made of stainless steel wire are soldered to and led subcutaneously from connectors that will later be cemented to the skull using dental acrylic (i.e., the headpiece). The insulation is removed from the end of the wires, which are sewn into muscle bellies for EMG. Bipolar cuff electrodes made of polymer (8) can also be placed around nerves

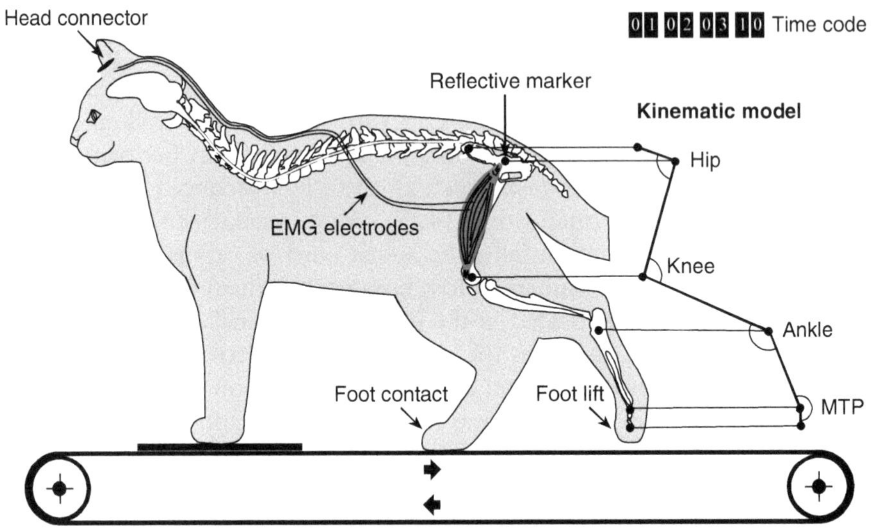

Fig. 1. General methodology for the study of locomotion in cats with spinal lesions. With a complete lesion of the spinal cord at T13, cats are positioned with their forelimbs standing on a platform fixed above the belt, while the hindlimbs are free to walk on the moving treadmill. With a partial lesion, the cat is free to walk with all four limbs on the treadmill. Pairs of electromyographic (EMG) electrodes are implanted into multiple muscles (only one pair is represented here). The multipin EMG connector is cemented to the skull using dental acrylic. Reflective markers are placed on bony prominent landmarks of the left hindlimb to reconstruct angular excursions. The swing and stance phases of each cycle are determined from foot contacts and foot lifts.

to evoke reflexes using electrical stimulation. To deliver drugs intrathecally, a cannula can be inserted in the subarachnoid space with a capped opening fixed to the headpiece. Drugs that cross the blood–brain-barrier can also be delivered intraperitoneally (i.p.). During and for a few days after surgery, animals are given an analgesic, such as buprenorphine. The animals are monitored daily by trained personnel for signs of discomfort.

A few days after implantation, baseline data (e.g., EMG, kinematics) are obtained while the cat walks on the treadmill. Kinematic data are obtained by placing reflective markers on bony landmarks and captured with a video camera to reconstruct angular excursions during locomotion (see Fig. 1). Foot lifts and contacts of the four limbs are determined off-line from video recordings using custom-made software. Following a few weeks of baseline recordings, a spinal lesion is performed. The spinal lesion, which can be partial or complete, is made by making an incision above the desired segments and by removing some tissue (e.g., muscle, connective tissue) to expose 1–2 vertebrae. In the case of a spinalization or dorsal lesion, the dorsal part of the vertebral segment is removed (i.e., a laminectomy) to expose the spinal cord. The dura is carefully cut with small surgical scissors, and lidocaine is applied topically and injected within the spinal cord. The spinal cord is then cut using surgical scissors and an absorbable hemostat is packed within the gap to prevent regrowth. Muscle and skin are then closed in anatomical layers before anesthesia is discontinued. If the SCI is complete, the bladder must be expressed daily by the experimenters for the duration of the study. Some animals with incomplete SCI also require bladder expression for the first few days following the lesion. Depending on the experimental protocol, the cat can then be retrained to recover hindlimb walking (see Sect. 5.2 for more details on locomotor training).

The chronically implanted cat has contributed substantially to the field of SCI research, from fundamental and clinical perspectives. It was key in demonstrating that the spinal cord, devoid of supraspinal inputs, can generate a well-organized locomotor pattern with full weight bearing, plantar foot placements, and activation of hindlimb muscles. It has also shown that the chronically injured spinal cord can continue to adapt to different conditions, such as treadmill speed, various stimuli, and further injury. Some of the key findings and insights derived from the chronically implanted cat are discussed later on.

3.2. The Decerebrate Unparalyzed Cat

Acute terminal experiments are valuable because they permit procedures that are more invasive than what is possible in freely behaving animals. In acute experiments, the cat is decerebrated, meaning that tissue from the brain is removed, primarily from the cortex and rostral brain. During decerebration, the cortex is removed without damaging the thalamus, basal ganglia, and brainstem, so that cardiorespiratory functions are unaffected. It can be done by suctioning brain tissue or by removing the blood supply to specific regions (i.e., anemic decerebration). Following decerebration, anesthesia is discontinued and neuronal excitability within the remaining CNS returns, sometimes in an elevated state. Decerebration is performed to remove sentience so that invasive measures can be made in an excitable system that is not depressed by anesthesia. The type of decerebration can vary by transecting the neuraxis at different levels (e.g., decorticate, premammillary, and postmammillary), which have varying effects on the interactions between supraspinal and spinal structures. For example, a premammillary decerebration spares a region of the brainstem that is involved in generating spontaneous locomotor activity, whereas with a postmammillary decerebration electrical stimulation of the brainstem is required to elicit locomotion (9). The different types of decerebration in the context of locomotor control are reviewed in detail by Whelan (9).

In acute terminal experiments, several procedures can be performed, including electrical stimulation of circumscribed regions of the brainstem (9, 10), spinal cord (11–14), and other brain structures (15, 16). Peripheral nerves, dorsal roots, and ventral roots can also be stimulated to evoke or condition motor responses (17, 18). The EMG of several muscles can be evaluated at rest and during locomotion using percutaneous intramuscular electrodes.

The limbs can also be immobilized and tendons can be attached to pullers and force transducers to stretch or vibrate specific muscles or to measure force generation (19, 20).

The decerebrate cat has been instrumental in our understanding of the descending control and sensory regulation of locomotion. Specifically, the decerebrate cat preparation established that walking was regulated by centers originating in the brainstem (reviewed by (9, 21, 22)). Complex interactions between cortical and subcortical structures shape the output from the brainstem to the spinal cord to initiate and regulate locomotion. Treadmill locomotion in the decerebrate cat was also used to demonstrate the importance of reflex pathways from group I ankle extensor afferents in regulating the duration of the stance phase and the stance-to-swing transition (23, 24).

3.3. The Decerebrate Paralyzed Cat

Some major breakthroughs in neuroscience were performed in the decerebrate paralyzed cat. The decerebration is the same as described in the previous section. Paralysis is achieved by injecting a drug intravenously, such as curare, which blocks transmission across the neuromuscular junction. As a result, no movement can occur and animals must be artificially ventilated to maintain respiratory function. Despite the absence of movement, locomotor-like activity can be recorded from muscle nerves, ventral roots, and/or by intracellular recordings from spinal interneurons or motoneurons. Recording locomotor-like activity in the absence of overt movement has been termed "fictive locomotion." Although the absence of movement eliminates phasic sensory feedback, which is important for regulating locomotor output, the effect of specific inputs on the spinal locomotor network can be evaluated in relative isolation. Fictive locomotion was first recorded in the adult cat (25).

Fictive motor patterns can occur spontaneously or can be triggered by various types of stimulations (i.e., tactile, electrical, or pharmacological). During fictive locomotion, the extension and flexion phases correspond approximately to the stance and swing phases of walking. Figure 2 shows the EMG and electroneurography (ENG) of several hindlimb muscles and nerves, respectively, during treadmill walking (Fig. 2a) and during an episode of spontaneous fictive locomotion (Fig. 2b) in adult cats. During both types of locomotion, the burst in the knee flexor/hip extensor semitendinosus (St) muscle or the combined posterior biceps-semitendinosus (PBSt) nerve is of short duration and begins slightly earlier than the burst in other flexor muscles or nerves, such as tibialis anterior (TA, ankle flexor) and sartorius (Srt, hip flexor/knee extensor) during treadmill walking or TA and extensor digitorum longus (EDL, ankle flexor/toe extensor) during fictive locomotion. The similarity in the sequential activation of various nerves and muscles during fictive and real locomotion, respectively, indicates that a large part of the pattern is centrally generated, although

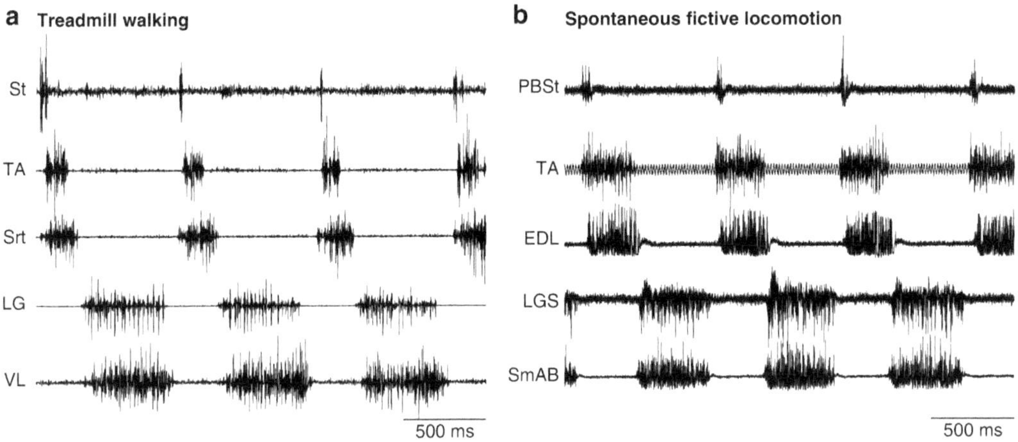

Fig. 2. Pattern of activation during real and fictive locomotion in adult cats. (**a**) The electromyography of several muscles of the left hindlimb during treadmill walking at a speed of 0.4 m/s in an adult cat. (**b**) The electroneurography of several nerves of the left hindlimb during spontaneous fictive locomotion in an adult cat. *St* semitendinosus, *TA* tibialis anterior, *Srt* sartorius, *LG* lateral gastrocnemius, *VL* vastus lateralis, *PBSt* posterior biceps-semitendinosus, *EDL* extensor digitorum longus, *LGS* lateral gastrocnemius-soleus, *SmAB* semimembranosus-anterior biceps.

supraspinal signals and sensory feedback certainly play a major role in shaping the final pattern during real movements through dynamic interactions with the spinal network (comprehensively reviewed in (26)).

Another considerable advantage of the paralyzed preparation is that intracellular recordings can be made from spinal neurons, which requires a great deal of recording stability (i.e., very small to no movement). The first intracellular recording of a mammalian neuron, a spinal motoneuron, was made in the cat (27). Since that time, many different types of neurons within the spinal cord of the cat have been identified and recorded at rest, during fictive locomotion, and fictive scratch. At present, the number of identified neurons in the spinal cord of the cat constitutes the largest body of work of its kind in a mammalian species (for detailed reviews see (28, 29)). Evaluating the cellular properties of the spinal cord before and after SCI is important for targeted therapies.

4. Lesion Types and Recovery of Locomotion

The cat has been important in detailing the recovery of locomotion, as well as the deficits incurred following complete spinal lesions (i.e., spinalization) and various types of partial spinal lesions. Some of the studies that have investigated the recovery of locomotion after various types of spinal lesions are indicated in Table 1, which are grouped by lesion type and level of injury. Different descending pathways within the spinal cord are involved in different aspects of locomotion (for reviews see (21, 22, 26)). For instance, the reticulospinal pathway has a role in initiating and modulating the locomotor pattern, while the vestibulospinal pathway is more concerned with postural control. Cortico- and rubrospinal pathways are involved in volitional and anticipatory aspects of locomotion. Moreover, descending pathways originating in the brainstem release neurotransmitters, such as serotonin (5-HT) and noradrenaline (NA) that can have potent effects on spinal neuronal excitability. The location of these pathways within the spinal cord of the cat are well known (30, 31). The dorsolateral spinal cord contains cortico- and rubrospinal tracts and part of the medullary reticulospinal pathway. Part of the serotoninergic pathway from the Raphe nucleus is also located dorsolaterally. The dorsal columns, which convey sensory information to supraspinal structures are located in the dorsomedial spinal cord.

The ventral spinal cord contains the pontine and medullary reticulospinal pathways and the lateral vestibulospinal tract. Part of the serotoninergic pathway from the Raphe nucleus and most of the noradrenergic pathway from the Locus Coeruleus are also located ventrally. Consequently, injury to specific regions or tracts of the spinal cord produces characteristic deficits, which have been well described in the adult cat for low thoracic injuries (i.e., T12–T13).

4.1. Complete Transection or Spinalization

A complete transection of the spinal cord, or spinalization, abolishes all descending inputs from structures rostral to the lesion (Fig. 3b). Complete SCI animal models have been instrumental in our understanding of the spinal control of locomotion. Numerous studies have investigated how the hindlimb locomotion recovers with time, training, and/or drug injections in kittens (32–37) and adult cats (38–56). With locomotor training, which is discussed later on, hindlimb walking (i.e., spinal locomotion) can recover after a minimum of 3–4 weeks. Figure 4 shows an example of kinematics and EMGs during treadmill locomotion at 0.4 m/s before (gray) and after (black) spinalization in the same cat (54). After spinalization, step length (Fig. 4a, b) and the duration of the locomotor cycle are generally reduced. In this cat, the duration of the cycle was 1,079 and 684 ms before and after spinalization, respectively. Thus, to maintain the same treadmill speed, the hindlimbs of the spinal cat must walk with a faster rhythm. Maximal angular excursions are generally reduced during spinal locomotion (Fig. 4c). The timing of most muscles is generally similar before and after spinalization, whereas the amplitude of the activity in flexor muscles typically increases while that of extensors decreases, although this varies across extensors and between animals (Fig. 4d). However, there are some notable changes in timing that are frequently observed. In particular, the delay in the onset of the knee flexor St activity with the hip flexor Srt is reduced, and both muscles can discharge simultaneously.

Table 1
Studies that have investigated the recovery of hindlimb locomotion following various types of spinal lesions. Studies are grouped according to the type of lesion and the level of transection. C, cervical; L, lumbar; T, thoracic

Type of lesion	Level of spinal lesion	Studies
Complete	T12–L1	Robinson and Goldberger (1986a, b); Lovely et al. (1986); Carter and Smith (1986); Barbeau and Rossignol (1987, 1990, 1991); Barbeau et al. (1987, 1993); Lovely et al. (1990); de Guzman et al. (1991); Howland et al. (1995); Bélanger et al. (1996); Chau et al. (1998); de Leon et al. (1998, 1999); Bouyer et al. (2001); Giroux et al. (2003); Bouyer and Rossignol (2003); Frigon and Rossignol (2008a, 2008b, 2009)
Partial—dorsal/dorsolateral	C2–C3	English (1980)
	T6	English (1985)
	T10–T13	English (1980, 1985); Zmyslowski et al. (1993); Gorska et al. (1993); Jiang and Drew (1996)
Partial—ventral/ ventrolateral	T10–L1	Gorska et al. (1990, 1993); Bem et al. (1995); Brustein and Rossignol (1998, 1999)
Partial—lateral hemisection	C2	Kato et al. (1983, 1984, 1985)
	T10–L3	Murray and Goldberger (1974); Kato et al. (1983, 1984, 1985); Bregman and Goldberger (1983); Eidelberg et al. (1986); Kato (1989, 1990); Helgren and Goldberger (1993); Kuhtz-Buschbeck et al. (1996); Tester et al. (2008); Barriere et al. (2008, 2010); Frigon et al. (2009)
	L4–S1	Kuhtz-Buschbeck et al. (1996)
Dual—serial lateral hemisections	C2 then T6; T12 then T6	Kato et al. (1983, 1984, 1985)
Dual—lateral hemisection + myelotomy	L2–L3 then caudal myelotomy	Kato (1989, 1990)
Dual—lateral hemisection + spinalization	T10–T11 then T12–T13	Barriere et al. (2008, 2010); Frigon et al. (2009)

4.2. Dorsal/ Dorsolateral Lesions

Dorsal lesions of the spinal cord disrupt the dorsal columns and the cortico- and rubrospinal tracts, depending on the extent and location of the lesion (Fig. 3c). Dorsal lesions of the spinal cord have been used in numerous studies to evaluate the recovery of quadrupedal locomotion (57–61). In adult cats, voluntary quadrupedal locomotion recovers within 10 days after dorsal lesions (61). Bilateral

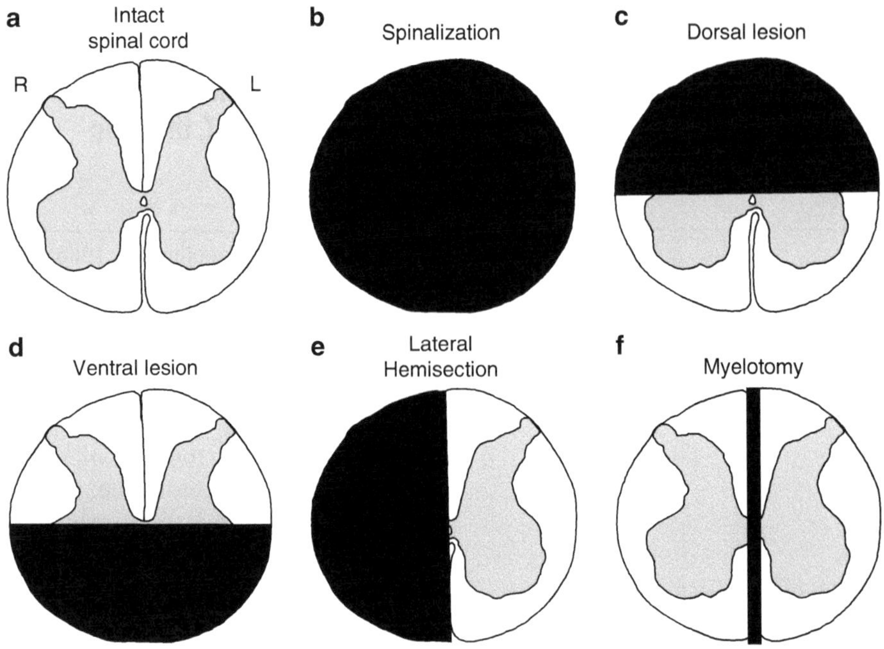

Fig. 3. Schematic illustration of different spinal lesions performed in the cat. (**a**) A transverse view of the spinal cord at low thoracic levels. (**b**) A transection or spinalization of the spinal cord, which completely sections a small portion of the spinal cord in the transverse plane, effectively isolating all structures located caudal to the lesion. (**c**) A bilateral dorsal lesion, which removes the dorsal half of the spinal cord. (**d**) A bilateral ventral lesion, which removes the ventral half of the spinal cord. (**e**) A lateral hemisection, which removes one side of the spinal cord. (**f**) A myelotomy, which splits the spinal cord along the midline disconnecting the right and left sides.

damage to the dorsal/dorsolateral spinal cord at low thoracic levels results in longer steps, paw drag during the early swing phase, and an impaired ability to step over obstacles. Paw drag can persist in cats with more extensive lesions. With large lesions, the duration of the cycle and of the swing phase can increase. Posture and the coupling between hindlimbs are only transiently impaired following a dorsal lesion. The delay in the onset of the knee flexor St activity with the hip flexor Srt is reduced and both muscles can discharge simultaneously, similar to what is observed following spinalization.

4.3. Ventral/Ventrolateral Lesions

Ventral lesions of the spinal cord disrupt vestibulospinal and reticulospinal tracts, as well as serotoninergic and noradrenergic pathways. Ventral lesions have been studied in fewer studies (62–66) because they are harder to perform selectively. Following damage to the ventral/ventrolateral spinal cord at low thoracic levels, there

Fig. 4. (continued) left foot contact. Stance is measured from left foot contact to left toe off. (**c**) Rectified, averaged, single burst EMG traces for selected muscles during intact (*gray*) and spinal (*black*) locomotion in a cat. Each trace is the average of 15–20 cycles, shown twice, normalized to left foot contact. *Srt* sartorius, *St* semitendinosus, *LG* lateral gastrocnemius, *VL* vastus lateralis.

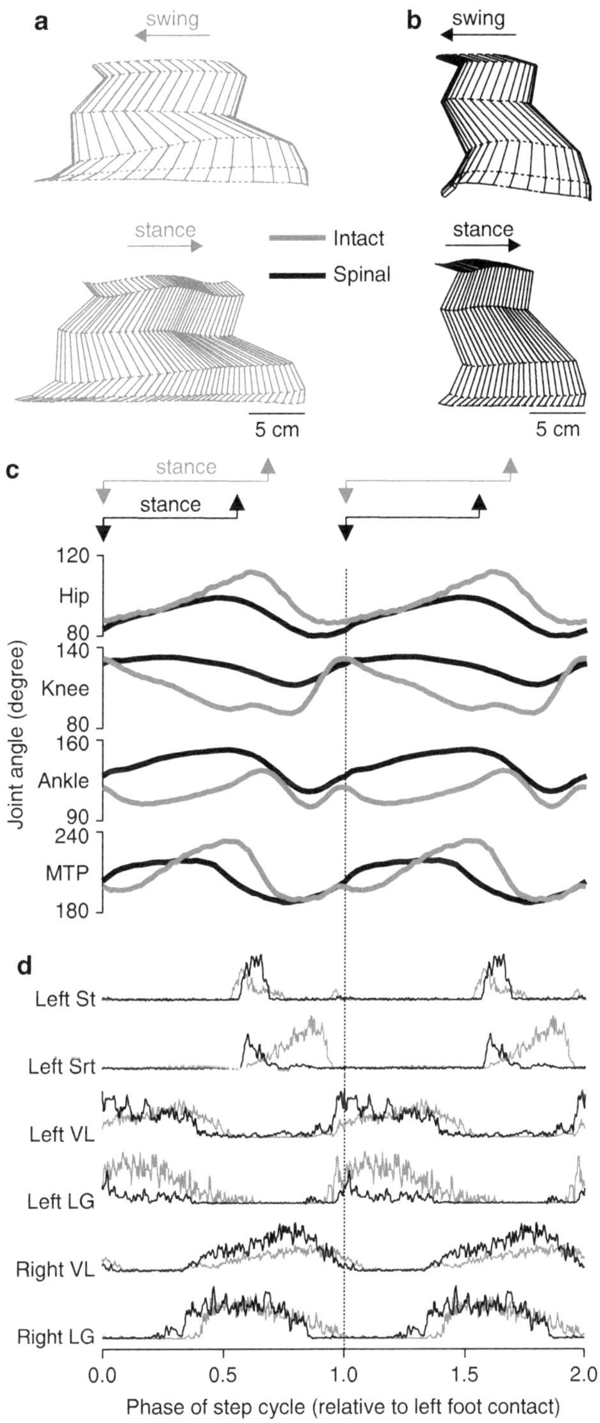

Fig. 4. Kinematic and electromyographic data of a cat during treadmill locomotion at a speed of 0.4 m/s before and after spinalization. (**a**) Reconstruction of the left hindlimb as it moves during locomotion. Note that to prevent overlap of the stick diagrams, each one is displaced by an amount equal to the displacement of the foot along the horizontal axis. (**b**) Angular excursions of the hip, knee, ankle, and metatarsophalangeal (MTP) joints of the left hindimb before (*gray*) and after (*black*) spinalization in the same cat, normalized to

is an initial period with loss of hindlimb walking and weight support that depends on the extent of the lesion (63–65). After a few days to a few weeks, hindlimb locomotion and weight support gradually recover. Despite the recovery of quadrupedal locomotion, lateral stability remains poor. Moreover, the coupling between the fore- and hindlimbs was altered following ventral lesions, as the frequency of the locomotor rhythm differed at both girdles in the most severely injured cats. In addition, the deficits were more pronounced during incline treadmill walking.

4.4. Lateral Hemisections

A lateral hemisection is made by sectioning half of the spinal cord on the left or right side at a specific segment, which is akin to a Brown-Séquard SCI (see Fig. 3e). The lateral hemisection is a popular model to investigate the recovery of hindlimb locomotion in cats (67–80). During the first few days after lateral hemisection, the hindlimb on the side of injury displays flaccid paresis and drags on a moving treadmill belt (70, 71, 75, 76, 80). If the lateral hemisection is performed at cervical levels, both the forelimb and hindlimb on the side of the lesion become paretic for approximately 1 week (70, 71). Animals require assistance for hindquarter support and body equilibrium during locomotion, although the deficits can vary between animals. Within 2 weeks, hindlimb weight bearing recovers and the affected hindlimb progresses from a passive crutch with minimal limb excursion to an active locomotor pattern (75, 76). With larger lesions, the coordination between the fore- and hindlimbs (i.e., intergirdle coordination) can be permanently lost, which is characterized by a different rhythm at the two girdles (80). The strategy used to adapt quadrupedal locomotion after lateral hemisection varies between cats and appears to depend on the presence or loss of intergirdle coordination.

4.5. Dual Spinal Lesion Paradigms

Three different types of dual lesion paradigms are illustrated in Fig. 5. A few studies in cats have used two spinal lesions separated in time (i.e., staggered) to determine what structures might be responsible for the recovery of hindlimb locomotion after SCI. For instance, Kato and colleagues (70, 71) performed a unilateral hemisection at C2 or T12, which was followed several weeks later by a second unilateral hemisection on the opposite side of the spinal cord at mid-thoracic levels (Fig. 5a). Such staggered lateral hemisections disrupt long supraspinal and propriospinal pathways. In all cats, a voluntary quadrupedal locomotion recovered within a few days to a few weeks. However, the coordination between the fore- and hindlimbs was permanently lost. Kato also performed a hemisection at L2–L3, which was followed a few weeks later by a caudal myelotomy (73, 74), which is a longitudinal split along the midline of the spinal cord (Figs. 3f and 5b). The hemisection followed by myelotomy isolates the spinal cord controlling one hindlimb. With such lesions, locomotion initially consists of a tripod gait involving

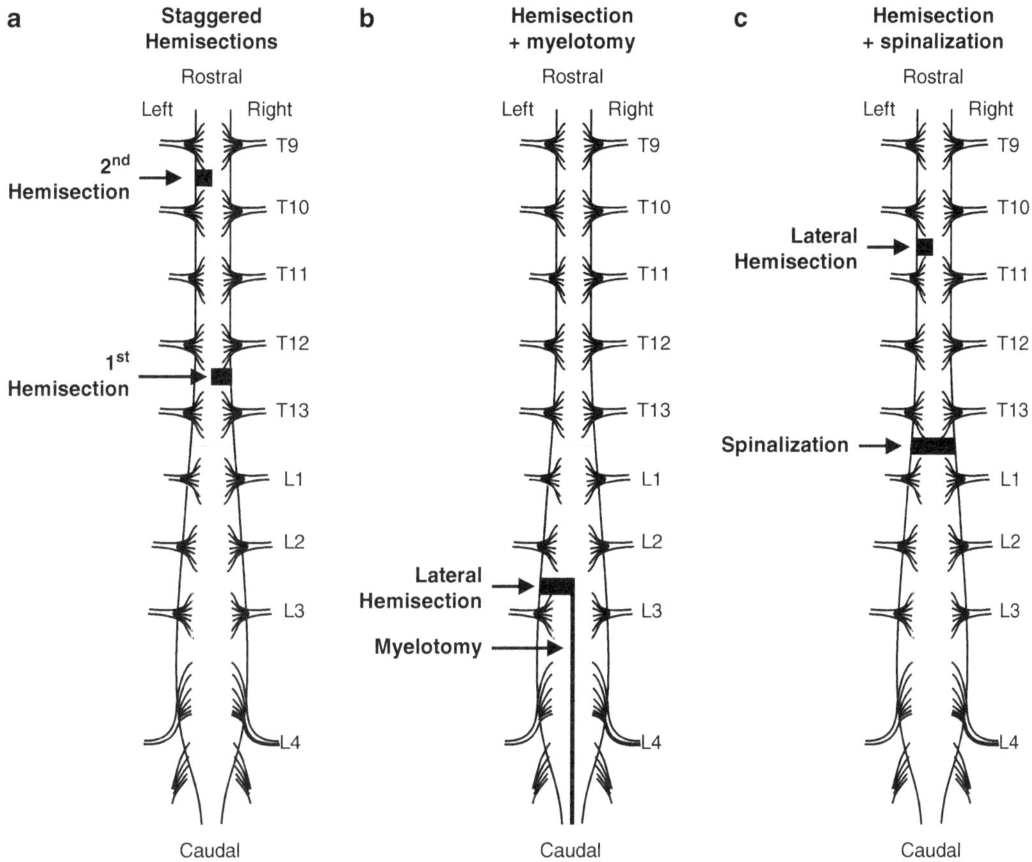

Fig. 5. Dual spinal lesion paradigms in the cat. (**a**) Staggered hemisections. In the example shown in the figure, a lateral hemisection was made at T12 on the right side followed several weeks later by a mid-thoracic hemisection on the left side (70, 71). (**b**) A lateral hemisection followed by a caudal myelotomy. A hemisection was made at L2–L3 and was followed several weeks later by longitudinally splitting the spinal cord along the midline from the hemisection to the end of the cauda equina (73, 74). (**c**) A lateral hemisection at T9–T10 followed several weeks later by a complete transection 2–3 segments below at T12–T13 (78–80).

the forelimbs and the nonisolated hindlimb. A few days to a few weeks later, the isolated hindlimb recovers its walking ability, although the coordination between hindlimbs is lost.

More recently, another type of dual spinal lesion paradigm was used (78–80), which consisted of a lateral hemisection at low thoracic levels followed several weeks later by a complete spinal transection (Fig. 5c). All cats recovered a voluntary quadrupedal locomotion following the hemisection. What was most remarkable was that following the spinalization, hindlimb locomotion spontaneously recovered within hours, a process that normally takes 3–4 weeks of intense treadmill training in chronic spinal cats. The underlying mechanisms involved in locomotor recovery with dual spinal lesion paradigms are discussed later on.

Therefore, different types of spinal lesions produce characteristic deficits that reflect in part damage to specific pathways. What is clear, however, is that no descending pathway is essential for the recovery of a voluntary quadrupedal locomotion following an incomplete SCI.

5. Key Findings

Many important findings for SCI research were first reported in the cat. The following sections summarize some of the seminal studies in the cat, highlighting the importance of their findings for SCI research.

5.1. The Spinal Origin of Locomotion: The Central Pattern Generator

Arguably, the most important discovery of the twentieth century for the control of walking was the demonstration that the spinal cord could generate the basic rhythm of locomotion by itself, which was first done in the adult cat. Thomas Graham Brown, a Scottish physiologist, transected the spinal cord of adult cats at the 12th thoracic (T12) segment and performed an extensive deafferentation of the hindlimbs by sectioning muscle nerves, thus abolishing supraspinal and peripheral inputs to the spinal cord (81). Only the nerves to the two test muscles, the gastrocnemius (ankle extensor) and tibialis anterior (*TA*, ankle flexor), were left intact, and sensory inputs from these muscles to the spinal cord were abolished by sectioning their dorsal roots. Despite the absence of inputs to the spinal cord, rhythmic alternating activity of the ankle extensor and flexor muscles occurred, which was evaluated by recording movement of the tendons of both muscles. This groundbreaking demonstration indicated that the generation of a rhythmic locomotor-like pattern is intrinsic to the spinal cord. The work by Brown is summarized in an excellent review article (82).

Despite what now seems like a clear demonstration for a spinal origin of locomotion, Brown's findings did not garner much support among his peers. The prevailing view that locomotion was of "reflex" origin (83), persisted into the 1960s. It was not until the work by Anders Lundberg and his colleagues in Sweden on spinalized cats (84–87) that Brown's findings resurfaced and became widely accepted. The assumption that an autonomous locomotor network also exists within the human spinal cord has greatly impacted therapeutic interventions in human SCI patients, although a definitive experimental demonstration of a spinal locomotor CPG in humans will never occur. For reviews discussing the existence of a spinal locomotor CPG in humans, see (88–90).

The spinal CPG for locomotion is also largely genetically determined, as kittens spinalized early after birth, before performing any walking movements, will express spinal locomotion (32–34). However, some elements of the spinal locomotor CPG are malleable and most likely contribute to the recovery of hindlimb walking after SCI (91). Experimental data from cats have been used to conceptualize or model the organization of the spinal locomotor CPG (92–96). At present, the functional organization of the mammalian locomotor CPG remains elusive, and extensive work in rodents and cats is ongoing. Therefore, a system of interconnected interneurons within the spinal cord organizes the basic pattern of locomotion, which can function autonomously from supraspinal structures following a complete SCI to restore walking ability. However, it should be emphasized that the CPG after complete SCI interacts with tonic and phasic sensory inputs from peripheral receptors located in joints, muscles, and skin. Thus, a better term is spinal "pattern generator (PG)" to denote an intrinsic spinal neuronal network that produces locomotor activity with inputs from the periphery. As discussed later on, the spinal locomotor PG is also critical in the recovery of locomotion following an incomplete SCI, even though supraspinal inputs can still access the spinal circuitry.

5.2. Locomotor Training

Locomotor training is based on the premise that providing sensory cues consistent with walking induces activity-dependent plasticity in the spinal network that generates locomotion (reviewed in (97, 98)). Locomotor training after a complete SCI was developed in adult cats during the 1980s concurrently in the laboratories of Serge Rossignol at the Université de Montréal (41) and of Victor "Reggie" Edgerton at the University of California at Los Angeles (38).

In the first few days after complete SCI, the legs of the cat are paretic, with little to no reflex movements or muscle spasms. During this period, the legs drag on the treadmill in an extended position with weak or absent spontaneous activity. Locomotor training consists of two experimenters moving the hindlimbs in a pattern consistent with walking and placing the plantar surface of the paws on the treadmill belt while it moves. The forelimbs are positioned on a stationary platform. After a few days, reflex movements and muscle spasms emerge and locomotor-like activity can be triggered by stimulating the perineal region (i.e., the skin located below the tail). At this point, the locomotor pattern is disorganized, with inappropriate contacts of the foot dorsum with the treadmill belt, short step lengths, inconsistent bouts of activity, and poor weight support. A Plexiglas separator is placed between the hindlimbs to prevent crossing of the legs. After a minimum of 3–4 weeks of locomotor training, 3–5 times a week, cats regain the ability to generate an organized hindlimb walking pattern with plantar foot contacts and weight support (38, 41, 42). However, postural control is permanently lost and the experimenter must hold the tail of the animal to provide equilibrium. The time course of recovery and the quality of spinal locomotion vary from one cat to another.

It was shown, in the adult cat, that functional recovery after complete SCI is specific to the type of training (99). For instance, cats that were trained to stand performed poorly at walking, whereas cats that were trained to walk had weak standing capacity. However, based on personal observations, a combination of stand and locomotor training works bests for locomotor recovery (unpublished observations). Combining stand and locomotor training appears to act synergistically, most likely by strengthening hindlimb extensor muscles, which is key to recovering weight support. The Edgerton group also showed that the spinal cord can "remember." For example, when spinal cats were locomotor-trained for 12 weeks and training was then stopped for 12 weeks, the capacity for hindlimb walking diminished considerably (99). However, if locomotor training was restarted, hindlimb walking returned much more rapidly than initially observed. Therefore, locomotor training can facilitate the recovery of hindlimb walking following complete SCI, training appears to be task-specific, and the spinal cord devoid of signals from supraspinal structures can "remember."

5.3. The Importance of Sensory Inputs for Locomotor Recovery

Although Sherrington's original hypothesis that locomotion was generated by a reflex mechanism was inaccurate, it is evident that inputs from reflex pathways are critical in shaping the locomotor pattern through dynamic interactions with the spinal locomotor CPG (comprehensively reviewed in (26)). Moreover, sensory feedback to the spinal cord after SCI is critical for functional recovery, which was first made evident in the cat. For instance, early after spinalization, stimulating the perineal region facilitates the emergence of locomotion in adult cats. The skin from the perineum is thought to provide general excitability to the spinal locomotor network. If sensory inputs to the spinal cord are abolished before SCI, either by sectioning muscle nerves (56, 100), cutaneous nerves (53), or dorsal roots (101), the recovery of spinal locomotion is compromised (e.g., poor weight support, disorganized locomotor pattern). Some animals denervated before spinalization will not recover spinal locomotion (56). It thus appears that the integrity of sensory pathways at the time of SCI is critical for functional recovery thereafter.

Some sensory cues have more important roles during locomotion in the cat than others. For example, the position of the hip at the end of stance (102) and the amount of loading on ankle extensors (23) influence the stance-to-swing transition. Proprioceptive inputs from hip flexors or ankle extensors regulate phase transitions and can entrain the rhythm during fictive locomotion in adult cats (103–111) and/or treadmill walking in unparalyzed decerebrate cats (112–117). Experimental studies in humans successfully translated some of the cat data, primarily for sensory inputs from ankle extensors (118–120), thus showing the importance of sensory cues during bipedal human walking. The importance of hip position and ankle loading is stressed during rehabilitative walking in human spinal cord-injured patients (98).

Inputs from the skin, particularly from the foot, also have specific functions during locomotion. For example, mechanically stimulating the dorsum of the foot during the swing phase evokes a coordinated response that removes the stimulated limb from the obstacle in intact and spinal cats (121, 122), as well as humans (123–125). Removing skin inputs from the foot by sectioning the five cutaneous nerves produces only transient deficits during treadmill locomotion in intact cats (126). However, after complete SCI, these same cats were unable to correctly place their paws on the plantar surface despite several weeks of locomotor training (53). In addition, if the cutaneous denervation was performed after spinalization, some locomotor deficits appeared, such as paw drag during swing, but only once all skin inputs from the foot were removed. Therefore, inputs from the skin of the foot become increasingly important for locomotion after SCI.

5.4. Pharmacological Activation of the Injured Spinal Cord

Spinal cord injury disrupts or abolishes the release of brainstem-derived neuromodulators, which normally plays a major role in regulating spinal neuronal excitability (22, 127–130). Several studies have shown that certain drugs can potently activate and/or modulate the spinal locomotor network after SCI in the cat.

Studies performed in the cat during the 1960s and 1970s laid the foundation for the pharmacological activation of the injured spinal cord. For instance, Lundberg and colleagues first showed that the acutely transected spinal cord generated locomotor-like discharges when DOPA was injected intravenously (i.v.), which is thought to trigger the release of noradrenaline and/or dopamine by intraspinal neurons (84–86). Subsequently, it was shown that i.v. injection of clonidine, an alpha-2 noradrenergic agonist, in decerebrate acutely spinalized (T12) cats elicited locomotion when the animals were placed on a treadmill (32). Naftchi then evaluated the effect of clonidine administration (i.v. and oral) on adult cats that received a severe contusion SCI at T3–T5 (131). All cats given clonidine recovered hindlimb walking, whereas control cats showed signs of paralysis and spasticity. It was then shown that injecting clonidine (i.p.) triggered hindlimb locomotion within the first 2 weeks after complete transection of the spinal cord (T12–T13) in adult cats (40). Moreover, combining locomotor training with intrathecal clonidine injections accelerated (~1 week instead of 3–4 weeks) the recovery of hindlimb walking after complete SCI in adult cats (49, 132). Other noradrenergic agonists, such as tizanidine and oxymetazoline, are also effective in the early stages after complete SCI in the cat. Thus, noradrenergic agonists can potently activate the spinal locomotor network in the cat after SCI.

However, blocking the noradrenergic system, by intrathecal injection of yohimbine, a noradrenergic antagonist, does not abolish hindlimb locomotion in the adult cat once it has recovered from spinalization (133). This indicates that spinal locomotion, in the cat, is maintained by other systems. To evaluate the involvement of the glutamatergic system, *N*-methyl-D-aspartate (NMDA) and/or 2-amino-5-phosphonovaleric acid (AP-5) were injected intrathecally after spinalization in adult cats (133, 134). NMDA is a glutamate receptor agonist, whereas AP-5 is an NMDA antagonist. Although NMDA failed to initiate treadmill locomotion in the first few days (3–5 days) after spinalization, it greatly improved the pattern when the cats were beginning to recover some hindlimb movements (7–8 days after spinalization) (134). Once spinal locomotion is firmly established, NMDA produces only minor effects on the pattern (52). At higher doses, NMDA temporarily blocks spinal locomotion, which recovers within 1 h. On the other hand, the NMDA antagonist, AP-5, completely blocks spinal locomotion, which can be restored by injecting NMDA. Therefore, the glutamatergic system is important for maintaining spinal locomotion.

Studies have also shown that manipulating inhibitory transmission within the injured spinal cord influenced locomotor recovery in chronic spinal cats. The main inhibitory neurotransmitters within the spinal cord are glycine and gamma-aminobutyric acid (GABA). The level of a GABA-synthesizing enzyme was shown to increase after a complete SCI in adult cats (135). Increased levels of inhibitory neurotransmitters (i.e., more inhibition) could depress neuronal excitability and impair specific spinal circuits. Interestingly, levels of the GABA-synthesizing enzyme were downregulated by locomotor treadmill training (i.e., less inhibition), but not by stand training (136). Moreover, administering glycinergic or GABAergic antagonists (i.e., less inhibition) improved hindlimb locomotion in chronic spinal cats that were not responding to treadmill training (36, 50). It thus appears that locomotor training and pharmacology modify the level of inhibition within spinal circuits, thus facilitating functional recovery.

In the chronic spinal cat, serotoninergic agonists do not initiate bouts of locomotor-like activity. Instead, serotonin modulates the pattern of locomotion but only once it has recovered through training or by pharmacologically activating the alpha-2 noradrenergic system (43). Namely, serotoninergic agonists increased the EMG activity of hindlimb muscles, particularly extensors, during locomotion in chronic spinal cats.

Despite striking results in the cat and other species, pharmacologically activating the injured spinal cord in humans has proven less than remarkable, even detrimental, for the recovery of walking (137, 138). There appears to be considerable interspecies variability in the types of drugs that can activate the spinal locomotor network and more work across several species, including nonhuman primates, is required to determine the effectiveness of various drugs and their mechanisms of action on spinal circuits after various types of SCIs.

5.5. Intrinsic Motoneuron Properties

As stated earlier, the first recording of a neuron in a mammal was in a cat motoneuron (27). One of the hallmarks of SCI is the loss of neuronal excitability in the early stages after injury. A major breakthrough performed in the paralyzed cat preparation was the identification of persistent inward currents (PICs) in spinal motoneurons (139), which has proven important in the context of SCI. When activated, PICs can considerably amplify synaptic inputs and sustain cell discharge. It is thought that after SCI motoneuronal excitability is depressed due to the loss of PICs, thus preventing the recovery of motor functions (reviewed in (128, 140)). With time, PICs recover in parallel with sensorimotor functions. Although, ongoing work in PICs is done primarily in rodents (e.g., (141)), it was thanks to the cat preparation that these important intrinsic neuronal properties were first identified.

5.6. The Role of the Intrinsic Spinal Locomotor Network After Spinal Cord Injury

After a complete SCI, the recovery of hindlimb walking must involve the locomotor network intrinsic to the spinal cord. In humans, most SCIs are incomplete, with varying amounts of descending and ascending connections between supraspinal structures and the spinal cord remaining. As a result, the recovery of hindlimb walking could involve changes in spared descending or ascending pathways to and from the spinal cord without necessitating much plasticity within the spinal locomotor network. However, recent studies showed that the locomotor circuitry intrinsic to the spinal cord is strongly involved in the recovery of hindlimb walking following partial SCI, which was made evident by the use of a dual spinal lesion paradigm in adult cats (78). In these experiments, a lateral hemisection was first made at spinal segments T10–T11 on the left side followed by several weeks of locomotor recovery. Once the animals made no more noticeable improvement in quadrupedal walking for several days, a complete spinal transection was made at T12–T13 (i.e., 2 segments below the hemisection). Remarkably, cats had a well-organized hindlimb walking pattern within hours of the complete SCI, a process that normally takes a minimum of 3–4 weeks of treadmill training without pharmacological aids. Therefore, considerable changes must have occurred over time within the spinal cord following the partial SCI, which made the spinal PG function more autonomously without the need for supraspinal inputs. From these studies in the cat, it is clear that the circuitry intrinsic to the spinal cord is involved in hindlimb locomotor recovery following complete and partial SCIs, perhaps through similar mechanisms. The role of descending signals from supraspinal structures in the recovery of hindlimb walking following an incomplete SCI still needs to be clarified.

Earlier work in the cat using serial hemisections also showed that the circuitry intrinsic to the spinal cord is involved in locomotor recovery following partial SCI. For instance, Kato and colleagues performed a unilateral hemisection at C2 or T12, which was followed several weeks later by a second unilateral hemisection on the opposite side of the spinal cord at mid-thoracic levels (70, 71). Such staggered hemisections effectively disrupt long supraspinal and propriospinal pathways to the lumbosacral spinal cord and abolished the coordination between the fore- and hindlimbs. The recovery of voluntary quadrupedal locomotion could be due to short de novo propriospinal pathways that span 1–2 segments, as also recently suggested in mice following similar staggered hemisections (142). What is clear is that functional changes within intrinsic spinal circuits are strongly implicated in the recovery of locomotion after incomplete SCI.

6. Concluding Remarks

In summary, several important discoveries for SCI research were first made or developed in the cat. The notion of an intrinsic spinal locomotor network that can be activated using various types of stimulation was an important step in developing targeted rehabilitative strategies for spinal cord-injured humans. The work performed in relatively large mammalian species, such as the cat, is vital to facilitate the translation of results to humans.

References

1. Beloozerova IN, Sirota MG (1986) Activity of motosensory cortex neurons in the cat during natural walking and stepping over obstacles. Neurophysiologia 18:546–549
2. McFadyen BJ, Lavoie S, Drew T (1999) Kinetic and energetic patterns for hindlimb obstacle avoidance during cat locomotion. Exp Brain Res 125:502–510
3. McVea DA, Pearson KG (2006) Long-lasting memories of obstacles guide leg movements in the walking cat. J Neurosci 26:1175–1178
4. Amos A, Armstrong DM, Marple-Horvat DE (1987) A ladder paradigm for studying skilled and adaptive locomotion in the cat. J Neurosci Methods 20:323–340
5. Amos A, Armstrong DM, Marple-Horvat DE (1987) Four mechanisms used to study skilled and adaptive ladder locomotion in the cat. J Physiol 382:11P
6. Gorassini MA, Prochazka A, Hiebert GW, Gauthier MJA (1994) Corrective responses to loss of ground support during walking I. Intact cats. J Neurophysiol 71:603–609
7. Hiebert GW, Gorassini MA, Jiang W, Prochazka A, Pearson KG (1994) Corrective responses to loss of ground support during walking II. Comparison of intact and chronic spinal cats. J Neurophysiol 71:611–622
8. Julien C, Rossignol S (1982) Electroneurographic recordings with polymer cuff electrodes in paralyzed cats. J Neurosci Methods 5:267–272
9. Whelan PJ (1996) Control of locomotion in the decerebrate cat. Prog Neurobiol 49: 481–515
10. Shik ML, Severin FV, Orlovsky GN (1966) Control of walking and running by means of electrical stimulation of the mid-brain. Biophysics 11:756–765
11. Gerasimenko YP, Avelev VD, Nikitin OA, Lavrov IA (2003) Initiation of locomotor activity in spinal cats by epidural stimulation of the spinal cord. Neurosci Behav Physiol 33:247–254
12. Barthelemy D, Leblond H, Provencher J, Rossignol S (2006) Nonlocomotor and locomotor hindlimb responses evoked by electrical microstimulation of the lumbar cord in spinalized cats. J Neurophysiol 96:3273–3292
13. Barthelemy D, Leblond H, Rossignol S (2007) Characteristics and mechanisms of locomotion induced by intraspinal microstimulation

and dorsal root stimulation in spinal cats. J Neurophysiol 97:1986–2000
14. Musienko PE, Bogacheva IN, Gerasimenko YP (2007) Significance of peripheral feedback in the generation of stepping movements during epidural stimulation of the spinal cord. Neurosci Behav Physiol 37:181–190
15. Mori S, Matsui T, Kuze B, Asanome M, Nakajima K, Matsuyama K (1999) Stimulation of a restricted region in the midline cerebellar white matter evokes coordinated quadrupedal locomotion in the decerebrate cat. J Neurophysiol 82:290–300
16. Matsuyama K, Mori F, Nakajima K, Drew T, Aoki M, Mori S (2004) Locomotor role of the corticoreticular-reticulospinal-spinal interneuronal system. Prog Brain Res 143:239–249
17. Hultborn H, Malmsten J (1983) Changes in segmental reflexes following chronic spinal cord hemisection in the cat. I. Increased monosynaptic and polysynaptic ventral root discharges. Acta Physiol Scand 119:405–422
18. Hultborn H, Malmsten J (1983) Changes in segmental reflexes following chronic spinal cord hemisection in the cat. II. Conditioned monosynaptic test reflexes. Acta Physiol Scand 119:423–433
19. Donelan JM, Pearson KG (2004) Contribution of force feedback to ankle extensor activity in decerebrate walking cats. J Neurophysiol 92: 2093–2104
20. Donelan JM, Pearson KG (2004) Contribution of sensory feedback to ongoing ankle extensor activity during the stance phase of walking. Can J Physiol Pharmacol 82:589–598
21. Armstrong DM (1986) Supraspinal contributions to the initiation and control of locomotion in the cat. Prog Neurobiol 26:273–361
22. Jordan LM, Liu J, Hedlund PB, Akay T, Pearson KG (2008) Descending command systems for the initiation of locomotion in mammals. Brain Res Rev 57:183–191
23. Duysens J, Pearson KG (1980) Inhibition of flexor burst generation by loading ankle extensor muscles in walking cats. Brain Res 187:321–332
24. Pearson KG (2008) Role of sensory feedback in the control of stance duration in walking cats. Brain Res Rev 57:222–227
25. Grillner S, Zangger P (1979) On the central generation of locomotion in the low spinal cat. Exp Brain Res 34:241–261
26. Rossignol S, Dubuc R, Gossard JP (2006) Dynamic sensorimotor interactions in locomotion. Physiol Rev 86:89–154
27. Eccles JC, Eccles RM, Lundberg A (1957) The convergence of monosynaptic excitatory afferents on to many different species of alpha motoneurones. J Physiol 137:22–50
28. Jankowska E (1992) Interneuronal relay in spinal pathways from proprioceptors. Prog Neurobiol 38:335–378
29. Jankowska E, Maxwell DJ, Bannatyne BA (2007) On coupling and decoupling of spinal interneuronal networks. Arch Ital Biol 145: 235–250
30. Petras JM (1967) Cortical, tectal and segmental fiber connections in the spinal cord of the cat. Brain Res 6:275–324
31. Holstege JC, Kuypers HGJM (1987) Brainstem projections to spinal motoneurons: an update. Neuroscience 23:809–821
32. Forssberg H, Grillner S (1973) The locomotion of the acute spinal cat injected with clonidine i.v. Brain Res 50:184–186
33. Forssberg H, Grillner S, Halbertsma J (1980) The locomotion of the low spinal cat. I. Coordination within a hindlimb. Acta Physiol Scand 108:269–281
34. Forssberg H, Grillner S, Halbertsma J, Rossignol S (1980) The locomotion of the low spinal cat: II. Interlimb coordination. Acta Physiol Scand 108:283–295
35. Robinson GA, Goldberger ME (1986) The development and recovery of motor function in spinal cats. I. The infant lesion effect. Exp Brain Res 62:373–386
36. Robinson GA, Goldberger ME (1986) The development and recovery of motor function in spinal cats. II. Pharmacological enhancement of recovery. Exp Brain Res 62:387–400
37. Howland DR, Bregman BS, Tessler A, Goldberger ME (1995) Development of locomotor behavior in the spinal kitten. Exp Neurol 135:108–122
38. Lovely RG, Gregor RJ, Roy RR, Edgerton VR (1986) Effects of training on the recovery of full-weight-bearing stepping in the adult spinal cat. Exp Neurol 92:421–435
39. Carter MC, Smith JL (1986) Simultaneous control of two rhythmical behaviors. II. Hindlimb walking with paw-shake response in spinal cat. J Neurophysiol 56:184–195
40. Barbeau H, Julien C, Rossignol S (1987) The effects of clonidine and yohimbine on locomotion and cutaneous reflexes in the adult chronic spinal cat. Brain Res 437:83–96
41. Barbeau H, Rossignol S (1987) Recovery of locomotion after chronic spinalization in the adult cat. Brain Res 412:84–95
42. Lovely RG, Gregor RJ, Roy RR, Edgerton VR (1990) Weight-bearing hindlimb stepping in treadmill-exercised adult spinal cat. Brain Res 514:206–218

43. Barbeau H, Rossignol S (1990) The effects of serotonergic drugs on the locomotor pattern and on cutaneous reflexes of the adult chronic spinal cat. Brain Res 514:55–67
44. Barbeau H, Rossignol S (1991) Initiation and modulation of the locomotor pattern in the adult chronic spinal cat by noradrenergic, serotonergic and dopaminergic drugs. Brain Res 546:250–260
45. de Guzman CP, Roy RR, Hodgson JA, Edgerton VR (1991) Coordination of motor pools controlling the ankle musculature in adult spinal cats during treadmill walking. Brain Res 555:202–214
46. Barbeau H, Chau C, Rossignol S (1993) Noradrenergic agonists and locomotor training affect locomotor recovery after cord transection in adult cats. Brain Res Bull 30: 387–393
47. Bélanger M, Drew T, Provencher J, Rossignol S (1996) A comparison of treadmill locomotion in adult cats before and after spinal transection. J Neurophysiol 76:471–491
48. de Leon RD, Hodgson JA, Roy RR, Edgerton VR (1998) Locomotor capacity attributable to step training versus spontaneous recovery after spinalization in adult cats. J Neurophysiol 79:1329–1340
49. Chau C, Barbeau H, Rossignol S (1998) Early locomotor training with clonidine in spinal cats. J Neurophysiol 79:392–409
50. de Leon RD, Tamaki H, Hodgson JA, Roy RR, Edgerton VR (1999) Hindlimb locomotor and postural training modulates glycinergic inhibition in the spinal cord of the adult spinal cat. J Neurophysiol 82:359–369
51. Bouyer LJG, Whelan P, Pearson KG, Rossignol S (2001) Adaptive locomotor plasticity in chronic spinal cats after ankle extensors neurectomy. J Neurosci 21:3531–3541
52. Giroux N, Chau C, Barbeau H, Reader TA, Rossignol S (2003) Effects of intrathecal glutamatergic drugs on locomotion. II. NMDA and AP-5 in intact and late spinal cats. J Neurophysiol 90:1027–1045
53. Bouyer LJG, Rossignol S (2003) Contribution of cutaneous inputs from the hindpaw to the control of locomotion: 2. Spinal cats. J Neurophysiol 90:3640–3653
54. Frigon A, Rossignol S (2008) Adaptive changes of the locomotor pattern and cutaneous reflexes during locomotion studied in the same cats before and after spinalization. J Physiol 586:2927–2945
55. Frigon A, Rossignol S (2008) Locomotor and reflex adaptation after partial denervation of ankle extensors in chronic spinal cats. J Neurophysiol 100:1513–1522
56. Frigon A, Rossignol S (2009) Partial denervation of ankle extensors prior to spinalization in cats impacts the expression of locomotion and the phasic modulation of reflexes. Neuroscience 158:1675–1690
57. English AW (1980) Interlimb coordination during stepping in the cat: effects of dorsal column section. J Neurophysiol 44:270–279
58. English AW (1985) Interlimb coordination during stepping in the cat. The role of the dorsal spinocerebellar tract. Exp Neurol 87:96–108
59. Zmyslowski W, Gorska T, Majczynski H, Bem T (1993) Hindlimb muscle activity during unrestrained walking in cats with lesions of the lateral funiculi. Acta Neurobiol Exp 53: 143–153
60. Gorska T, Majczynski H, Bem T, Zmyslowski W (1993) Hindlimb swing, stance and step relationships during unrestrained walking in cats with lateral funicular lesion. Acta Neurobiol Exp 53:133–142
61. Jiang W, Drew T (1996) Effects of bilateral lesions of the dorsolateral funiculi and dorsal columns at the level of the low thoracic spinal cord on the control of locomotion in the adult cat: I. Treadmill walking. J Neurophysiol 76:849–866
62. Gorska T, Bem T, Majczynski H (1990) Locomotion in cats with ventral spinal lesions: support patterns and duration of support phases during unrestrained walking. Acta Neurobiol Exp 50:191–200
63. Gorska T, Bem T, Majczynski H, Zmyslowski W (1993) Unrestrained walking in cats with partial spinal lesions. Brain Res Bull 32: 241–249
64. Bem T, Gorska T, Majczynski H, Zmyslowski W (1995) Different patterns of fore-hindlimb coordination during overground locomotion in cats with ventral and lateral spinal lesions. Exp Brain Res 104:70–80
65. Brustein E, Rossignol S (1998) Recovery of locomotion after ventral and ventrolateral spinal lesions in the cat. I. Deficits and adaptive mechanisms. J Neurophysiol 80:1245–1267
66. Brustein E, Rossignol S (1999) Recovery of locomotion after ventral and ventrolateral spinal lesions in the cat. II. Effects of noradrenergic and serotoninergic drugs. J Neurophysiol 81:1513–1530
67. Murray M, Goldberger ME (1974) Restitution of function and collateral sprouting in the cat spinal cord: the partially hemisected animal. J Comp Neurol 158:19–36
68. Bregman BS, Goldberger ME (1983) Infant lesion effect: I. Development of motor behavior following neonatal spinal cord damage in cats. Brain Res 9:103–117

69. Kato M, Murakami S, Yasuda K, Hirayama H, Hikino K (1983) Increase of extensor tonus of forelimbs in chronic cats with bilateral serial hemisections of the spinal cord at different levels. Neurosci Lett 41:289–293
70. Kato M, Murakami S, Yasuda K, Hirayama H (1984) Disruption of fore-and hindlimb coordination during overground locomotion in cats with bilateral serial hemisection of the spinal cord. Neurosci Res 2:27–47
71. Kato M, Murakami S, Hirayama H, Hikino K (1985) Recovery of postural control following chronic bilateral hemisections at different spinal cord levels in adult cats. Exp Neurol 90:350–364
72. Eidelberg E, Nguyen LH, Deza LD (1986) Recovery of locomotor function after hemisection of the spinal cord in cats. Brain Res Bull 16:507–515
73. Kato M (1989) Chronically isolated lumbar half spinal cord produced by hemisection and longitudinal myelotomy generates locomotor activities of the ipsilateral hindlimb of the cat. Neurosci Lett 98:149–153
74. Kato M (1990) Chronically isolated lumbar half spinal cord generates locomotor activities in the ipsilateral hindlimb of the cat. Neurosci Res 9(1):22–34
75. Helgren ME, Goldberger ME (1993) The recovery of postural reflexes and locomotion following low thoracic hemisection in adult cats involves compensation by undamaged primary afferent pathways. Exp Neurol 123:17–34
76. Kuhtz-Buschbeck JP, Boczek-Funcke A, Mautes A, Nacimiento W, Weinhardt C (1996) Recovery of locomotion after spinal cord hemisection: an X-ray study of the cat hindlimb. Exp Neurol 137:212–224
77. Tester NJ, Howland DR (2008) Chondroitinase ABC improves basic and skilled locomotion in spinal cord injured cats. Exp Neurol 209:483–496
78. Barriere G, Leblond H, Provencher J, Rossignol S (2008) Prominent role of the spinal central pattern generator in the recovery of locomotion after partial spinal cord injuries. J Neurosci 28:3976–3987
79. Frigon A, Barriere G, Leblond H, Rossignol S (2009) Asymmetric changes in cutaneous reflexes after a partial spinal lesion and retention following spinalization during locomotion in the cat. J Neurophysiol 102:2667–2680
80. Barriere G, Frigon A, Leblond H, Provencher J, Rossignol S (2010) Dual spinal lesion paradigm in the cat: evolution of the kinematic locomotor pattern. J Neurophysiol 104(2):1119–1133
81. Brown TG (1911) The intrinsic factors in the act of progression in the mammal. Proc Roy Soc Lond B 84:308–319
82. Heckmann CJ, Gorassini MA, Bennett DJ (2005) Persistent inward currents in motoneuron dendrites: implications for motor output. Muscle Nerve 31:135–156
83. Stuart DG, Hultborn H (2008) Thomas Graham Brown (1882–1965), Anders Lundberg (1920-), and the neural control of stepping. Brain Res Rev 59:74–95
84. Sherrington CS (1910) Flexion-reflex of the limb, crossed extension-reflex, and reflex stepping and standing. J Physiol 40:28–121
85. Anden NE, Jukes MGM, Lundberg A (1966) The effect of DOPA on the spinal cord. 2. A pharmacological analysis. Acta Physiol Scand 67:387–397
86. Jankowska E, Jukes MGM, Lund S, Lundberg A (1967) The effect of DOPA on the spinal cord. 5. Reciprocal organization of pathways transmitting excitatory action to alpha motoneurones of flexors and extensors. Acta Physiol Scand 70:369–388
87. Jankowska E, Jukes MGM, Lund S, Lundberg A (1967) The effects of DOPA on the spinal cord. 6. Half centre organization of interneurones transmitting effects from the flexor reflex afferents. Acta Physiol Scand 70:389–402
88. Lundberg A (1981) Half-centres revisited. In: Szentagothai J, Palkovits M, Hamori J (eds) Regulatory functions of the CNS. Principles of motion and organization, vol 1, Advances in physiological science. Pergamon Press, Budapest, pp 155–167
89. Dimitrijevic MR, Gerasimenko Y, Pinter MM (1998) Evidence for a spinal central pattern generator in humans. Ann N Y Acad Sci 860:360–376
90. Duysens J, Van de Crommert HW (1998) Neural control of locomotion; the central pattern generator from cats to humans. Gait Posture 7:131–141
91. MacKay-Lyons M (2002) Central pattern generation of locomotion: a review of the evidence. Phys Ther 82:69–83
92. Rossignol S, Bouyer L, Barthélemy D, Langlet C, Leblond H (2002) Recovery of locomotion in the cat following spinal cord lesions. Brain Res Rev 40:257–266
93. Grillner S (1981) Control of locomotion in bipeds, tetrapods, and fish. In: Brookhart JM, Mountcastle VB (eds) Handbook of physiology. The nervous system II. American Physiological Society, Bethesda, MD, pp 1179–1236
94. Perret C (1983) Centrally generated pattern of motoneuron activity during locomotion in

the cat. In: Roberts A, Roberts BL (eds) Neural origin of rhythmic movements, vol 37, Society for experimental biology symposium. Cambridge University Press, Cambridge, pp 405–422
95. Rybak IA, Shevtsova NA, Lafreniere-Roula M, McCrea DA (2006) Modelling spinal circuitry involved in locomotor pattern generation: insights from deletions during fictive locomotion. J Physiol 577:617–639
96. Rybak IA, Stecina K, Shevtsova NA, McCrea DA (2006) Modelling spinal circuitry involved in locomotor pattern generation: insights from the effects of afferent stimulation. J Physiol 577:641–658
97. McCrea DA, Rybak IA (2008) Organization of mammalian locomotor rhythm and pattern generation. Brain Res Rev 57:134–146
98. Edgerton VR, Leon RD, Harkema SJ, Hodgson JA, London N, Reinkensmeyer DJ, Roy RR, Talmadge RJ, Tillakaratne NJ, Timoszyk W, Tobin A (2001) Retraining the injured spinal cord. J Physiol 533:15–22
99. Harkema SJ (2001) Neural plasticity after human spinal cord injury: application of locomotor training to the rehabilitation of walking. Neuroscientist 7:455–468
100. de Leon RD, Hodgson JA, Roy RR, Edgerton VR (1999) Retention of hindlimb stepping ability in adult spinal cats after the cessation of step training. J Neurophysiol 81:85–94
101. Carrier L, Brustein L, Rossignol S (1997) Locomotion of the hindlimbs after neurectomy of ankle flexors in intact and spinal cats: model for the study of locomotor plasticity. J Neurophysiol 77:1979–1993
102. Goldberger ME (1977) Locomotor recovery after unilateral hindlimb deafferentation in cats. Brain Res 123:59–74
103. Grillner S, Rossignol S (1978) On the initiation of the swing phase of locomotion in chronic spinal cats. Brain Res 146:269–277
104. Andersson O, Forssberg H, Grillner S, Wallén P (1981) Peripheral feedback mechanisms acting on the central pattern generators for locomotion in fish and cat. Can J Physiol Pharmacol 59:713–26
105. Andersson O, Grillner S (1981) Peripheral control of the cat's step cycle. I. Phase dependent effects of ramp-movements of the hip during "fictive locomotion". Acta Physiol Scand 113:89–101
106. Andersson O, Grillner S (1983) Peripheral control of the cat's step cycle. II. Entrainment of the central pattern generators for locomotion by sinusoidal hip movements during "fictive locomotion". Acta Physiol Scand 118: 229–239
107. Conway BA, Hultborn H, Kiehn O (1987) Proprioceptive input resets central locomotor rhythm in the spinal cat. Exp Brain Res 68: 643–656
108. Kriellaars DJ, Brownstone RM, Noga BR, Jordan LM (1994) Mechanical entrainment of fictive locomotion in the decerebrate cat. J Neurophysiol 71:2074–2086
109. Guertin P, Angel MJ, Perreault M-C, McCrea DA (1995) Ankle extensor group I afferents excite extensors throughout the hindlimb during fictive locomotion in the cat. J Physiol 487:197–209
110. Perreault M-C, Angel MJ, Guertin P, McCrea DA (1995) Effects of stimulation of hindlimb flexor group II afferents during fictive locomotion in the cat. J Physiol 487:211–220
111. Stecina K, Quevedo J, McCrea DA (2005) Parallel reflex pathways from flexor muscle afferents evoking resetting and flexion enhancement during fictive locomotion and scratch in the cat. J Physiol 569:275–290
112. Frigon A, Sirois J, Gossard J-P (2010) The effects of ankle and hip muscle afferent inputs on rhythm generation during fictive locomotion. J Neurophysiol 103:1591–1605
113. Pearson KG, Ramirez JM, Jiang W (1992) Entrainment of the locomotor rhythm by group Ib afferents from ankle extensor muscles in spinal cats. Exp Brain Res 90: 557–566
114. Whelan PJ, Hiebert GW, Pearson KG (1995) Stimulation of the group I extensor afferents prolongs the stance phase in walking cats. Exp Brain Res 103:20–30
115. Hiebert GW, Whelan PJ, Prochazka A, Pearson KG (1996) Contribution of hind limb flexor muscle afferents to the timing of phase transitions in the cat step cycle. J Neurophysiol 75:1126–1137
116. Whelan PJ, Pearson KG (1997) Comparison of the effects of stimulating extensor group I afferents on cycle period during walking in conscious and decerebrate cats. Exp Brain Res 117:444–452
117. Lam T, Pearson KG (2001) Proprioceptive modulation of hip flexor activity during the swing phase of locomotion in decerebrate cats. J Neurophysiol 86:1321–1332
118. Lam T, Pearson KG (2002) Sartorius muscle afferents influence the amplitude and timing of flexor activity in walking decerebrate cats. Exp Brain Res 147:175–185
119. Duysens J, Clarac F, Cruse H (2000) Load-regulating mechanisms in gait and posture: comparative aspects. Physiol Rev 80:83–133
120. af Klint R, Cronin NJ, Ishikawa M, Sinkjaer T, Grey MJ (2010) Afferent contribution to

locomotor muscle activity during unconstrained overground human walking: an analysis of triceps surae muscle fascicles. J Neurophysiol 103:1262–1274

121. af Klint R, Mazzaro N, Nielsen JB, Sinkjaer T, Grey MJ (2010) Load rather than length sensitive feedback contributes to soleus muscle activity during human treadmill walking. J Neurophysiol 103:2747–2756
122. McVea DA, Pearson KG (2007). Long-lasting, context-dependent modification of stepping in the cat after repeated stumbling-corrective responses. J Neurophysiol 97: 659–69
123. Forssberg H (1979) Stumbling corrective reaction: a phase-dependent compensatory reaction during locomotion. J Neurophysiol 42:936–953
124. Schillings AM, Van Wezel BMH, Duysens J (1996) Mechanically induced stumbling during human treadmill walking. J Neurosci Methods 67:11–17
125. Zehr EP, Stein RB (1999) What functions do reflexes serve during human locomotion? Prog Neurobiol 58:185–205
126. Lam T, Wolstenholme C, van der Linden M, Pang MY, Yang JF (2003) Stumbling corrective responses during treadmill-elicited stepping in human infants. J Physiol 553: 319–331
127. Bouyer LJG, Rossignol S (2003) Contribution of cutaneous inputs from the hindpaw to the control of locomotion: 1. Intact cats. J Neurophysiol 90:3625–3639
128. Heckman CJ, Lee RH, Brownstone RM (2003) Hyperexcitable dendrites in motoneurons and their neuromodulatory control during motor behavior. Trends Neurosci 26:688–695
129. Naftchi NE (1982) Functional restoration of the traumatically injured spinal cord in cats by clonidine. Science 217:1042–1044
130. Heckman CJ, Hyngstrom AS, Johnson MD (2008) Active properties of motoneurone dendrites: diffuse descending neuromodulation, focused local inhibition. J Physiol 586:1225–1231
131. Heckman CJ, Mottram C, Quinlan K, Theiss R, Schuster J (2009) Motoneuron excitability: the importance of neuromodulatory inputs. Clin Neurophysiol 120:2040–2054
132. Chau C, Barbeau H, Rossignol S (1998) Effects of intrathecal a_1- and a_2-noradrenergic agonists and norepinephrine on locomotion in chronic spinal cats. J Neurophysiol 79: 2941–2963
133. Giroux N, Reader TA, Rossignol S (2001) Comparison of the effect of intrathecal administration of clonidine and yohimbine on the locomotion of intact and spinal cats. J Neurophysiol 85:2516–2536
134. Chau C, Giroux N, Barbeau H, Jordan LM, Rossignol S (2002) Effects of intrathecal glutamatergic drugs on locomotion. I. NMDA in short-term spinal cats. J Neurophysiol 88: 3032–3045
135. Tillakaratne NJ, Mouria M, Ziv NB, Roy RR, Edgerton VR, Tobin AJ (2000) Increased expression of glutamate decarboxylase (GAD(67)) in feline lumbar spinal cord after complete thoracic spinal cord transection. J Neurosci Res 60:219–230
136. Tillakaratne NJ, de Leon RD, Hoang TX, Roy RR, Edgerton VR, Tobin AJ (2002) Use-dependent modulation of inhibitory capacity in the feline lumbar spinal cord. J Neurosci 22:3130–3143
137. Dietz V, Colombo G, Jensen L, Baumgartner L (1995) Locomotor capacity of spinal cord in paraplegic patients. Ann Neurol 37:574–582
138. Remy-Neris O, Barbeau H, Daniel O, Boiteau F, Bussel B (1999) Effects of intrathecal clonidine injection on spinal reflexes and human locomotion in incomplete paraplegic subjects. Exp Brain Res 129:433–440
139. Schwindt PC, Crill WE (1980) Properties of a persistent inward current in normal and TEA-injected motoneurons. J Neurophysiol 43: 1700–1724
140. Hultborn H (2003) Changes in neuronal properties and spinal reflexes during development of spasticity following spinal cord lesions and stroke: studies in animal models and patients. J Rehabil Med 41 Suppl: 46–55
141. Murray KC, Nakae A, Stephens MJ, Rank M, D'Amico J, Harvey PJ et al (2010) Recovery of motoneuron and locomotor function after spinal cord injury depends on constitutive activity in 5-HT2C receptors. Nat Med 16:694–700
142. Courtine G, Song B, Roy RR, Zhong H, Herrmann JE, Ao Y et al (2008) Recovery of supraspinal control of stepping via indirect propriospinal relay connections after spinal cord injury. Nat Med 14:69–74

Chapter 9

Using Naturally Occurring Spinal Cord Injury in Domestic Dogs to Explore Novel Therapeutic Options

Nick D. Jeffery, Nicolas Granger, and Robin J.M. Franklin

Abstract

Modeling human disease in animals has traditionally been used to elucidate pathogenesis and test promising new approaches to treatment. This approach has now identified a plethora of interventions that ameliorate functional deficits associated with spinal cord injury (SCI) in rodents. A major current challenge is to translate these interventions into humans. However, there are several important differences between experimental SCI in rodents and the equivalent clinical condition in humans, chiefly that of lesion heterogeneity, which can result in failure to translate a statistically detectable effect in a laboratory into a benefit of sufficient magnitude to be of value to clinical patients. Here we show how veterinary clinical SCI cases can help in testing the feasibility and value of translating specific putative therapies from laboratory to clinic. Dogs commonly present with SCI, undergo similar diagnostic and therapeutic procedures as their human counterparts and have a similar prognosis. We show how effects on spinal cord function can be objectively and precisely quantified in clinical canine patients and discuss how this type of data provides a unique viewpoint with which to evaluate promising experimental interventions prior to initiating human clinical trials.

Key words: Spinal cord injury, Translation, Transplantation, Dog, Olfactory ensheathing cell, Veterinary

1. Introduction

Traditionally, animal models in medical research are used chiefly to dissect the pathogenesis of specific aspects of human disease and, subsequently, to assess the efficacy of novel therapies. This approach has many facets, such as modeling of known disease mechanisms (including spinal cord contusion), but is epitomized by studies on genetically modified mice, in which the effects of perturbation of a specific gene can be studied in detail in the absence of confounding factors.

Modern research into spinal cord injury (SCI) has become primarily focused on rodent models, because they provide homogenous

Håkan Aldskogius (ed.), *Animal Models of Spinal Cord Repair*, Neuromethods, vol. 76,
DOI 10.1007/978-1-62703-197-4_9, © Springer Science+Business Media, LLC 2013

populations of genetically characterized lines of animals. There are also fewer economic and ethical constraints with rodent models. However, some of the earliest investigations into SCI used dogs in the same manner. For instance, Allen demonstrated in 1911 the importance of secondary damage in the injured spinal cord using a canine "weight drop" model (1). Used in this way, experimental dogs have the advantage of their (relative) large size and availability of in-bred strains (notably research beagles) but current *mores* in Europe, Japan, and the USA dictate that experiments on dogs are less acceptable, on both financial and ethical grounds. Therefore, this use of dogs in SCI research has become uncommon in Japan and the Western world.

This traditional approach to the use of dogs in SCI research can be crudely caricatured as thinking of dogs as "large rodents," in which the same experiments that could be done in rodents are instead carried out in dogs. A different approach—the emphasis of this chapter—is to study disease that occurs spontaneously in domestic dogs, and the effects of novel therapeutic interventions on that disease. Here we use the term "domestic dog" to refer to pet dogs whose owners seek veterinary care when clinical signs of disease are displayed. This population of dogs can therefore become a powerful translational "model" for investigation of the effects of novel treatments prior to their use in human patients. Dogs also benefit from this approach if the treatment under investigation is efficacious. This approach can therefore be characterized informally as treating pet dogs as "surrogate human" patients, in which therapies can be trialed more cheaply and with no risk to human patients. This is made possible because pet dogs are frequently and spontaneously affected by diseases that closely resemble their human counterparts (see for example (2)) and for which, similar to human patients, there may be no effective treatment. SCI is one such notable example (3).

In summary therefore, the current use of rodents in investigation of new therapeutic approaches is mainly for "proof-of-principle" experiments to determine whether there is a demonstrable *statistical difference* (improvement) in outcome following an intervention; in contrast, cohorts of pet dogs presenting with SCI can be used to determine whether the intervention has a *clinically useful* effect (i.e., are the effects sufficiently large and robust to be of medical value?).

Generally, for the domestic dog population to play such a role it is important that the disease under study should be well characterized and of sufficient prevalence. Furthermore, in common with human clinical trials, there is a need for well-defined, preferably numerical, outcome measures with which to define efficacy. In this chapter we describe how these features apply to SCI in the domestic dog population. From the human medical perspective, the value of trials in domestic dogs is primarily to identify interventions that have *sufficient demonstrable clinical benefit despite heterogeneity of patient presentation* to make human trials worthwhile. From the point of view of potential funders, such trials are orders of magnitude less expensive than human trials and allow for small "wrinkles" in technique to be ironed out before human trials commence. From the point of view of veterinarians, their clients and patients, such trials aid translation of advances in laboratory science directly into the veterinary clinic.

2. Background

2.1. History of the Use of Dogs for SCI Study

Dogs were the experimental subjects of the first attempt to model SCI (1): following laminectomy and placement of a metal interface device on the exposed dura mater a weight was dropped a specified distance to cause a contusion lesion. Subsequently the neurologic outcome was correlated with the force (weight x distance) and extent of histologic damage to the spinal cord. The model has since been refined, by use of computer-controlled feedback devices, and is now predominantly used in rodents, but remains essentially the same. Presumably dogs were selected for experiments because they were relatively large and, at the time, both accessible and acceptable subjects for medical research of this nature.

In subsequent years, dogs and cats were used for many experiments perturbing the CNS and in the 1950s Tarlov and coworkers published a series of experiments in which grading of neurologic deficits after SCI in animals was first attempted (4–6). The grading was dependent on the severity of pelvic limb paresis that occurred after thoracolumbar SCI and the experiments also documented how much functional improvement could occur and over what time period. These experiments established that with long-term follow-up, even severe neurologic deficits could be reversed, although there was an upper limit to the severity from which recovery of locomotion could be expected.

Within the last two decades rodent models have almost entirely replaced dogs and cats in SCI research, and functional deficit can be defined with precision in these species (7, 8). Dogs have continued to be used for some spinal experiments but predominantly those concerned with the orthopedics of spinal injury. For instance, Al-Mefty and coworkers implanted plastic screws into the vertebral bodies of dogs to model chronic spinal cord compression and the effects of decompression (9, 10). This experimental use of dogs for spinal disease investigation is therefore based on their relatively large size. In general, experimental work in dogs has become ethically unacceptable in Japan and the Western world (although it is still possible to conduct SCI experiments on dogs in the USA—for example see (11) or (12)). In contrast, in Korea and China there are many current experiments investigating the effects of various stem cell transplants on outcome after experimental SCI in dogs (13–17). Such use of dogs is an example of experiments testing a "proof-of-principle," in contrast to studies in canine clinical cases aiming directly to determine translational potential.

2.2. "Natural" Spinal Cord Injury in Dogs: The Veterinary Aspect

Domestic dogs suffering from naturally occurring SCI have been treated by veterinarians for at least a century and the pathogenesis of intervertebral disk disease, the most common cause of SCI in dogs, was elucidated in detail about 60 years ago (18, 19). There are two major causes of SCI in domestic dogs: external trauma (usually road traffic accidents) causing vertebral fracture or luxation (20), and intervertebral disk nuclear extrusion occurring as a consequence of intervertebral disk degeneration (IVDD) (21). Intervertebral disk degeneration is a very common condition, especially in certain small, chondrodystrophic breeds, notably the Dachshund, in which prevalence is generally estimated to be ~19%, and possibly much higher in specific family lines (22). Intervertebral disk disease is particularly prevalent in many small dogs because they are chondrodystrophic and thereby at increased risk of chondroid degeneration of the IVD nucleus (19). Mostly as a result of this, SCI is very frequently encountered in veterinary practice; there are many hundreds of centers in Western countries in which such cases are treated and each veterinary neurosurgical center can expect to treat between 200 and 500 cases per year. As a result of this extensive experience over 50 years the standard of care for dogs with SCI has greatly improved so that dogs with SCI now receive very similar diagnostic tests, treatment options, prognostication and after-care as their human SCI patient counterparts (23). This large pool of canine SCI patients therefore provides a largely under-exploited population in which putative therapeutics can be tested.

2.3. Pathological Findings

The acute pathological consequences of SCI share many similarities between rodents, dogs, cats and humans. In all species progressively more severe contusive injuries are associated with a centrifugally directed progression of tissue destruction (24–27) and it has also been demonstrated that small diameter axons are less susceptible to destruction following contusive injuries than larger axons (23, 26). However, some discrepancies are found between humans, cats, and dogs on the one hand and rodent models on the other. For example, Nashmi and Fehlings (28) found an increase in diameter of surviving axons in the dorsal columns of the injured rat spinal cord as opposed to that of the cat (24).

Experimental SCI in rodents is most commonly induced using an impact applied to the dorsal aspect of the cord, whereas in both humans and dogs the inciting injury usually originates from a ventral (anterior) direction, because of the common causes (fracture-luxation or extruded IVD nucleus) (Fig. 1). This causes different

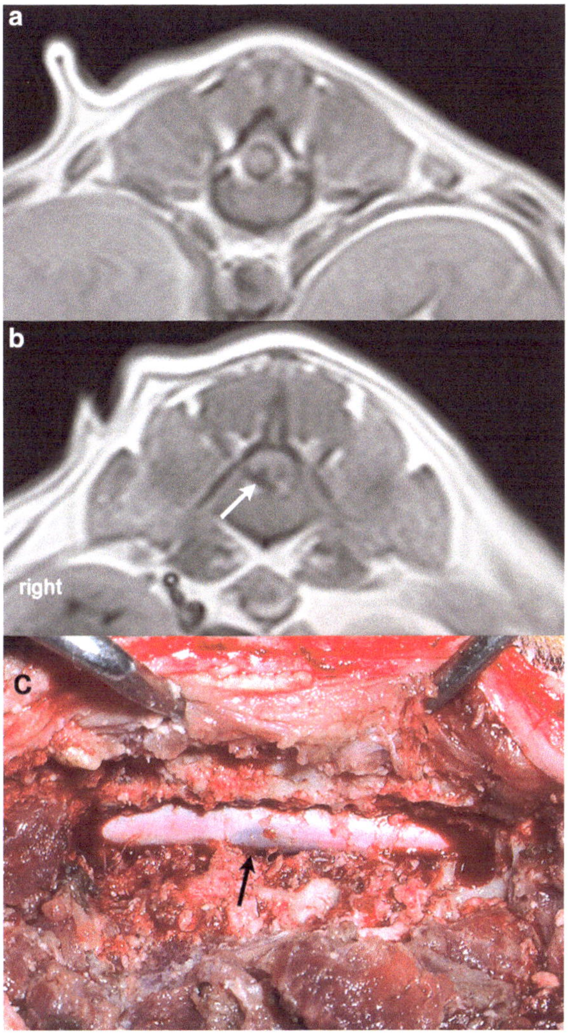

Fig. 1. Clinical aspects of canine spinal cord injury. (**a**) T1-weighted transverse MRI scan of the spinal cord close to the thoracolumbar junction in a normal dog; (**b**) comparable level transverse T1W MRI scan in a dog with right sided intervertebral disk extrusion causing compression of the spinal cord (*white arrow*); (**c**) surgical appearance of the spinal cord during surgical decompression, hematoma of the spinal cord parenchyma is visible through the dura mater (*black arrow*) (**c**: Hidalgo A, with permission).

patterns of dorsal–ventral lesion asymmetry that has functional significance because dorsal lesions (which affect sensory pathways more predominantly) are associated with better locomotion than ventral lesions (which cause more destruction to motor pathways) (29). Also, small differences in macroscopic vascular patterns may affect precise patterns of tissue injury: in dogs, as in humans, there are two dorsolateral arteries and one ventral midline artery (30), whereas in rats, there are two small anastomotic chains of arteries forming irregular loops of vessels (31).

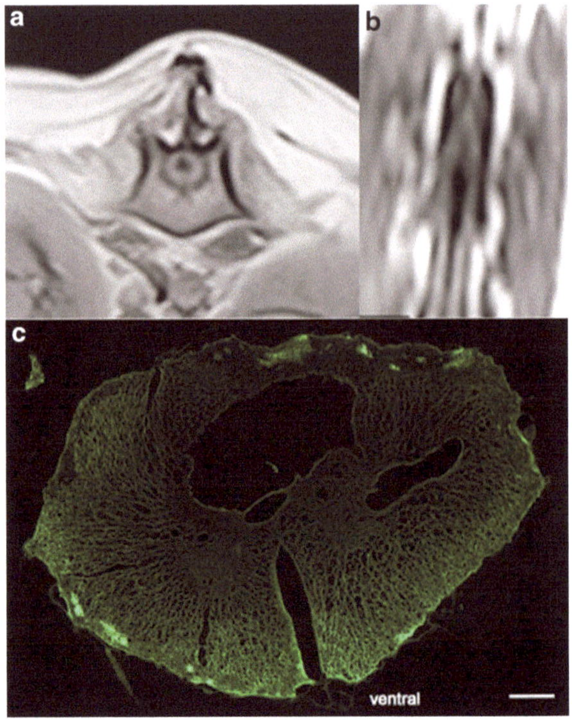

Fig. 2. Post-injury spinal cord cavitation. (**a**) Transverse T1W MRI scan of spinal cord immediately caudal to the injury level showing a cavity (which appears *black*) within the parenchyma; (**b**) same lesion viewed in dorsal plane T1W MRI scan—note that the volume of the lesion can be derived from images (**a**) and (**b**); (**c**) transverse section of a post-injury cavitation following immunohistochemical reaction for glial fibrillary acidic protein (*green*) (12 μm section). Scale bar: 500 μm.

In contrast to the relatively high degree of homogeneity of lesion appearance between individuals in the acute stages after contusive injury, pathological heterogeneity, especially with regard to the extent of demyelination–remyelination and formation of cystic structures or syringomyelia (Fig. 2), is readily apparent in more chronic stages of clinical SCI in both dogs and humans. In terms of potential for limiting functional recovery, intraspinal cavitation is important because it may have profound effects on both axonal sprouting and the ability to remyelinate (32). The chronic phase of the injury also highlights a key problem with modeling SCI in the laboratory, because the post-injury scarring process in the mouse spinal cord differs greatly from that in the rat, dog and human. In mice progressive necrosis and cavity formation does not occur as it does in other species. Instead there is formation of a continuous connective tissue matrix, much as occurs in non-neural tissue (33). This phenomenon is likely to be specific to mice because it does not occur in rats or hamsters (34). Thus, genetically modified mice, while useful to answer specific research questions, may not be appropriate for examination of all aspects of pathogenesis or gene activation patterns following SCI. In contrast, canine clinical cases model accurately the patterns of chronic sequelae that occur in humans and which may represent obstacles to successful translation of strategies from rodent to human.

3. Concepts Underlying Use of Veterinary Patients in SCI Research

The study of SCI in the laboratory has identified numerous interventions that are successful in improving functional outcome in experimental rodents, but there is now a pressing need to determine how best to translate successful strategies into clinically applicable treatments for human patients. Because of the factors outlined above, spontaneous SCI in domestic dogs constitutes an excellent intermediate model for this purpose.

To illustrate this specific point, use of dogs as experimental models for proof-of-principle experiments is rarely of great value, since there is little additional information that can be obtained over that available from rodent experiments, which are cheaper, more tightly controllable and ethically less complicated. Instead, domestic dogs that sustain naturally occurring SCI in the same way as humans provide excellent models for *translational* research, because they mirror so well the *clinical presentation* in human patients.

Proof-of-principle laboratory studies on experimental animals rely on establishing homogenous injuries within a homogenous population in which the effect of an intervention can be most readily observed. For instance, typically, the nature of the injury will be stereotypical (e.g., computer controlled impact device) in a tightly controlled sample of rats of the same weight, gender and genetic background. This implies a poor degree of "generalizability" of the investigations to other situations including, most importantly, human patient populations. In contrast, but in common with human patients, domestic dogs with SCI have natural injuries, in which the precise nature of the injury varies between individuals (e.g., varying components of contusion and compression, resulting from fractures, luxations, disk extrusions), there is a variable genetic background and there will be varying degrees of functional loss even within defined patient sub-groups. Further, rodent models of SCI also tend to minimize anatomical disruption of the spinal cord and surrounding structures, which favors axonal regrowth following an intervention (35), whereas structural integrity is frequently compromised in clinical injuries to dogs or humans. Similarly, dogs that sustain SCI because of a road traffic accident will, like human patients, present as poly-traumatized cases—a complicating factor that is never reproduced in experimental animals. It is likely that poor perfusion and oxygenation of the spinal cord as a result of multiple organ damage will exacerbate histologic injury and be detrimental to recovery.

Veterinarians are increasingly aware of the possibilities for translating laboratory research into the veterinary clinic and dog owners not only demand state-of-the-art treatment for their pets but also are well informed (via the internet) about possible novel treatments. These factors now have conspired to produce a large pool of domestic dogs that have severe SCI, in which the lesion can be well characterized using state-of-the-art imaging technology, and could form part of multicenter trials.

In summary, the central concept is that pet dogs form a "surrogate human" population of clinical patients in which to test the efficacy of potential treatments for SCI. Specific advantages of using naturally occurring SCI in domestic dogs as a model for investigating interventions are: (1) patterns of SCI are very similar to those in human patients; (2) diagnosis and patient categorization are made using the same tools; (3) similar methods can be used for functional assessment; (4) there is a large pool of chronically injured patients in both species; and, (5) domestic dogs and humans share a similar environment and can receive similar methods of rehabilitation.

4. Advantages of this Model for Exploring Specific Aspects of Spinal Cord Injury

4.1. Study of Treatment "Dose" in Dogs

The translation of therapies from laboratory animals to human patients necessitates defining the dose of a given treatment required to obtain the expected effect in a different, much larger, species. Defining the dose may be a problem in itself although may be done by simply scaling up the rat dose for the body weight, or body surface area, of a human patient (which may be an adequate approach for drug administration). Another method, which may be a better approach for cell therapies, would be to scale up the dose for the size of the spinal cord or by correcting the dose for the volume of the lesion measured by magnetic resonance imaging (MRI) (Fig. 2) (36). For example, the normal spinal cord diameter of Dachshund dogs in the caudal thoracic region (T9) is 4.0 ± 0.5 mm and 3.0 ± 0.2 mm at a lesion site (T12, T13 or L1) 3 months after an injury ($n = 10$ dogs with a mean body weight of 6.2 ± 1.8 kg; unpublished data; measurement from MR images). In comparison, the rat normal spinal cord diameter measured on MRI varies between 1 and 3 mm (36), and the gross human spinal cord diameter is around 7 mm in the caudal thoracic region (37).

However, because human or dog spinal cord lesions are longer and wider than lesions obtained in laboratory animals, not only the number of cells but the migratory properties of the cells tested, as well as the optimal method of delivery of the cells to populate the entire lesion must also be considered. The key difference between cell and drug therapies in this regard is that, although the calculated number of cells may be appropriate, it will also be critical that the cells are able to arrive at a lesion site within a specified amount of time (i.e., when the environment is responsive to their effects). The size of a lesion may present an obstacle in this respect because the time that is required for a cell to migrate through a large lesion may delay it sufficiently to make it arrive too late to be of benefit. Therefore, the dog forms a good model in which treatment doses and timing could be tested for their flexibility in development toward human use, since spinal cord lesions are larger than those in rats.

4.2. Histopathological Study of the Spinal Cord after Intervention

The availability of human spinal cord pathological specimens is very limited and when available, poor tissue preservation due to delayed fixation can limit acquisition of high quality data. Dogs have a shorter lifespan that humans, and may live only between a few months to several years following SCI and so are more commonly available for *post mortem* examination including, in some cases, tissue prepared by perfusion-fixation. Owners of dogs that have taken part in exploratory studies on novel therapeutic interventions are highly motivated by the research and commonly allow *post mortem* examination.

4.3. Environmental Factors

Recovery following a SCI is highly influenced by the activities of the patient, either performed autonomously (if still possible) or via a rehabilitation program. For example, treadmill training can have a major impact on plasticity of the intrinsic spinal cord circuitry in humans, cats and rats (38). It has also recently been found that behaviors can compete. For example, environmental enrichment in rats (which favors non-specific "self-training") can cancel the benefits obtained by combining an intervention with specific task training (39). Following SCI, domestic dogs are treated in directly comparable environments as human patients and may also benefit from rehabilitation, such as physiotherapy and hydrotherapy, in which dogs engage very readily. Thus, domestic dogs could also be used to investigate how best to integrate medical interventions and rehabilitation programs.

5. Practicalities of Studying SCI Interventions in Dogs

5.1. Lesion Types

All types of lesion found in humans, including pure contusion, mixed contusion-compression, chronic compression, dynamic compression and vascular lesions have been recognized in dogs, and are readily identifiable on advanced imaging studies. However, penetrating trauma, such as that resulting from knife or gunshot wounds, is uncommon.

Domestic dogs are therefore available for modeling many different aspects of SCI although, as in humans, the most common clinical lesion results from acute mixed compressive and contusive events, such as that associated with fracture-luxation or intervertebral disk nuclear extrusion. This sub-population of dog patients is large, because the incidence rate of nuclear IVD extrusion is very high, especially within specific sub-populations.

5.2. Lesion Sites

SCI can occur at any site in dogs, but is most common adjacent to the thoracolumbar junction (TLJ), probably because of the changing nature of the mechanics and stabilizing influences within this region (19). In dogs, the TLJ contains the spinal cord (which extends into L5 vertebra) and therefore injuries at this level model well the effect of spinal cord lesions in the thoracic part of the human vertebral column. Effects of the lesion on brain control of the distal extremities, including bladder and limbs, can therefore be examined, with the exception of the control of sexual function (which would be difficult to study in dogs).

5.3. Survival Periods and Complications

The shorter general lifespan of dogs implies that canine veterinary patients will become available for *post mortem* studies much sooner after an intervention than their human counterparts. Nevertheless, dog SCI patients will usually survive many years after their initial injury, providing an opportunity to study the effects of an intervention over a longer follow-up period than is possible in rodents. On the other hand, dogs that have become paraplegic following SCI are susceptible to the same secondary problems as human paraplegics, such as ascending urinary tract infections, providing another opportunity for assessment of the impact of interventions in prehuman clinical trials.

5.4. Monitoring Function After Injury

Routine methods of clinical analysis and classification of function in canine SCI patients are very similar to those available for human patients, relying on simple, long-established, scoring systems. However, domestic dogs are highly amenable to sophisticated analysis of movement and the quadrupedal gait can be exploited to provide detailed quantifiable data on the coordination of limb movements between thoracic and pelvic limb girdles. It is particularly helpful in this respect that the most common site of injury in dogs is between T3 and L3 spinal cord segments (and by far most commonly between T12 and L2 (dog spinal cord has 13 thoracic and 7 lumbar segments)) and so lesions at that site disrupt communication between thoracic and pelvic girdles. The quadrupedal gait permits assessment of this coordination in a way that would not be feasible in human subjects and allows precise quantification of deficits and the response to therapeutic interventions (see below).

5.5. Routes for Delivery of Therapeutic Agents

Any of the feasible approaches for drug delivery in human patients can be used in dogs, including intravenous, intrathecal or even intraparenchymal routes of delivery of both cells and drugs. There are regulations in most countries regarding the ethics of interventions in domestic dogs, but these are not so restrictive as to prevent clinical trials of novel therapies, similar to those that would be permitted in human patients, so long as there is adequately documented scientific justification for the intervention. Regulation of interventions in veterinary patients varies throughout the world and frequently differs from the regulation of experimental procedures carried out in laboratory animals. One obstacle in the UK is the specific prohibition of placebos in clinical trials. This can make it difficult to properly evaluate in a double-blinded study (i.e., when both veterinary investigator and owner are blinded) the implications and treatment efficacy of a novel intervention versus sham therapy. However, for severely injured animals there is (of course) no currently effective therapy, meaning that it is feasible to choose to compare two different plausible interventions based on data obtained in laboratory rodents. Furthermore, there are frequently patient groups that can be identified and act as reasonable, albeit imperfect, controls (for instance paraplegic dogs whose owners choose not to take part in a clinical treatment trial). The specific prohibition of placebos places obstacles on surgical interventions in particular, since it would be very difficult for an Ethical Committee to accept sham surgery for a paralyzed dog.

6. Outcome Measures

Until recently, a potential drawback to the use of domestic dogs in SCI research was the lack of well-defined outcome measures of the effect of an intervention. Such a limitation also applies to some degree to human patients, since routine clinical outcome scores are coarse estimates of function and subjective (40) and therefore may differ from 1 day to another despite a stable neurologic deficit. Therefore the recent development of outcome measure tools for canine patients (41, 42) suggests that it might well be possible to detect changes in function with greater subtlety than in human patients.

In general, the *clinical* scoring instruments that are used to measure the neurological deficits and the function in dogs with SCI are identical to those used for human patients, allowing direct translation of this type of outcome data from dogs to human patients.

6.1. Clinical Score

The severity of functional loss following SCI in dogs has been subject to clinical scoring since the 1950s, using Tarlov scales, or that implemented by Hoerlein (43). Recognizing the need for more precise tools, Olby and others (44) developed a scoring scheme similar to the BBB score used in rodents (7) for use in dogs. Although providing considerably more subtlety, the scoring scheme is not linear, which limits accurate and flexible statistical analysis over the range of possible scores. Furthermore, subjective scores remain susceptible to bias and are often incomplete because they do not take into account urinary and fecal incontinence; also, assessment of fine sensation, which is a component of similar schemes in human patients, is difficult in dogs.

6.2. Spinal Cord Imaging

Advanced imaging techniques of the spinal cord such as high field MRI and 16-slice computed tomography (CT) are increasingly available for veterinary patients (45), although dogs must be anesthetized during such examinations. These techniques allow detailed analysis of spinal cord integrity similar to that available in routine clinical practice in humans and can be used to recognize post-traumatic complications of SCI such as syringomyelia (46). Extended regions of hyperintensity on T2-weighted MRI scans are now known to be correlated with poor recovery (45, 47). Diffusion tensor imaging and other less routine MRI techniques, although not commonly used in dogs currently, could also be applied with access to the appropriate software (23).

6.3. Electrophysiology

Electrodiagnostic equipment used for dogs is identical to that used for human patients and similar recordings as can be made in human patients can be obtained from dog patients, providing information on integrity of impulse transmission through the damaged region of spinal cord (48–50). This includes:

- *Transcranial magnetic motor evoked potentials*: compound muscle action potentials recorded in limb muscles following magnetic stimulation of the head or spine. Motor evoked potentials are used to study long-distance descending motor pathways;
- *Somatosensory evoked potentials*: field or ascending potentials recorded over the vertebral column or skull following stimulation of a peripheral nerve, spinal cord (spinal-evoked potential) or skin (dermatomal somatosensory evoked potential). Somatosensory evoked potentials are used to study ascending sensory pathways;
- *Late waves* recorded from peripheral nerves, such as the Hoffman reflex (H-reflex), reflect motor neuron excitability following SCI (51) and can assess plasticity of the spinal cord. The H-reflex can be recorded in paraplegic dogs (Fig. 3).

There is extensive veterinary experience with these techniques and somatosensory evoked potentials and spinal-evoked potentials have been shown to be clinically relevant to grade SCI in dogs (49, 52). Transcranial magnetic motor evoked potentials (TMMEPs) can be difficult to record in the acute phase of SCI in non-ambulatory dogs (53), but recent investigations have shown that the severity of conduction deficits following TMMEP in dogs with cervical spondylomyelopathy (a chronic compressive lesion of

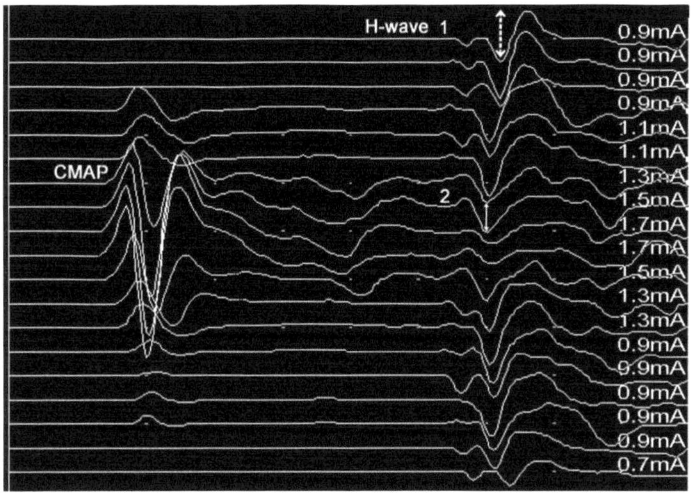

Fig. 3. H-wave recording in a dog following thoracolumbar spinal cord injury. The distal tibial nerve is stimulated and the recording obtained from the interosseous muscles; the H-wave amplitude is higher during sub-maximal stimulation of the tibial nerve (1, *dashed line* with *double arrowheads*) compared to supra-maximal stimulation (2, *plain line double arrowheads*); the presence of the H-wave demonstrates the integrity of the lumbar spinal cord below the lesion site; CMAP: compound muscle action potential corresponding to the direct interosseous muscle response; H-wave: late potential.

the cervical spinal cord) correlate with the severity of the clinical signs (54). Somatosensory evoked potentials have not been recorded in dogs with thoracolumbar disk herniation and loss of pain sensation caudal to the lesion, but measurements have been made in the acute phase of SCI (49). Recently, we have recorded SSEPs in the chronic phase of SCI in patients that have absent or equivocal pain perception in the hindquarters (41).

6.4. Urodynamic Testing

Similar to human patients with SCI (55), dogs with SCI will present neurogenic bladder dysfunction such as reflex voiding and detrusor over-reactivity complicated by vesico-sphincterial dysynergia, or detrusor atony (41). Urodynamic equipment used for analysis of human continence can be used in dogs (56) to obtain cystometric, urethral profilometric, and sphincter electromyogram data. This information is directly relevant to understanding possible effects of novel interventions in human patients—for instance, routine urodynamic assessments of function in human SCI patients (57, 58), such as bladder compliance, number of involuntary detrusor contractions, sphincter activity during bladder filling can readily be obtained to define the same aspects of continence in dogs with SCI.

6.5. Gait Analysis

Disruption of quadrupedal coordination after SCI provides an enticing opportunity for functional analysis in dogs. Following severe thoracolumbar SCI, there is complete loss of pelvic limb movement, but often "spinal walking"—defined as stepping by the pelvic limbs in the absence of clinically detectable spinal cord transmission across

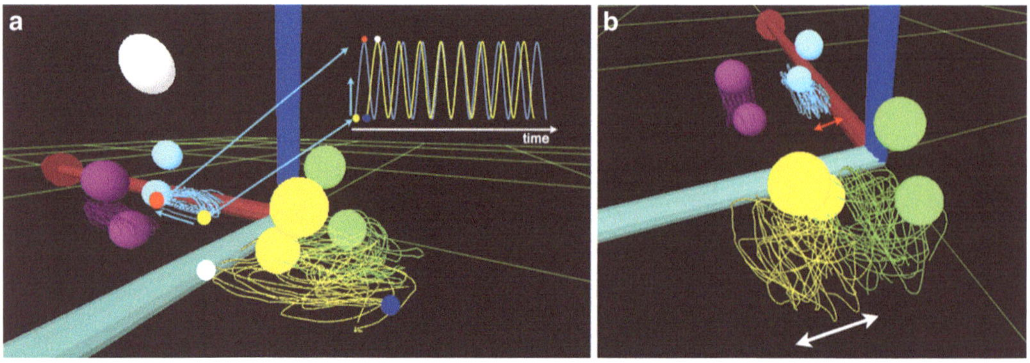

Fig. 4. 3D sketch diagram of paw movements as a SCI dog walks on a treadmill. (**a**) The *light blue* and *purple balls* associated with traces represent the dog's right and left thoracic limb paws and the *green* and *yellow balls* associated with traces represent the dog's right and left pelvic limb paws. The other markers are placed on the elbows and stifles. Note the erratic motion of the pelvic limbs paws compared to those of the (normal) thoracic limb. *Inset*: during forward motion the markers describe step cycles (for example the stance phase is the part of the cycle that goes from the *yellow dot* to the *red dot* on the right paw). Plotted against time, each limb step cycle can be represented as a sine wave. The time relationship between the thoracic and pelvic limb curves (for example between the maxima of those curves: *red* and *white dots*) can be used to quantify coordination between the thoracic and the pelvic limbs (the interval between the *red* and the *white dots* changes between each step cycle in this example); (**b**) Caudolateral view of the same dog demonstrating great variability in positioning of the pelvic limb paws (*white line* with *double arrowheads*) compared to the thoracic limbs (*red line* with *double arrowheads*). This variability provides a numerical quantification of lateral instability.

the lesion—can appear within a few months in dogs (59). A characteristic feature of "spinal walking" is the lack of coordination between thoracic and pelvic limb movements, and there is also a lack of pelvic limb lateral stability. Computerized gait analysis provides methods to quantify both these aspects of spinal walking, meaning that it is possible to attribute a numerical score to the changes in coordination that occur during either spontaneous or intervention-induced recovery (Fig. 4) (for a review see (41)). Using these methods we have shown that normal dogs have extremely rigid coordination between thoracic and pelvic limb motion, and also that this pattern is recovered following mild SCI (41, 60, 61). We are currently using these techniques to evaluate the outcome after intraspinal transplantation of olfactory ensheathing cells (OECs).

6.6. Sensory Testing

Von Frey filament testing can be applied as easily in dogs as in rodents, although this method of skin sensitivity testing is not very reliable, perhaps because of the variation in hair coat texture and depth, or simply because information from the skin is readily ignored in dogs.

6.7. Necessity of Combining Outcome Measures

The various outcome measures are complementary to one another and it is only by obtaining a combination of data that deficits following SCI can be fully defined. Each outcome measure will assess a limited selection of spinal cord pathways; for example TMMEPs likely reflect function of long descending motor tracts, whereas coordination between the thoracic and pelvic limbs in a dog probably also reflects the function of short distance propriospinal tracts in the thoracolumbar region. Further, each clinical lesion has unique anatomical characteristics, often with spared intact spinal cord tissue, even if there is clinically complete loss of all function. Tissue spared following trauma may only contribute to one function, say bladder continence but not locomotion, and therefore clinical patients expressing specific patterns of deficits could, in future, be selected as appropriate candidates (because of their perceived higher likelihood of a positive response) to receive specific novel interventions. For example, it is known that only ~10% of descending axons in the thoracolumbar spinal cord are required for locomotion in cats (62) but it is unlikely that all cases with 10% of spared spinal cord will *only* have spared descending motor axons. An intervention that aims to restore locomotion will thus have better chances if trialed on cases that have many spared motor axons rather than spared sensory axons (and these could potentially be identified through examination of MEPs and SEPs).

Diffusion MRI provides an alternative, rapidly developing, method to obtain a complete representation of spared tissue after SCI. Such tractography, using diffusion tensor imaging, can selectively highlight the different white matter tracts in cats and characterize disruption associated with SCI (63). This technique has yet to be applied in dogs, but should be feasible because of their larger size.

7. Limitations of the Dog Clinical SCI Model

7.1. Non-verbal

The most obvious limitation of dog patients as models of human patients is their inability to verbalize their impression of their disease. Although this does not prevent objective measurement of outcomes that have direct relevance to human patients, such as functional evidence of conduction through the damaged spinal cord, it does prevent assessment of subtle changes in sensation or aspects of the pain experienced by human patients. However, it remains possible to indirectly assess the dog's quality of life by means of questionnaires answered by the owner on visual analogue scales for example.

7.2. Anatomical Considerations

Another prominent drawback to use of dogs as a SCI model is that they do not have a well-developed corticospinal tract (CST). Functional disturbances following SCI in dogs therefore reflect injuries to different pathways than those that are thought to be critical in people. There are two aspects of SCI in which CST function is important: first, because of the dominance of the CST, it is thought that its regeneration, or an adequate compensating plastic response, is a prerequisite for satisfactory restoration of many motor functions in human patients; second, the CST is a long pathway and therefore presents a greater obstacle to regeneration (since segmental bridging may not adequately reproduce its function). In dogs there are many other long tract pathways (specifically the rubrospinal tract) that may also be studied for their plasticity and regeneration and may aid in illustrating general principles, although they do not fulfill the same functions as the CST in humans. Therefore, modeling of SCI interventions in primates, with their large and important CST, provides essential complementary information to that available from treatment applied to canine clinical patients. In comparison to rodent models, dogs do not have great disadvantages in investigating effects of an intervention on the CST (since the rodent CST is considerably less well developed than in humans and similarly influences motor neurons indirectly (64)), apart from the lack of well-defined hand movements.

7.3. Lack of Canine-Specific Laboratory Reagents

Even if the ethical issues were to be overcome, dogs would still be inferior experimental animals compared to rodents for many aspects of proof-of-principle SCI research because there are fewer laboratory reagents available for detailed study of cellular responses (such as specific antibodies).

7.4. Lesion Heterogeneity

Lesion heterogeneity has a double-edged effect. On the one hand, each clinical spinal cord lesion has, to some degree, "unique" characteristics in terms of precise localization, severity and speed of onset, which could be thought of as an obstacle to applicability as a "model" for research purposes. However, on the other hand, such lesion heterogeneity is a feature of the clinical target population—humans with SCI. In both species it is possible, using clinically routine techniques to classify patients into broadly similar groups. Therefore, the dog SCI population mimics closely that of the target human population meaning that data from clinical canine studies has good generalizability to similar human patients. Nevertheless, it implies the need for careful selection of the patients enrolled in a study.

8. Key Findings from SCI Research in Clinical Dog Patients

Previous data obtained using dogs has predominantly been from experimental animals (in which they have been used as "large rats") but there is a small body of work, albeit not high quality randomized clinical trials, using domestic dogs as models to examine effects of various therapies for SCI. For instance, Borgens used domestic dogs to investigate the effects of electric currents pulsed across the damaged region of the spinal cord (65) and Blight tested the effect of 4-aminopyridine in chronic SCI in dogs in a phase I trial (66). Both studies yielded promising results, but because of the non-randomized nature of the trials it is difficult to draw robust conclusions regarding efficacy. More recently, however, a randomized, blinded, placebo-controlled clinical trial of *N*-acetylcysteine to improve function after SCI has been conducted (that showed no effect) (67) and many further studies are planned or currently underway (68).

OECs have long been thought to be a promising method for treatment of SCI in humans and the properties of canine OECs have been extensively investigated during the last decade (69–72). We have confirmed the safety of transplanting OECs to the spinal cord of dogs with severe SCI using an open surgical approach (72); percutaneous fluoroscopic-guided intraspinal injection of cells (see (14)) may be preferable because it is less invasive, although ascertainment of the precise site of injection is less straightforward. Currently we are engaged in a trial to measure the efficacy of OEC transplantation on restoring coordination of the thoracic and pelvic limbs compared to the effects of cell transport medium (containing various growth factors) alone. The results of such studies will have implications for future studies in humans regarding the methods of using the intervention, the design of outcome measures and whether to proceed or hold back on pursuing this particular intervention in human patient trials. Similar studies on other interventions can serve as a prelude prior to proceeding with much more expensive human trials.

It is likely that results from clinical trials in domestic dogs will prompt further questions that will need to be solved through laboratory experiments, such as optimal methods of treatment delivery, and such developments will then aid design of new clinical trials. We envisage a scenario in which the dog model can be used routinely to screen putative therapies for their translational potential and, in turn, use the results to define subjects for future laboratory analysis (e.g., investigating cell delivery methods).

9. Added Value for Understanding Human Spinal Cord Injury

9.1. Translation

Insights gained from using the domestic dog "model" consist mainly in understanding translational issues, such as the methods needed for administering treatment and methods of analysis rather than for understanding fundamental aspects of SCI in humans. However, exceptions may occur in which further insights may be gained, such as observations made in dogs regarding effects of an intervention on specific aspects of functional loss or amelioration. For instance a specific effect of an intervention in dogs might be the restoration of normal bladder function rather than restoration of normal locomotion. Such a conclusion is biologically plausible (because of the different diameter of the relevant axons and the nature of the intra-cord pathways) and could easily be overlooked in rodent experiments.

9.2. Modeling Chronic Lesions

It is unlikely that human patients will receive treatment aimed at regenerating or otherwise repairing the spinal cord (for example a cell therapy) immediately after trauma. First, clinicians must determine the severity of injury and, by implication, the extent to which spontaneous functional recovery may occur, either following conventional treatment or simply through elapse of time. Second, cell therapies may not be effective in the acute phase of the injury because of the inhospitable host environment at that stage (73). For example, intraspinal transplantation of umbilical cord blood cells in rats improves neurologic outcome, possibly via anti-inflammatory effects, but the cells are not found in the spinal cord at 6 weeks post transplant (74). Therefore there is a need for models of the chronic phase of the injury and for several reasons this is difficult to model in experimental rodents. Dogs with chronic SCI (i.e., more than 3 months after initiation) are readily available, since not all cases currently respond to conventional therapy, and so can be used for studies on novel interventions at this stage.

9.3. Detailed Functional Outcome Analysis

Detailed gait analysis (which is currently not widely available in rodent SCI models) might also be able to dissect effects of an intervention on different CNS pathways. We have previously highlighted this possibility in our experiments on gait analysis in dogs (60, 61) in which we can differentiate coordination between pelvic and thoracic limbs in the sagittal plane from that required to achieve lateral stability during locomotion. We hypothesize that sagittal coordination between thoracic and pelvic limbs might perhaps (especially after a severe SCI) be mediated via propriospinal pathways whereas coordination mediated by brainstem nuclei will be required for lateral stability.

9.4. Histopathological Outcome Data

A further specific advantage is the availability of pathological material from SCI patients—especially those that have been enrolled in treatment trials—at a much earlier point than would be possible for human patients. Thus, we may be able to deduce that specific methods of delivery are inappropriate when translating from the rat to the dog—with implication that further problems could be anticipated when translating to human patients.

References

1. Allen AR (1911) Surgery of experimental lesion of spinal cord equivalent to crush injury of fracture dislocation of spinal column: a preliminary report. J Am Med Assoc 57:878–880
2. Paoloni M, Khanna C (2008) Translation of new cancer treatments from pet dogs to humans. Nat Rev Cancer 8:147–156
3. Jeffery ND, Smith PM, Lakatos A, Ibanez C, Ito D, Franklin RJ (2006) Clinical canine spinal cord injury provides an opportunity to examine the issues in translating laboratory techniques into practical therapy. Spinal Cord 44:584–593
4. Tarlov IM, Klinger H, Vitale S (1953) Spinal cord compression studies. I. Experimental techniques to produce acute and gradual compression. AMA Arch Neurol Psychiatry 70: 813–819

5. Tarlov IM, Klinger H (1954) Spinal cord compression studies. II. Time limits for recovery after acute compression in dogs. AMA Arch Neurol Psychiatry 71:271–290
6. Tarlov IM (1954) Spinal cord compression studies. III. Time limits for recovery after gradual compression in dogs. AMA Arch Neurol Psychiatry 71:588–597
7. Bassso DM, Beattie MS, Bresnahan JC (1995) A sensitive and reliable locomotor rating scale for open field testing in rats. J Neurotrauma 12:1–21
8. Basso DM, Beattie MS, Bresnahan JC, Anderson DK, Faden AI, Gruner JA et al (1996) MASCIS evaluation of open field locomotor scores: effects of experience and teamwork on reliability. Multicenter Animal Spinal Cord Injury Study. J Neurotrauma 13:343–359
9. Al-Mefty O, Harkey HL, Marawi I, Haines DE, Peeler DF, Wilner HI et al (1993) Experimental chronic compressive cervical myelopathy. J Neurosurg 79:550–561
10. Harkey HL, Al-Mefty O, Marawi I, Peeler DF, Haines DE, Alexander LF (1995) Experimental chronic compressive cervical myelopathy: effects of decompression. J Neurosurg 83:336–341
11. Assina R, Sankar T, Theodore N, Javedan SP, Gibson AR, Horn KM et al (2008) Activated autologous macrophage implantation in a large-animal model of spinal cord injury. Neurosurg Focus 25:E3
12. Rabinowitz RS, Eck JC, Harper CM Jr, Larson DR, Jimenez MA, Parisi JE et al (2008) Urgent surgical decompression compared to methylprednisolone for the treatment of acute spinal cord injury: a randomized prospective study in beagle dogs. Spine (Phila Pa 1976) 33: 2260–2268
13. Kim BG, Kang YM, Phi JH, Kim YH, Hwang DH, Choi JY et al (2010) Implantation of polymer scaffolds seeded with neural stem cells in a canine spinal cord injury model. Cytotherapy 12:841–845
14. Lee JH, Chang HS, Kang EH, Chung DJ, Choi CB, Hwang SH et al (2009) Percutaneous transplantation of human umbilical cord blood-derived multipotent stem cells in a canine model of spinal cord injury. J Neurosurg Spine 11:749–757
15. Lee JH, Chung WH, Kang EH, Chung DJ, Choi CB, Chang HS et al (2011) Schwann cell-like remyelination following transplantation of human umbilical cord blood (hUCB)-derived mesenchymal stem cells in dogs with acute spinal cord injury. J Neurol Sci 300:86–96
16. Lim JH, Byeon YE, Ryu HH, Jeong YH, Lee YW, Kim WH et al (2007) Transplantation of canine umbilical cord blood-derived mesenchymal stem cells in experimentally induced spinal cord injured dogs. J Vet Sci 8:275–282
17. Ryu HH, Lim JH, Byeon YE, Park JR, Seo MS, Lee YW et al (2009) Functional recovery and neural differentiation after transplantation of allogenic adipose-derived stem cells in a canine model of acute spinal cord injury. J Vet Sci 10:273–284
18. Hansen HJ (1951) A pathologic-anatomical interpretation of disc degeneration in dogs. Acta Orthop Scand 20:280–293
19. Hansen HJ (1952) A pathologic-anatomical study on disc degeneration in dog, with special reference to the so-called enchondrosis intervertebralis. Acta Orthop Scand Suppl 11:1–117
20. Jeffery ND (2010) Vertebral fracture and luxation in small animals. Vet Clin North Am Small Anim Pract 40:809–828
21. Brisson BA (2010) Intervertebral disc disease in dogs. Vet Clin North Am Small Anim Pract 40:829–858
22. Ball MU, McGuire JA, Swaim SF, Hoerlein BF (1982) Patterns of occurrence of disk disease among registered dachshunds. J Am Vet Med Assoc 180:519–522
23. Webb AA, Ngan S, Fowler D (2010) Spinal cord injury II: Prognostic indicators, standards of care, and clinical trials. Can Vet J 51:598–604
24. Blight AR, Decrescito V (1986) Morphometric analysis of experimental spinal cord injury in the cat: the relation of injury intensity to survival of myelinated axons. Neuroscience 19: 321–341
25. Bunge RP, Puckett WR, Becerra JL, Marcillo A, Quencer RM (1993) Observations on the pathology of human spinal cord injury. A review and classification of 22 new cases with details from a case of chronic cord compression with extensive focal demyelination. Adv Neurol 59:75–89
26. Fehlings MG, Tator CH (1995) The relationships among the severity of spinal cord injury, residual neurological function, axon counts, and counts of retrogradely labeled neurons after experimental spinal cord injury. Exp Neurol 132:220–228
27. Smith PM, Lakatos A, Barnett SC, Jeffery ND, Franklin RJ (2002) Cryopreserved cells isolated from the adult canine olfactory bulb are capable of extensive remyelination following transplantation into the adult rat CNS. Exp Neurol 176:402–406
28. Nashmi R, Fehlings MG (2001) Changes in axonal physiology and morphology after chronic compressive injury of the rat thoracic spinal cord. Neuroscience 104:235–251

29. Schucht P, Raineteau O, Schwab ME, Fouad K (2002) Anatomical correlates of locomotor recovery following dorsal and ventral lesions of the rat spinal cord. Exp Neurol 176:143–153
30. Fletcher TF (1993) Spinal cord and meninges (Chapter 14). In: Evans EH (ed) Miller's anatomy of the dog, 3rd edn. WB Saunders Co, Philadelphia
31. Woolam DH, Millen JW (1955) The arterial supply of the spinal cord and its significance. J Neurol Neurosurg Psychiatry 18:97–102
32. Hagg T, Oudega M (2006) Degenerative and spontaneous regenerative processes after spinal cord injury. J Neurotrauma 23:264–280
33. Zhang Z, Fujiki M, Guth L, Steward O (1996) Genetic influences on cellular reactions to spinal cord injury: a wound-healing response present in normal mice is impaired in mice carrying a mutation (WldS) that causes delayed Wallerian degeneration. J Comp Neurol 371:485–495
34. Inman DM, Steward O (2003) Physical size does not determine the unique histopathological response seen in the injured mouse spinal cord. J Neurotrauma 20:33–42
35. Steward O, Zheng B, Tessier-Lavigne M (2003) False resurrections: distinguishing regenerated from spared axons in the injured central nervous system. J Comp Neurol 459:1–8
36. Behr VC, Weber T, Neuberger T, Vroemen M, Weidner N, Bogdahn U et al (2004) High-resolution MR imaging of the rat spinal cord in vivo in a wide-bore magnet at 17.6 Tesla. MAGMA 17:353–358
37. Ko HY, Park JH, Shin YB, Baek SY (2004) Gross quantitative measurements of spinal cord segments in human. Spinal Cord 42:35–40
38. Dietz V (2009) Body weight supported gait training: from laboratory to clinical setting. Brain Res Bull 78:I–VI
39. Garcia-Alias G, Barkhuysen S, Buckle M, Fawcett JW (2009) Chondroitinase ABC treatment opens a window of opportunity for task-specific rehabilitation. Nat Neurosci 12: 1145–1151
40. Savic G, Bergström EM, Frankel HL, Jamous MA, Jones PW (2007) Inter-rater reliability of motor and sensory examinations performed according to American Spinal Injury Association standards. Spinal Cord 45:444–451
41. Jeffery ND, Hamilton L, Granger N (2011) Designing clinical trials in canine spinal cord injury as a model to translate successful laboratory interventions into clinical practice. Vet Rec 168:102–107
42. Webb AA, Jeffery ND, Olby NJ, Muir GD (2004) Behavioural analysis of the efficacy of treatments for injuries to the spinal cord in animals. Vet Rec 155:225–230
43. Hoerlein BF (1953) Intervertebral disc protrusions in the dog. I. Incidence and pathological lesions. Am J Vet Res 14:260–269
44. Olby NJ, De Risio L, Muñana KR, Wosar MA, Skeen TM, Sharp NJ et al (2001) Development of a functional scoring system in dogs with acute spinal cord injuries. Am J Vet Res 62:1624–1628
45. Levine JM, Fosgate GT, Chen AV, Rushing R, Nghiem PP, Platt SR et al (2009) Magnetic resonance imaging in dogs with neurologic impairment due to acute thoracic and lumbar intervertebral disk herniation. J Vet Intern Med 23:1220–1226
46. Granger N, Jeffery ND (2009) MRI findings in severe chronic spinal cord injury of dogs. In: British Small Animal Veterinary Association Congress Proceeding, Brimingham
47. Ito D, Matsunaga S, Jeffery ND, Sasaki N, Nishimura R, Mochizuki M et al (2005) Prognostic value of magnetic resonance imaging in dogs with paraplegia caused by thoracolumbar intervertebral disk extrusion: 77 cases (2000–2003). J Am Vet Med Assoc 227:1454–1460
48. Holliday TA (1992) Electrodiagnostic examination. Somatosensory evoked potentials and electromyography. Vet Clin North Am Small Anim Pract 22:833–857
49. Poncelet L, Michaux C, Balligand M (1993) Somatosensory potentials in dogs with naturally acquired thoracolumbar spinal cord disease. Am J Vet Res 54:1935–1941
50. Nollet H, Van Ham L, Deprez P, Vanderstraeten G (2003) Transcranial magnetic stimulation: review of the technique, basic principles and applications. Vet J 166:28–42
51. Hiersemenzel LP, Curt A, Dietz V (2000) From spinal shock to spasticity: neuronal adaptations to a spinal cord injury. Neurology 54:1574–1582
52. Shores A, Redding RW, Knecht CD (1987) Spinal-evoked potentials in dogs with acute compressive thoracolumbar spinal cord disease. Am J Vet Res 48:1525–1530
53. Sylvestre AM, Cockshutt JR, Parent JM, Brooke JD, Holmberg DL, Partlow GD (1993) Magnetic motor evoked potentials for assessing spinal cord integrity in dogs with intervertebral disc disease. Vet Surg 22:5–10
54. Da Costa RC, Poma R, Parent JM, Partlow G, Monteith G (2006) Correlation of motor evoked potentials with magnetic resonance imaging and neurologic findings in Doberman Pinschers with and without signs of cervical spondylomyelopathy. Am J Vet Res 67:1613–1620
55. Fowler CJ, Griffiths D, De Groat WC (2008) The neural control of micturition. Nat Rev Neurosci 9:453–466

56. Goldstein RE, Westropp JL (2005) Urodynamic testing in the diagnosis of small animal micturition disorders. Clin Tech Small Anim Pract 20:65–72
57. Biering-Sorensen F, Craggs M, Kennelly M, Schick E, Wyndaele JJ (2008) International lower urinary tract function basic spinal cord injury data set. Spinal Cord 46:325–330
58. Ho CH, Linsenmeyer TA, Millis SR (2000) The reproducibility of urodynamic studies of neurogenic bladders in spinal cord injury. J Spinal Cord Med 23:276–283
59. Handa Y, Naito A, Watababe S, Komatsu S, Shimizu Y (1986) Functional recovery of locomotive behavior in the adult spinal dog. Tohoku J Exp Med 148:373–384
60. Hamilton L, Franklin RJ, Jeffery ND (2007) Development of a universal measure of quadrupedal forelimb-hindlimb coordination using digital motion capture and computerised analysis. BMC Neurosci 8:77
61. Hamilton L, Franklin RJ, Jeffery ND (2008) Quantification of deficits in lateral paw positioning after spinal cord injury in dogs. BMC Vet Res 4:47
62. Blight AR (1983) Cellular morphology of chronic spinal cord injury in the cat: analysis of myelinated axons by line-sampling. Neuroscience 10:521–543
63. Cohen-Adac J, Benali H, Hoge RD, Rossignol S (2008) In vivo DTI of the healthy and injured cat spinal cord at high spatial and angular resolution. Neuroimage 40:685–697
64. Yang HW, Lemon RN (2003) An electron microscopic examination of the corticospinal projection to the cervical spinal cord in the rat: lack of evidence for cortico-motoneuronal synapses. Exp Brain Res 149:458–469
65. Borgens RB, Toombs JP, Breur G, Widmer WR, Waters D, Harbath AM et al (1999) An imposed oscillating electrical field improves the recovery of function in neurologically complete paraplegic dogs. J Neurotrauma 16:639–657
66. Blight AR, Toombs JP, Bauer MS, Widmer WR (1991) The effects of 4-aminopyridine on neurological deficits in chronic cases of traumatic spinal cord injury in dogs: a phase I clinical trial. J Neurotrauma 8:103–119
67. Baltzer WI, McMichael MA, Hosgood GL, Kerwin SC, Levine JM, Steiner JM et al (2008) Randomized, blinded, placebo-controlled clinical trial of N-acetylcysteine in dogs with spinal cord trauma from acute intervertebral disc disease. Spine (Phila Pa 1976) 33:1397–1402
68. Olby N (2010) The pathogenesis and treatment of acute spinal cord injuries in dogs. Vet Clin North Am Small Anim Pract 40:791–807
69. Skinner AP, Pachnicke S, Lakatos A, Franklin RJ, Jeffery ND (2005) Nasal and frontal sinus mucosa of the adult dog contain numerous olfactory sensory neurons and ensheathing glia. Res Vet Sci 78:9–15
70. Bock P, Beineke A, Techangamsuwan S, Baumgartner W, Wetzer K (2007) Differential expression of HNK-1 and p75(NTR) in adult canine Schwann cells and olfactory ensheathing cells in situ but not in vitro. J Comp Neurol 505:572–585
71. Techangamsuwan S, Kreutzer R, Kreutzer M, Imbschweiler I, Rohn K, Wewetzer K et al (2009) Transfection of adult canine Schwann cells and olfactory ensheathing cells at early and late passage with human TERT differentially affects growth factor responsiveness and in vitro growth. J Neurosci Methods 176:112–120
72. Jeffery ND, Lakatos A, Franklin RJ (2005) Autologous olfactory glial cell transplantation is reliable and safe in naturally occurring canine spinal cord injury. J Neurotrauma 22:1282–1293
73. Woodhouse A, Vincent AJ, Kozel MA, Chung RS, Waite PM, Vickers JC et al (2005) Spinal cord tissue affects ensheathing cell proliferation and apoptosis. Neuroreport 16:737–740
74. Chua SJ, Bielecki R, Yamanaka N, Fehlings MG, Rogers IM, Casper RF (2010) The effect of umbilical cord blood cells on outcomes after experimental traumatic spinal cord injury. Spine (Phila Pa 1976) 35:1520–1526

Chapter 10

Primate Models of Spinal Repair

Corinna Darian-Smith

Abstract

Nonhuman primate models (NHPs) of spinal injury and repair play an increasingly important role in the translation of promising rodent data to the spinal injured patient. This chapter discusses why NHP models of spinal injury are important and provides a description of each of the different lesion models that are currently being used. The pros and cons of using each type of lesion are outlined, as are the species options. A dorsal laminectomy, which is common to all spinal lesions, is also described.

Key words: Spinal injury, Monkey, Dorsal rhizotomy, Dorsal column lesion, Spinal hemisection, Contusion injury

1. Introduction

The nonhuman primate (NHP) model of spinal injury is an important clinical link between data attained in the rodent and the spinal injured patient. This chapter will focus exclusively on NHP models that are used to examine neuronal repair following spinal injury. Most but not all models focus on cervical injury and the effects of the respective lesion on hand/arm function, since hand use and the underlying pathways in the monkey (i.e., the macaque) are so similar to humans, and even small improvements in hand use can greatly improve the quality of life in spinal injured patients. Only a few groups use a lumbar model which has particular implications for autonomic dysfunction as well as hindlimb sensorimotor recovery. Overall, however, there are relatively few groups worldwide using NHP models of spinal repair, which is a reflection of the many challenges associated with using such animals, including regulatory requirements and financial constraints (see discussion below). An increasing appreciation of the many significant differences between rodents and primates, as well as an increased pressure to push basic science findings to clinical trial at a faster rate, is

Håkan Aldskogius (ed.), *Animal Models of Spinal Cord Repair*, Neuromethods, vol. 76,
DOI 10.1007/978-1-62703-197-4_10, © Springer Science+Business Media, LLC 2013

now fuelling a need for nonhuman clinical models of spinal injury. My focus will be on the kinds of questions that are being addressed by each of the specific models. The intention is to provide a general overview of the NHP models of spinal repair currently in use, so that interested readers will be aware of some of the specific challenges associated with each one.

2. Using Nonhuman Primates for the Study of Spinal Injury and Repair

2.1. Why Use a Primate Rather Than a Rodent Model?

Rodent models (in addition to other models described in this book) play a critical role in the study of cellular, molecular, and biochemical mechanisms mediating CNS (central nervous system) plasticity and repair following central spinal and spinal nerve injuries. Furthermore, the many genetic strains and molecular/genetic techniques available in rodents far outreach what can currently be used in the primate. Rodents can be used in greater numbers to perfect and adapt techniques that can then be used to probe particular mechanisms of spinal repair in NHPs. However, rodent models do not adequately model potential clinical treatments, as the overwhelming failure rate of clinical trials testifies (1). And, rodent models do not provide a good direct clinical link for the study of sensorimotor function, either in the forelimb or hindlimb, following such injuries. Rats use their forelimbs and paws to reach for and grasp food objects quite effectively (2, 3), but they do not have the fractionated digit use, opposing digits and hand anatomy, or spinal and forebrain circuitry that have evolved independently over 60 to 70 million years to enable the sophisticated hand use that is synonymous with the macaque and human. Both rats and monkeys are quadrupeds, and in this respect, each has limitations for modeling hindlimb recovery and weight support that are key to human walking. In addition, rats have central pattern generator circuitry in the spinal cord that plays a key role in the apparent recovery of rhythmic hindlimb stepping following complete transection of the spinal cord (4–6). Following a spinal injury that leaves a permanent gap in their lumbar cord, the observation that rats can "walk again" (be it under very specific conditions) is interesting in its own right, and indicates that for basic hindlimb locomotion in the rodent, brainstem, and cortical circuitry are not critical for stereotypical hindlimb movements. However, this observation can be misleading if it is extrapolated to the clinical condition, where weight bearing (particularly bipedal), and a primate forebrain and spinal anatomy are brought into the equation. It is therefore important to always keep in mind that there are differences in the anatomy (and molecular biochemistry, etc.) between the rodent and primate, and these differences can play a major role in the recovery that can occur following spinal injury. For reviews and figures showing differences between rodent and primate spinal sensorimotor anatomical pathways see refs. (7–10). Examples of some of the major differences include the following:

1. The corticospinal tract is far larger and more prominent in the primate than in the rodent for volitional movements of the hand, forelimb, and hindlimb. In rodents the corticospinal tract occupies the ventral portion of the dorsal columns as well as the dorsolateral white matter in the spinal cord, while in primates it occupies the dorsolateral white matter (with a small ventral medial component). In primates the corticospinal tract is also far more complex in its cortical origins and spinal terminations than is observed in the rodent (see also refs. (11, 12)).
2. In rodents the rubrospinal tract is prominent and plays an important role in limb/paw function, while in the primate the tract is relatively small and inconsequential in the human.
3. In addition to anatomical differences, inflammatory and immune responses also differ between rodents and primates (13). These factors could differentially effect the extent of postinjury repair, such that successful treatments in rodents may be (and often are) ineffective in primates or humans.
4. Another important difference between rats and primates that is not always appreciated is that adult rats do not appear to have a comparable capacity for pathway reorganization following certain types of spinal injury as is found in the NHP. For example, in the rat cortex, when a dorsal column lesion deactivates hindlimb and lower trunk representations, there is no reactivation of this cortical region by any part of the body (including spared hindlimb fibers) in the following several months (14). In contrast, a similar deafferentation in the monkey leads to extensive reorganization of the initially deactivated cortical and subcortical regions (see ref. (15) for review).

Another example of this is seen in a differing capacity for corticospinal plasticity following a hemisection injury. In the rat, the corticospinal tract has only been shown to recover 3–5% of its prelesion fiber density following a cervical hemisection (16, 17) compared to a reported 60% in the macaque (18).

Clearly, there are important reasons for using NHP models of spinal repair as a translational step to clinical trials, particularly as a means of enhancing the treatment efficacy and the safety of the clinical trial. However, primate spinal research should only be done in institutions with appropriate facilities and expertise, infrastructure, regulatory oversight, and veterinary care, and for these and other reasons it is never an easy undertaking. Macaque monkeys in particular are expensive, and per diem costs and overheads at some universities are extremely high. Despite this, the financial cost of even a single human clinical trial would be sufficient to support a number of NHP studies that could lead to safer, greater, and ultimately more rapid advances in our development of treatments for SCI (10).

2.2. Advantages and Disadvantages of Commonly Used NHP Species

Some monkey species are more suitable than others for addressing particular questions. For example, the macaque has a hand and supporting sensorimotor circuitry that are very similar to the human. The squirrel monkey, in contrast, has only a pseudo-opposing thumb, comparatively crude dexterity and a small brain with few sulci. Thus, if the impairment and recovery of hand function is key to a spinal injury/repair study, the macaque is clearly the preferred species. If, however, the study is designed to address cellular mechanisms and the measurement of subtle changes in hand function is not so critical, then a new world species is more easily justified, both in terms of handling, availability, ease of housing and enrichment, and overall expenditure. The lack of cortical sulci also provides easier access for certain imaging techniques where cortical changes are assessed (e.g., ref. (19)).

The major species currently being used to study spinal injury are the Old World macaques, and the New World squirrel monkeys and marmosets. Of course in all NHP work, care should be taken to use colony bred animals purchased from known, reputable and certified sources, so that no burden is placed on the wild populations.

Macaque monkeys are the most commonly used species for NHP spinal research (e.g., *Macaca mulatta, rhesus; M. cynomulgus also known as M. fascicularis*), and have several advantages over the New World species. *First*, there is a rich literature on their neuroanatomy and physiology—this provides a strong base for the interpretation of data. *Second*, as already mentioned, their hands and associated sensorimotor pathways are very similar to the human—this makes them an excellent translational model for arm and hand function. *Third*, macaque monkeys are also easily trained to perform quite sophisticated sensorimotor tasks that test tactile sensation and fine motor control of the digits. However, the more complex the task, the longer it takes to train individuals, so there may be tradeoffs between task complexity and training time requirements. The simplest task (i.e., reaching and grasping) can take 1–2 months of training while a complex task (i.e., a task requiring computer cues) with multiple training steps can take as long as 6 to 12 months or even longer. Given the associated challenges and daily costs of macaque research, simpler tasks that test the deficit directly that can be easily repeated across animals are usually best, both for the science and for the animal. However, macaque monkeys can also have drawbacks as a spinal injury model. Adults are large and males have muscle bulk on their shoulders that can be challenging for the surgeon and problematic for the monkey following surgery. This can be addressed by using young adults or females, which do not have muscle bulk on the upper back, though the latter are not always readily available from breeding colonies.

New World monkey species that have been used to address changes following spinal injury include squirrel monkeys (e.g., *Saimiri boliviensis* and *S. sciureus*) and marmosets, though aspects of pathway reorganization have also been studied in owl monkeys (genus *Aotus*). New World species (especially squirrel and marmoset monkeys) are small and relatively easy to breed for research purposes. These species are starting to be investigated as a potential parallel or alternate species to the macaque, especially in studies where hand function and brain size are not such an issue. The small size of these species (approximately 1 kg or less, vs. 3–12 kg in macaques), means they are easily housed in enriched environments and to some extent can be handled more readily, especially for post-operative rehabilitation. They do not carry herpes B (which is endemic to macaques), which can be fatal if transmitted to humans, so handling them carries fewer hazards with respect to human health. Due to their lower costs relative to the macaque, studies requiring a greater sample size are also more feasible with New World species. However, New World monkeys are not as phylogenetically close to the human as the macaques, and far less is known of their anatomical, physiological, and cellular/molecular organization.

3. What Are the Specific NHP Spinal Injury Models Currently in Use?

The location and extent of a spinal injury, the pathways effected, and the cellular and molecular microenvironment present both pre- and postinjury, all contribute to postinjury reorganization and repair, and thus the behavioral recovery and outcome of the animal. In order to determine the relative involvement of specific tracts in postinjury recovery, spinal injury models typically target major sensorimotor pathways (e.g., dorsal column, corticospinal tract) or portions of the cord encompassing major tracts, so that experiments can be reproduced and interpreted across relatively small number of animals. The following models are categorized according to the location and type of the spinal injury, and whether it lies outside the cord, or directly produces a central injury. All of the models described involve the disruption of the dura and thus the blood–brain barrier.

3.1. Spinal Nerve Lesions

3.1.1. Dorsal Root Lesion (Dorsal Rhizotomy)

Dorsal root lesions (DRLs) permanently block all sensory input (i.e., mechano-, proprio-, nociceptors), from entering the cord, but leave the motor output intact. Large DRLs (i.e., cervical segments C4–T4 or C3–T3) completely deafferent the hand and forearm of the macaque (or human), abolish voluntary manipulative movements (20, 21), and cause considerable neuronal reorganization at different levels of the neuraxis (8, 9, 22–25). Brachial plexus injuries in humans similarly remove sensory feedback from the hand and cause a severe loss of voluntary hand movements (26). But what happens after less devastating dorsal root injuries? To investigate functional recovery along with associated neuronal and cellular changes following dorsal root injury, my lab uses a restricted DRL model that is readily reproducible across animals and that can be altered in its extent to test the limits of reorganization and recovery. This model has made it possible to investigate acute, chronic, and long-term changes (>8 months) in the macaque monkey that accompany dorsal rhizotomies of a specific size. It is also a lesion that is easily performed in a rodent for parallel cellular/molecular studies. In the monkey model, only dorsal rootlets innervating the first two to three digits of one hand are cut, so voluntary hand movements and related neuronal reorganization can be assessed over a pre- and postlesion period of weeks or months. Importantly, this lesion model causes minimal distress to the monkey and requires only localized (often temporary) anesthesia of several digits of one hand. Since motor pathways remain intact, monkeys show only a very subtle impairment of digit placement immediately after the lesion. This model allows us to study mechanisms of recovery and reorganization that can be related to more serious clinical injuries, such as the dorsal root or brachial plexus lesions in humans that are typically more extensive and debilitating.

Making the DRL in our monkeys is relatively straightforward surgically, and has been described in detail elsewhere (7, 9, 24), as well as at the end of this chapter. For DRLs, young males (2–4 years) are anesthetized and secured in a stereotaxic frame with head flexed to extend the cervical spine. The C8 spine is larger than surrounding spines and provides a landmark for placement of the laminectomy, which exposes dorsal rootlets entering spinal segments C5–T1. Extracellular recordings are made of the discharge of one to a few axons at variable depths and locations in each fascicle as has been described in detail elsewhere (24, 27). Ten or more cutaneous receptive fields (RFs) are mapped from each rootlet to produce what we call a microdermatome map, which is then used to guide the selection of rootlets for cutting. The lesion is permanent, and cut centrally directed axons do not regrow into the dorsal root entry zone (DREZ). For the period of our postlesion analysis (up to 8 months), the peripheral axons and cell bodies (in dorsal root ganglia) of cut neurons also remain largely intact. On completion of the electrophysiological mapping of the microdermatomes and the transection of the rootlets, the overlying tissues are sutured in layers. We do not re-sew the dura as this can create unwanted adhesions with the cord. The monkeys in this procedure recover remarkably quickly, and are up and alert and eating within a few hours.

Immediately after a DRL, there is no electrophysiologically detectable input from the experimentally deafferented digits, and the corresponding spinal dermatomal (24), brainstem cuneate nucleus (25), and somatosensory cortex (24) maps are inactivated or functionally unresponsive to stimulation of the deafferented digits. These silent zones persist for many weeks at all levels of the pathway and correspond to what can be an initially severe behavioral deficit in the affected hand (28). Over the course of 2–3 months, and with regular hand use, there is some recovery of digit use, and as this occurs there is a corresponding reorganization demonstrable at each level of the neuraxis. In the spared dorsal root fascicles immediately adjacent to the lesion, cutaneous responses can be elicited where this was not previously possible (24). Significant somatotopic reorganization occurs in the cuneate nucleus of the brainstem and there is partial reemergence of input from the initially deafferented digits (25). There is also partial or complete reactivation of the somatosensory cortical digit map, depending on both the initial size of the lesion and the length of the postlesion time period (24, 25). Anatomical studies (27) have shown that a small population of spared afferent fibers continue to innervate the deafferented digits. These enter the cord adjacent to the lesion site, and are initially too weak in their signal response to drive any level of the pathway. Over the postlesion weeks and months, the spared primary afferent fibers sprout within the spinal dorsal horn and cuneate nucleus and presumably form additional connections with input-deprived postsynaptic target neurons. As such they drive reorganization at the spinal and higher levels of the neuraxis. These studies show a limit to the extent of reorganization possible but also demonstrate that even very small numbers of spared afferents can enable extensive postlesion reorganization at different levels of the neuraxis. Recent studies in our lab also show extensive changes to the motor pathways indirectly affected by this injury. For example, corticospinal projections originating in the primary motor cortex in the region of digit representation sprout within the dorsal horn in the cervical cord following the DRL (Darian-Smith et al., unpublished). Optimizing the facilitation of these spared fibers is a major focus for future study.

In addition to the physiological and anatomical responses to the dorsal root injury, we have also found adult neurogenesis within both the spinal cord and sensorimotor cortex. Adult neurogenesis occurs only in initially inactivated and subsequently reorganized regions of the spinal cord and cortex (29, 30), which means that it is very specifically injury induced. It occurs in both rats (and mice, unpublished), and to a greater extent in monkeys, which means that this is a conserved mammalian injury-induced mechanism. Its specific link with injury, even when the injury itself occurs a considerable distance away (outside the spinal cord), points to a postinjury microenvironment that is not only permissive but also inductive to the formation of new neurons. The presence of these neurons, which are mostly GABAergic inhibitory interneurons, suggests that they play a role in the remodeling of the neuronal circuitry. However, much remains to be determined.

We know that phenotypically these cells mature, and that a significant number of them survive for at least 2 months postlesion in the monkey. We have also shown that the dendritic morphology of a few short range excitatory pyramids observed within the cortex is indistinguishable from surrounding pyramids, suggesting integration with the local circuitry (30). We also know that there is astrocytic and oligodrocytic (Darian-Smith and Irvine, unpublished), proliferation within the spinal cord and cortex where there is injury-induced neurogenesis, but this is at a very low level compared with proliferation observed following a central injury. We (and other groups) do not observe neurogenesis within the cord following a central injury (e.g., a dorsal column lesion). This appears to be due to the massive gliosis or glial scarring that occurs following central injury and the inhibitory molecules produced during this complex inflammatory response (30–32). Understanding what makes the spinal microenvironment permissive versus nonpermissive to mechanisms of repair (such as axon growth, myelination or neurogenesis) currently forms a major focus of study in SCI research.

At the synaptic level, we see a significant increase in the number of GABAergic profiles within the dorsal horn 3–4 months following a DRL (33). This fits with an increase in GABAergic neurons in the region, but may also indicate that spared interneurons that have had their normal input removed are also sprouting and forming new and additional local synaptic connections. Intriguingly, following either a peripheral nerve or central injury, there is typically a downregulation of GABAergic connections within the spinal cord or brainstem (34–37). Additional studies are required to address these seemingly dichotomous results from peripheral nerve, spinal nerve, and central spinal injuries.

3.2. Central Spinal Injuries

3.2.1. Dorsal Column Transection

Dorsal column lesion or transection (DCL) has been studied for many decades, not so much as a spinal injury model per se, but to examine the specific function of this major ascending somatosensory fiber tract. A dorsal column lesion is relatively easily made via a laminectomy (see end of chapter), whereby the dura is opened to expose the cord, and can be reproduced, and reconstructed postmortem to establish its extent.

The effects of dorsal column lesions on hand and digit function have been investigated by several groups over the last several decades (see refs. (9) and (15), for reviews). Reported findings have been inconsistent with supposedly similar lesions producing deficits ranging from subtle to chronic and debilitating (38–41), but inconsistencies are explained by variable lesion size and a lack of systematic lesion reconstruction across studies. Vierck and colleagues, who were particularly systematic in their analyses, found that the cuneate fasciculus regulates finger positioning, and complete transection of this tract therefore disrupts fine motor control of the digits. Dorsal column transection also initially impairs a monkey's ability to distinguish the frequency or duration of tactile stimulation (21, 42), with some monkeys recovering this ability over several months of testing. These monkeys likely learned to use alternate cues and pathways to assess and distinguish textures. While Vierck and colleagues used descriptive measures of digit use (41), they also tested very specific parameters of hand function. For example, in one study, monkeys were trained to place their hand in a mechanical arm with a digit splint to test the monkey's ability to position digits after transection (40). The monkey then watched a computer screen and moved the index finger (shown by a cursor) into a target region on the screen, as a specified load was applied to the finger splint in order to perturb digit movement. When the cuneate fasciculus lesion was complete monkeys failed to adjust their finger position, but when the lesion was incomplete, spared fibers were able to reorganize and compensate for the lost input. In another study from the same group, macaques were trained to perform a key press task with individual digits to obtain a food reward, so that fractionated digit movements could be examined following a dorsal column lesion (38). That investigation showed that monkeys lost their ability to fractionate and instead used a subtle compensatory strategy to perform the key press task. Importantly, such observations required a very specific task to demonstrate the deficit, and this kind of subtlety could not have been demonstrated with New World NHP species that have less sophisticated hand function (e.g., squirrel monkeys or marmosets).

Cortical and subcortical changes have also been examined extensively following a DCL in the NHP (see ref. (15) for review). Briefly, complete section of the dorsal columns at a high cervical level blocks the ascending input from the forelimb and all parts of the body below the forelimb. Thus, areas 3b and 1 are initially extensively deactivated in these corresponding areas. Jain and colleagues (43) observed that after 6–8 months of recovery, deactivated regions of cortex appropriated input from surrounding sources. For example, the former hand region came to be activated by spared input from the arm, and input from the intact chin infiltrated the hand, trunk, and hindlimb regions. These reactivations were at least in part mediated by axonal sprouting occurring at the level of cuneate and trigeminal nuclei in the brainstem, though reorganization of the circuitry almost certainly occurred at the spinal, brainstem, thalamic, and cortical levels.

Since there is now a substantial body of behavioral and system wide physiological data about the effects of a cervical DCL, this is a particularly useful central spinal injury model for assessing therapeutic treatments postinjury. It is also important that, as with a DRL SCI model, this injury has little detectable effect on the daily behavior of the monkey.

3.2.2. Hemisection

Hemisection lesions have been used in macaques to look at the role of the corticospinal tract (CST), in addition to other ascending and descending pathways, in the recovery of voluntary limb movements. The CST comprises the single most important descending motor tract involved in distal forelimb muscle innervation and the execution of fine voluntary hand and digit movements and ~90% of its axons project contralaterally in the dorsolateral spinal cord. Kuypers and colleagues used brainstem lesions to target the CST in a series of famous studies during the 1960s (see ref. (44)), but they did not look at CST injury as part of a spinal hemisection. In the 1990s Galea and colleagues looked at this directly by examining the behavioral effects of unilateral hemisection at cervical level C3–C4 (45). Monkeys initially presented with a severe hemiparesis on the side of the lesion, but within a month, and irrespective of age, they regained the ability to pick up objects with the impaired hand and were able to perform a reach–grasp–retrieval task. The recovery was quite dramatic, but it was never fully complete (up to 3 years after the injury). The reach–retrieval task used in these spinal hemisection experiments was relatively simple, and monkeys had to reach to a clamp and remove a target object held at a set force. Despite some visual input, the initial reach phase of the task proved to be a significant parameter to measure, indicating that a persistent behavioral deficit involved a less direct trajectory and preshaping of the hand during the reach.

Corticospinal pathway transection also led to a persistent weakness of the opposing digits, which was indirectly assessed with the reach–retrieval task, as digit strength was required in the removal of the target at higher resistive forces. There was also a persistent increase in contact times at higher resistive forces, which may have been due to the loss of primary afferent input in addition to a residual weakness of the digits. The somatosensory component of the task was also impaired, and hand representation maps in the primary somatosensory cortex showed reorganization of the deprived region of cortex and a reduced hand map with abnormally large cutaneous receptive fields. Studies from this group (45, 46) suggested that previously existing "crossover" corticospinal projections were responsible, at least in part, for mediating behavioral recovery. The authors also proposed that the sprouting of these or other preexisting pathways played a role in the recovery, though this could not be demonstrated directly. More recent studies of cervical hemisection in the macaque (C7–8) now directly support these ideas. Rosenzweig and colleagues (47) reconstructed individual axons in non-lesioned macaques and found that CST axons branch and terminate bilaterally far more extensively than was previously appreciated. The same group then observed extensive spontaneous compensatory plasticity of spared corticospinal axons in a macaque cervical hemisection (C7) model (18). This was measured in terms of axon density below the level of the lesion.

They found that at 2 weeks following the lesion the density of axons originating from the motor cortex was 75% less than that observed prelesion and at 24 weeks it was only 25% less than prelesion levels. Local sprouting or neurite outgrowth of spared corticospinal axons therefore seems to play an important role in the recovery of hand function observed following a unilateral cervical hemisection. The work from both groups shows that a certain amount of spontaneous recovery is possible following a partial transection injury. This form of recovery is not observed in an equivalent model in the rodent.

Despite some capacity for spontaneous growth in the adult CNS, the extent and range of it is typically quite limited, so that significant attention has been paid in recent years to the identification and understanding of factors limiting central neuronal regeneration. Determining how such factors can be blocked or manipulated to optimize recovery is of obvious clinical interest for its treatment potential. While most of this work is conducted in rodents, promising findings have also led to translational studies in NHPs.

One such group in Switzerland has used the unilateral hemisection model to study the role of the neurite growth inhibitor Nogo-A in the postinjury recovery of hand function in the monkey (48–50). Nogo-A is among the better known myelin-associated growth inhibitors and is localized in myelin (51). Their work in macaque monkeys follows directly from rodent investigations that reported aberrant rubrospinal sprouting into the cervical ventral horn following mAB IN-1 treatment that blocks Nogo-A (52). In the monkey, they also found that with IN-1 treatment postinjury, there is enhanced regenerative/compensatory sprouting of corticospinal axon terminals within the spinal cord, and a greater improvement of hand function compared with control animals (50, 53). Unfortunately, while these studies are promising, the segmental placement of the lesions (at C7–8) at such a caudal level (i.e., hand innervation in the macaque spans C5–C8 with some inter-animal variation) spared much of the corticospinal projection to the affected hand, which means that the behavioral data cannot be easily interpreted.

Future studies are needed to identify cellular/molecular mechanisms in the NHP that enable the spontaneous growth of spared CST axons following SCI, so that these mechanisms can be enhanced therapeutically following partial spinal injury.

3.2.3. Complete Transection

Complete transection models have been used in rodents for many years, not only to look at pattern generators in locomotion within spinal cord circuitry, but because they reproduce many of the irreversible sensorimotor and autonomic dysfunctions experienced in para- and quadriplegic patients (54). Creating a complete transection model in the NHP is understandably rare (55, 56), given the health problems and daily challenges of managing individual animals.

It is not surprising then that only one group (in Spain) has attempted such a model in a NHP in recent years (57, 58). In that study, two macaque monkeys underwent complete transection at T8–9 and these animals were maintained for more than a year using human health care protocols and adapted cages. Though the study provides invaluable information regarding the appropriate care of these animals over a protracted period, the intensive daily regimen and handling required to maintain them at a good level of health makes it extremely unlikely that this will ever become a viable NHP SCI model in the foreseeable future.

3.2.4. Contusion

To date, few groups have attempted the development of a NHP spinal contusion model where the animals are kept for weeks to months postinjury ((59), in India; (60), in Japan). Bresnahan (61) used spinal contusions in rhesus macaques in 1978 to describe the ultrastructure of anterograde degeneration over time, but the details of lesions and postinjury impairments were not described and only three animals were kept beyond one postinjury week. Contusion injuries are relatively easily created with a laminectomy, and cord exposure similar to that described in this chapter for other SCI models. A device (similar to that used in rats) is then used (e.g., Allen weight drop technique, or modified NYU device), whereby a weight is dropped with controlled impact force and depth typically only on one side of the cord. Though a contusion injury reproduces what might be considered a typical clinical injury more accurately, which is the motivation driving the development of these models, the resulting impairment can be severe. This means that the practical and ethical issues surrounding this type of NHP model are not insubstantial. It is also extremely difficult to precisely reproduce injuries of this type across individuals (though this aspect is improving), which makes the assessment of behavioral and functional repair, and the contribution of different fiber tracts following injury and treatment difficult, especially given the small numbers of animals generally used in such investigations. Both Babu and colleagues (59) in the bonnet macaque, and Iwanami and colleagues in the marmoset (60), have recently developed spinal contusion models which they grade into three categories of injury severity. Iwanami and colleagues use C5 level injuries in marmosets, and have been able to reproduce their findings across animals. This group has since successfully used their model for translational neural stem/progenitor transplantation studies to follow up on promising findings in the rodent (60, 62). Though behavioral measures in a marmoset are relatively crude, this group has used a combination of spontaneous climbing and walking movements, hand grip power, and treadmill walking to assess functional recovery.

Though desirable for translational studies prior to clinical trial, the widespread use of a NHP contusion model such as the one being used by Okano and colleagues in Japan, seems unlikely in the near term. Loss of function following contusion injury can be severe, and the costs/benefits of using this type of lesion model must be carefully weighed against using a more controlled partial central injury such as a dorsal column lesion or hemisection.

4. Making a Cervical Laminectomy

The surgical laminectomy is common to all spinal injury models described above. In our experiments, we use young male *Macaca fascicularis* (2.5 to 3.5 kg), because they do not yet have the back muscle bulk of older males. This means an easier surgery, with less muscle damage than in a larger male, and we see a quick recovery post-surgery, with surprisingly little if any demonstrable back discomfort. *M. fascicularis*, it should be noted, are typically smaller and less bulky than *M. mulatta* or *M. nemestrina* males, so they are perhaps the best macaque species for spinal laminectomy. Muscle bulk is not a problem in any of the New World species. Monkeys are anesthetized (gaseous O_2/isoflurane) and positioned in a primate stereotaxic frame with the head flexed to extend the neck and cervical spine. The C8 spine is prominent in the macaque as well as the squirrel monkey (and possibly other NHPs), and serves as a landmark (the only landmark) for making an initial incision. Note that the cord segmental location will vary relative to this vertebral landmark, such that in younger animals the cord fills a greater portion of the rostrocaudal vertebral space, and as the animal torso lengthens the cord sits more rostrally. There is also animal to animal variation in the total length of the spinal cord so it is important to always verify where you are segmentally before making a lesion. After making the initial incision, the underlying fascia is blunt dissected and spines and dorsal vertebrae removed over the desired segments. In our dorsal rhizotomy experiments, we expose dorsal rootlets entering spinal segments C5–T1. In the macaque, striated muscle lies over the cord surface and this must be removed to expose the dura mater overlying the cord. In the squirrel monkey, this muscle layer is not clearly present so be aware of this when removing the overlying bone. We cut the dura mid-sagittally and retract the dura to expose the underlying rootlets. The dura is anchored with silk sutures, and digital photographs are taken and printed so that we can identify specific rootlets when we record from them. Retracting the dura is not necessary for all lesions. However, if the precise placement of a lesion at a lower cervical level is important, I would advise exposure of the rootlets. This is because the C8–T1 border is relatively easy to identify (in macaques and squirrel monkeys so presumably across primate species) as long as the dorsal root fascicles can be seen. Rootlets in segments C5–C8 enter the cord at ~90° to the cord, while T1 rootlets are finer in caliber and angle back at ~60–70°. This makes the segmental placement of any of the lesions described above predictable and reproducible. After making the lesion, whether this means cutting selected rootlets or cutting the cord directly as with a dorsal column or hemisection lesion (or something in between), the overlying tissues are sutured back together in layers. Because suturing can lead to the formation of unwanted adhesions over or around the cord, we find that the dura is best left alone and repositioned but not sutured. Subdermal and then dermal sutures are used for final closure.

5. The Future of Using NHP Models of Spinal Injury and Repair

It is increasingly clear that the safest and most efficient way to take therapeutically promising basic science to the clinic, is through the appropriate use of NHP SCI models. A great deal can be determined at the molecular and behavioral level in the rodent, but the gap between the rodent and human primate remains large, and even a relatively small (but highly controlled) study in the NHP could greatly improve the probability of clinical trial success. And in the end, this is likely to hasten therapeutic development rather than slow it. If nothing else, one could argue ethically that primate models of SCI should be used to test issues of safety in NHPs, before therapies with potentially harmful side effects (e.g., stem cell or pluripotent precursor transplantations), are tested in humans. NHP SCI translational investigations are still limited worldwide but this is likely to change as SCI clinical trials increase in number. In the USA, the only SCI clinical trial that has been undertaken took rodent data directly to the human (see Geron's clinical trial Web site—their rodent-based approach uses hESC-derived oligodendrocyte progenitor cells in acute SCI patients to remyelinate damaged axons). Though this trial was stopped in 2012 for business reasons, it was an example of a clinical trial that did not use a NHP model before testing in humans. Such work can have enormous potential, but sidestepping NHP translational studies can result in serious risk to the patient, and to the future of SCI clinical trials (should Stage 1 trials prove unsuccessful).

As new therapeutic approaches emerge, be they targeting axon growth inhibitor molecules (e.g., chondroitin sulphate proteoglycans, brevicans, and ephrins) or astrocytic activation, or myelin regeneration (see ref. (31) for review), NHP models of spinal injury and repair are likely to, and arguably should, play an increasingly important translational role.

References

1. DiMasi JA, Feldman L, Seckler A, Wilson A (2010) Trends in risks associated with new drug development: success rates for investigational drugs. Clin Pharmacol Ther 87:272–277
2. Anderson KD, Sharp KG, Hofstadter M, Irvine KA, Murray M, Steward O (2009) Forelimb locomotor assessment scale (FLAS): novel assessment of forelimb dysfunction after cervical spinal cord injury. Exp Neurol 220:23–33
3. Irvine KA, Ferguson AR, Mitchell KD, Beattie SB, Beattie MS, Bresnahan JC (2010) A novel method for assessing proximal and distal forelimb function in the rat: the Irvine, Beatties and Bresnahan (IBB) Forelimb Scale. J Vis Exp :(46). pii: 2246
4. Wolpaw JR, Tennissen AM (2001) Activity-dependent spinal cord plasticity in health and disease. Annu Rev Neurosci 24:807–843
5. Edgerton VR, Tillakaratne NJ, Bigbee AJ, de Leon RD, Roy RR (2004) Plasticity of the spinal neural circuitry after injury. Annu Rev Neurosci 27:145–167
6. Cha J, Heng C, Reinkensmeyer DJ, Roy RR, Edgerton VR, De Leon RD (2007) Locomotor ability in spinal rats is dependent on the amount of activity imposed on the hindlimbs during treadmill training. J Neurotrauma 24:1000–1012
7. Darian-Smith C (2007) Monkey models of recovery of voluntary hand movement after spinal cord and dorsal root injury. ILAR J 48: 396–410
8. Darian-Smith C (2009) Plasticity of somatosensory function during learning, disease and injury. In: Gardner E, Kaas JH (eds) The senses: a comprehensive reference. Somatosensation, vol 6. Academic, San Diego, pp 259–298
9. Darian-Smith C (2009) Synaptic plasticity, neurogenesis, and functional recovery after spinal cord injury. Neuroscientist 15:149–165
10. Courtine G, Bunge MB, Fawcett JW, Grossman RG, Kaas JH, Lemon R et al (2007) Can experiments in nonhuman primates expedite the translation of treatments for spinal cord injury in humans? Nat Med 13:561–566
11. Lemon RN (2008) Descending pathways in motor control. Annu Rev Neurosci 31: 195–218
12. Lemon RN, Griffiths J (2005) Comparing the function of the corticospinal system in different species: organizational differences for motor specialization? Muscle Nerve 32:261–279
13. Tuszynski MH, Grill R, Jones LL, McKay HM, Blesch A (2002) Spontaneous and augmented growth of axons in the primate spinal cord: effects of local injury and nerve growth factor-secreting cell grafts. J Comp Neurol 449:88–101
14. Jain N, Florence SL, Kaas JH (1995) Limits on plasticity in somatosensory cortex of adult rats: hindlimb cortex is not reactivated after dorsal column section. J Neurophysiol 73:1537–1546
15. Kaas JH, Qi H-X, Burish MJ, Gharbawie OA, Onifer SM, Massey JM (2008) Cortical and subcortical plasticity in the brains of humans, primates, and rats after damage to sensory afferents in the dorsal columns of the spinal cord. Exp Neurol 209:407–416
16. Weidner N, Ner A, Salimi N, Tuszynski MH (2001) Spontaneous corticospinal axonal plasticity and functional recovery after adult central nervous system injury. Proc Natl Acad U S A 98:3513–3518
17. Ghosh A, Sydekum E, Haiss F, Peduzzi S, Zörner B, Schneider R et al (2009) Functional and anatomical reorganization of the sensory-motor cortex after incomplete spinal cord injury in adult rats. J Neurosci 29:12210–12219
18. Rosenzweig ES, Courtine G, Jindrich DL, Brock JH, Ferguson AR, Strand SC et al (2010) Extensive spontaneous plasticity of corticospinal projections after primate spinal cord injury. Nat Neurosci 13:1505–1510
19. Chen LM, Friedman RM, Ramsden BM, LaMotte RH, Roe AW (2001) Fine-scale organization of SI (area 3b) in the squirrel monkey revealed with intrinsic optical imaging. J Neurophysiol 86:3011–3029
20. Mott FW, Sherrington CS (1895) Experiments upon the influence of sensory nerves upon movement and nutrition of the limbs. Proc R Soc Lond B 57:481–488
21. Vierck CJ (1982) Comparison of the effects of dorsal rhizotomy or dorsal column transection on motor performance of monkeys. Exp Neurol 75:566–575
22. Pons TP, Garraghty PE, Ommaya AK, Kaas JH, Taub E, Mishkin M (1991) Massive cortical reorganization after sensory deafferentation in adult macaques. Science 252:857–860
23. Woods TM, Cusick CG, Pons TP, Taub E, Jones EG (2000) Progressive transneuronal changes in the brainstem and thalamus after long-term dorsal rhizotomies in adult macaque monkeys. J Neurosci 20:3884–3899
24. Darian-Smith C, Brown S (2000) Functional changes at periphery and cortex following dorsal root lesions in adult monkeys. Nat Neurosci 3:476–481
25. Darian-Smith C, Ciferri M (2006) Cuneate nucleus reorganization following a cervical

dorsal rhizotomy in the macaque monkey. J Comp Neurol 498:552–565
26. Nagano A (1998) Treatment of brachial plexus injury. J Orthop Sci 3:71–80
27. Darian-Smith C (2004) Primary afferent terminal sprouting following a cervical dorsal root section in the macaque monkey. J Comp Neurol 470:134–150
28. Darian-Smith C, Ciferri M (2005) Loss and recovery of voluntary hand movements in the macaque following a cervical dorsal rhizotomy. J Comp Neurol 49:27–45
29. Vessal M, Aycock A, Garton A, Ciferri M, Darian-Smith C (2007) Adult neurogenesis in primate and rodent spinal cord: comparing a cervical dorsal rhizotomy with a dorsal column transection. Eur J Neurosci 26:2777–2794
30. Vessal M, Darian-Smith C (2010) Adult neurogenesis occurs in primate sensorimotor cortex following cervical dorsal rhizotomy. J Neurosci 30:8613–8623
31. Fawcett J (2009) Molecular control of brain plasticity and repair. Prog Brain Res 179:501–509
32. Silver J, Miller JH (2004) Regeneration beyond the glial scar. Nat Rev Neurosci 5:146–156
33. Darian-Smith C, Hopkins S, Ralston HJ 3rd (2010) Changes in synaptic populations in the spinal dorsal horn following dorsal rhizotomy in the monkey. J Comp Neurol 518:103–117
34. Ralston HJ 3rd, Ralston DD (1994) Medial lemniscal and spinal projections to the macaque thalamus: an electon microscopic study of differing GABAergic circuitry serving thalamic somatosensory mechanisms. J Neurosci 14: 2485–2502
35. Ralston HJ 3rd, Ohara PT, Meng XW, Wells J, Ralston DD (1996) Transneuronal changes of the inhibitory circuitry in the macaque somatosensory thalamus following lesions of the dorsal column nuclei. J Comp Neurol 371:325–335
36. Moore KA, Kohno T, Karchewski LA, Scholz J, Baba H, Woolf CJ (2002) Partial peripheral nerve injury promotes a selective loss of GABAergic inhibition in the superficial dorsal horn of the spinal cord. J Neurosci 22:6724–6731
37. Somers DL, Clemente FR (2002) Dorsal horn synaptosomal content of aspartate, glutamate, glycine and GABA are differentially altered following chronic constriction injury to the rat sciatic nerve. Neurosci Lett 323:171–174
38. Cooper BY, Glendinning DS, Vierck C (1993) Finger movement deficits in the stumptail macaque following lesions of the fasciculus cuneatus. Somatosens Mot Res 10:17–29
39. Glendinning DS, Cooper BY, Vierck CJ Jr, Leonard CM (1992) Altered precision grasping in stumptail macaques after fasciculus cuneatus lesions. Somatosens Mot Res 9:61–73
40. Glendinning DS, Vierck CJ Jr, Cooper BY (1993) The effect of fasciculus cuneatus lesions on finger positioning and long-latency reflexes in monkeys. Exp Brain Res 93:104–116
41. Leonard CM, Glendinning DS, Wilfong T, Cooper BY, Vierck CJ Jr (1992) Alterations of natural hand movements after interruption of fasciculus cuneatus in the macaque. Somatosens Mot Res 9:75–89
42. Vierck CJ, Cooper BY (1998) Cutaneous texture discrimination following transection of the dorsal spinal column in monkeys. Somatosens Mot Res 15:309–315
43. Jain N, Catania KC, Kaas JH (1997) Deactivation and reactivation of somatosensory cortex after dorsal spinal cord injury. Nature 386:495–498
44. Kuypers HGJM (1981) Anatomy of the descending pathways. In: Brooks VBV, Brookhart JM, Mountcastle VB (eds) Handbook of physiology, Section 1. The nervous system, vol 2. American Physiological Society, Bethesda, pp 597–666
45. Galea MP, Darian-Smith I (1997) Manual dexterity and corticospinal connectivity following unilateral section of the cervical spinal cord in the macaque monkey. J Comp Neurol 381: 307–319
46. Galea MP, Darian-Smith I (1997) Corticospinal projection patterns following unilateral section of the cervical spinal cord in the newborn and juvenile macaque monkey. J Comp Neurol 381:282–306
47. Rosenzweig ES, Brock JH, Culbertson MD, Lu P, Moseanko R, Edgerton VR et al (2009) Extensive spinal decussation and bilateral termination of cervical corticospinal projections in rhesus monkeys. J Comp Neurol 513:151–163
48. Fouad K, Klusman I, Schwab ME (2004) Regenerating corticospinal fibers in the Marmoset (*Callitrix jacchus*) after spinal cord lesion and treatment with the anti-Nogo-A antibody IN-1. Eur J Neurosci 20(9):2479–2482
49. Freund P, Wannier T, Schmidlin E, Bloch J, Mir A, Schwab ME et al (2007) Anti-Nogo-A antibody treatment enhances sprouting of corticospinal axons rostral to a unilateral cervical spinal cord lesion in adult macaque monkey. J Comp Neurol 502:644–659
50. Freund P, Schmidlin E, Wannier T, Bloch J, Mir A. Schwab ME et al (2009) Anti-Nogo-A antibody treatment promotes recovery of manual dexterity after unilateral cervical lesion in adult primates–re-examination and extension of behavioral data. Eur J Neurosci 29:983–996
51. Maier IC, Schwab ME (2006) Sprouting, regeneration and circuit formation in the injured spinal cord: factors and activity. Philos Trans R Soc Lond B Biol Sci 361:1611–1634

52. Raineteau O, Fouad K, Bareyre FM, Schwab ME (2002) Reorganization of descending motor tracts in the rat spinal cord. Eur J Neurosci 16:1761–1771
53. Wannier-Morino P, Schmidlin E, Freund P, Belhaj-Saif A, Bloch J, Mir A et al (2008) Fate of rubrospinal neurons after unilateral section of the cervical spinal cord in adult macaque monkeys: effects of an antibody treatment neutralizing Nogo-A. Brain Res 1217:96–109
54. Kirshblum S, Campagnolo DI, DeLisa JA (2002) Spinal cord medicine. Lippincott Williams & Wilkins, Philadelphia
55. Sherrington C (1899) On the spinal animal. Med Clin Trans 82:1–31
56. Fulton JF, Sherrington CS (1932) State of the flexor reflex in paraplegic dog and monkey respectively. J Physiol 75:17–22
57. Piedras MJ, Hernández-Laín A, Cavada C (2010) Clinical care and evolution of paraplegic monkeys (*Macaca mulatta*) over fourteen months post-lesion. Neurosci Res 69: 135–143
58. Hernández-Laín A, Piedras MJ, Cavada C (2011) Functional evaluation of paraplegic monkeys (*Macaca mulatta*) over fourteen months post-lesion. Neurosci Res 69:144–153
59. Babu RS, Periasamy P, Varadamurthy S, Sethuraman OS, Namasivayam A (2007) Locomotor behavior of bonnet macaques after spinal cord injury. Motor Control 11:71–85
60. Iwanami A, Yamane J, Katoh H, Nakamura M, Momoshima S, Ishii H et al (2005) Establishment of graded spinal cord injury model in a nonhuman primate: the common marmoset. J Neurosci Res 80:172–181
61. Bresnahan J (1978) An electron-microscopic analysis of axonal alterations following blunt contusion of the spinal cord of the rhesus monkey (*Macaca mulatta*). J Neurol Sci 37:59–82
62. Yamane J, Nakamura M, Iwanami A, Sakaguchi M, Katoh H, Yamada M et al (2010) Transplantation of galectin-1-expressing human neural stem cells into the injured spinal cord of adult common marmosets. J Neurosci Res 88:1394–1405

Chapter 11

Animal Models of Spinal Cord Ischemia

Hamdy Awad, Haytham Elgharably, and Phillip Popovich

Abstract

A serious complication of aortic cross clamping during thoracic and thoracoabdominal aortic aneurysm repair surgery is ischemic spinal cord injury (ISCI) and subsequent postoperative paraplegia. Several experimental models have been developed in various species that reproduce this clinical problem. This chapter provides a comprehensive summary of these experimental models and provides specific details regarding model reproducibility, the techniques used to induce ischemia and the predominant clinical and histopathological outcomes that are associated with these different models.

Key words: Animal models, Spinal cord, Ischemia, Aortic cross clamping, Aortic aneurysm repair

Abbreviations

ARM	Artery of Adamkiewicz, also known as the arteria radicularis magna
AXC	Aortic cross clamping
ISCI	Ischemic spinal cord injury
LSA	Left subclavian artery
SCI	Spinal cord injury
TAAA	Thoracoabdominal aortic aneurysm

1. Introduction

Spinal cord ischemia is an unavoidable consequence of aortic cross clamping (AXC) during thoracic and thoracoabdominal aortic aneurysm (TAAA) surgery. Interruption of the spinal cord blood supply causes ischemia and a cascade of injurious cellular and molecular processes that culminate in neuronal cell death. Postoperative paralysis can be immediate, but is often delayed, with the reported incidence (0–40%) varying among vascular centers as a function of the surgery and Crawford classification (1). Those patients not only suffer from severe long-term physical disability, but also have shorter survival rates. The 5-year survival rate for

Håkan Aldskogius (ed.), *Animal Models of Spinal Cord Repair*, Neuromethods, vol. 76,
DOI 10.1007/978-1-62703-197-4_11, © Springer Science+Business Media, LLC 2013

patients with spinal cord injury (SCI) after TAAA repair has been reported to be 44%, whereas it was 62% for those without spinal cord injury in a retrospective clinical study (1). Despite marked advances in surgical technique and the use of neuroprotective adjunct therapies (e.g., hypothermia, cerebrospinal fluid (CSF) drainage) that have been introduced to the spinal cord protection protocol during aortic aneurysm repair, postoperative paraplegia remains as one of the most devastating complications of TAAA surgery.

It is expected that animal models will be useful tools for studying neuronal ischemic injury mechanisms, and ultimately, knowledge gained from these models should lead to the development of more effective neuroprotective interventions that will prevent postoperative paraplegia. Different species of animals, including: mice, rats, dogs, rabbits, sheep, pigs and monkeys have been used to simulate the ischemic events that occur during aortic aneurysm surgery. However, each species offers specific advantages and poses unique challenges in an experimental setting. Most advantages and limitations are related to anatomical variations of the spinal cord blood supply and the technical approaches that are used to produce ischemia, both of which affect the reproducibility of ischemic spinal cord injury (ISCI) and subsequent neurological impairment.

In order to appreciate the subtleties of species' variations in spinal cord vasculature and how these may affect the variability of pre-clinical models, an overview of human spinal cord vascular anatomy is helpful.

2. Overview of Human Spinal Cord Blood Supply

The anterior spinal cord artery and the two posterior spinal arteries that arise from the vertebral arteries at the foramen magnum are the primary sources of blood to the human spinal cord. Blood flow from segmental spinal arteries helps reinforce flow to individual spinal segments (Fig. 1). Arteriolar sized vessels from the anterior spinal and both posterior spinal arteries encircle the spinal cord forming a fine, pial plexus. The segmental spinal arteries give rise to anterior and posterior radicular arteries, which supply the anterior and posterior spinal arteries along the spinal canal (14).

The number, size, and location of the radicular arteries varies. At each level of the spinal column, a radicular artery enters the intervertebral foramen to supply the spinal cord. At the thoracic and lumbar regions of the cord, segmental spinal arteries originate from the posterior branches of the intercostal and lumbar arteries, respectively. However, the origin of these vessels in the cervical and sacral regions is variable. In the cervical region they arise from the vertebral artery or the subclavian artery (via the ascending cervical branch of the thyrocervical trunk or the deep cervical branch of the

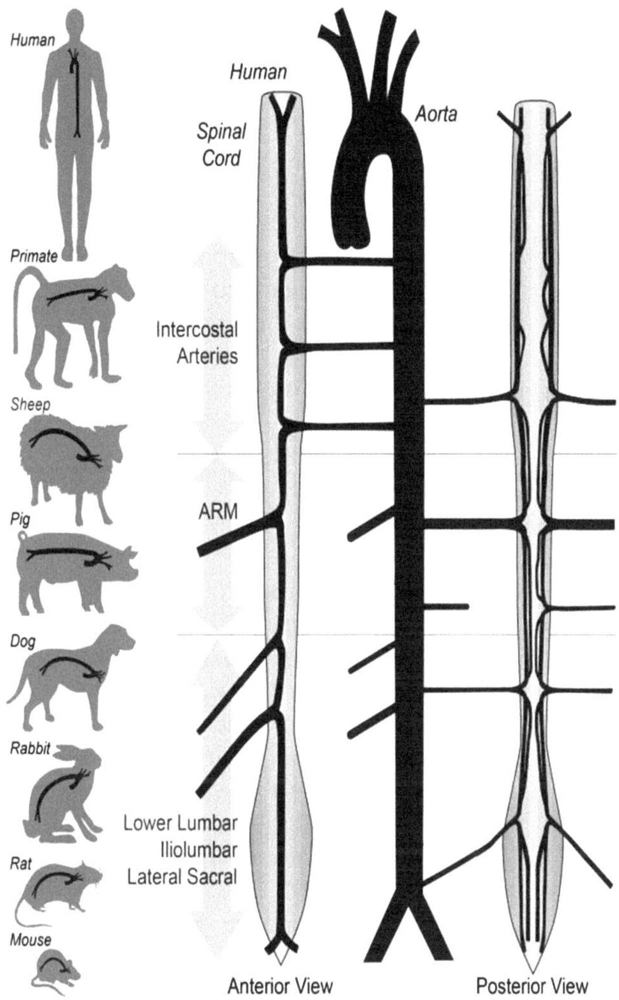

Fig. 1. A comparative overview of spinal cord blood vessels often targeted when modeling ischemic spinal cord injury caused by aortic cross clamping. In humans, the anterior spinal artery and two smaller posterior spinal arteries are the main vessels supplying the spinal cord. These latter spinal arteries receive radicular branches from the intercostal arteries in the thoracic region, while the arteria radicularis magna (ARM) is the primary arterial blood supply to the lower two-thirds of the spinal cord (usually between T11–L3). The cauda equina is supplied by lower lumbar, iliolumbar and lateral sacral arteries. Differences between spinal cord blood supply in humans and other species: *Mice*: similar to human (2); *Rat*: hetero-segmental aorta, ARM origin at T8–L2 (3–5); *Rabbit*: Uniform homosegmental blood supply; main blood supply to spinal cord is from segmental branches of the abdominal aorta (6); *Dogs*: ARM arises from the fifth lumbar artery (7–10); *Pigs*: Internal thoracic and subscapular arteries are larger than in humans – providing extensive collateral flow to the spinal cord. Segmental thoracic and lumbar arteries are larger than in humans (11); *Sheep*: No details about the spinal cord blood supply in sheep; *Primates*: Virtually identical to that in humans. The anterior spinal artery is continuous in man and baboon, which was not true in rabbits, dogs or pigs (12, 13).

costocervical region). The sacral segmental spinal arteries most frequently arise from the lateral sacral artery, but can originate from the iliolumbar, middle sacral, fifth lumbar or obturator artery (14).

The Artery of Adamkiewicz (also known as the arteria radicularis magna or ARM) is the largest of the radicular arteries and is the principle supply of arterial blood to the thoracoabdominal region of the spinal cord (Fig. 1; usually found between T11–L3). The Artery of Adamkiewicz is of great clinical importance, especially in regard to aortic aneurysm surgery. Preoperative localization of the Adamkiewicz artery and its segmental supplier before thoracic and thoracoabdominal aortic aneurysm repair can help to prevent postoperative paraplegia (15).

3. Animals Models of ISCI

3.1. Mouse ISCI Models

Generally, the blood supply of the mouse spinal cord is similar to that of humans. The ASA originates from the vertebral arteries and <10 anterior radicular arteries join it. Two posterior spinal arteries are supplied by several radicular arteries, which form large anastomoses around the spinal cord (2).

To date, four different models of mouse ISCI have been published (Table 1); three used an open surgical approach with different locations of AXC (Fig. 2), and the fourth induced ischemia via photochemical thrombosis of the dorsal spinal cord blood supply (16). The Lang-Lazdunski et al. and Wang et al. models cause immediate paralysis, while the Awad et al. model causes delayed paralysis with reproducible spinal cord pathology. The application of systemic hypothermia is thought to be responsible for promoting survival for months after ischemia/reperfusion (18). This low mortality and long-term survival allows scientists the opportunity to study the consequences of ischemia/reperfusion for many weeks or months after the insult. Mouse models of ISCI also offer an additional advantage, i.e., the ability to manipulate genetics to understand specific mechanisms that cause or contribute to ischemic neuronal injury.

3.2. Rat ISCI Models

Tveten compared the arrangement of spinal cord blood vessels in humans and rats, and noted some similarities, but also several differences (3, 4). For example, blood supplied by radicular arteries to the lower cervical and upper thoracic segments is increased in rat models compared to humans. Tventen also concluded that rats possess an ARM; however, its location was less variable, located slightly rostral to where it is found in humans. Also, with few exceptions, anastomoses between the ventral and the dorsal spinal arteries did not occur in rats. In contrast, Schievink et al. concluded that a human equivalent of the ARM does not exist in rats (5).

Table 1
Mouse models of ischemic spinal cord injury

References	Anatomical location of induced ischemia	Technique and outcome of induced ischemia
Lang-Lazdunski et al. 2000 (2)	Left internal mammary artery, the aortic arch between the left common carotid artery and the left subclavian artery (LSA) and the LSA at its origin	• *Approach*: Cross clamping the three vessels through a ventral midline cervico-thoracic incision. • *Ischemia time*: 9 or 11 min followed by perfusion for 24 or 48 h. • *Temperature*: 36 °C. • *Clinical outcome*: All animals developed immediate paralysis. Recovery from paralysis was dependent on the duration of ischemia. Mortality was 21% after 11 min of ischemia, while no mortality was reported after 9 min of ischemia.
Gaviria et al. 2006 (16)	T8 vertebrae	• *Approach*: Photochemical spinal cord ischemia by intravenous injection of Rose Bengal dye with 8 min irradiation of exposed spinal cord. • Post-ischemic changes were recorded by ex vivo magnetic resonance imaging and histological analysis. • *Clinical outcome*: After 4 h of injury, alterations of the spinal cord dorsal and lateral white matter were reflected by a progressive loss of white/grey matter contrasted with further ventral extension by 24 h.
Wang et al. 2010 (17)	Descending aorta at T8	• *Approach*: Aortic cross clamping (AXC) through left thoracotomy along the eighth rib. • *Ischemia time*: 0, 8, 10, 12 min. • *Clinical outcome*: All mice had acute weakness of the hind limb at 1 h post-ischemia and 100% motor deficit within 24 h with 30–50% recovery of normal motor functions by 7 days. Eighty percent of mice survived 28 days and all of had severe motor deficits immediately after injury.
Awad et al. 2010 (18)	Descending aorta 1 mm distal to LSA origin	• *Approach*: AXC through left thoracotomy between second and third ribs. • *Ischemia time*: 3–11 min @ 33 °C, 35 °C, or 36 °C. • *Clinical outcome*: The onset of paralysis was delayed for 24–36 h after reperfusion and 70% of mice developed severe hind limb paralysis. More than 95% of mice survived up to 9 weeks after surgery.

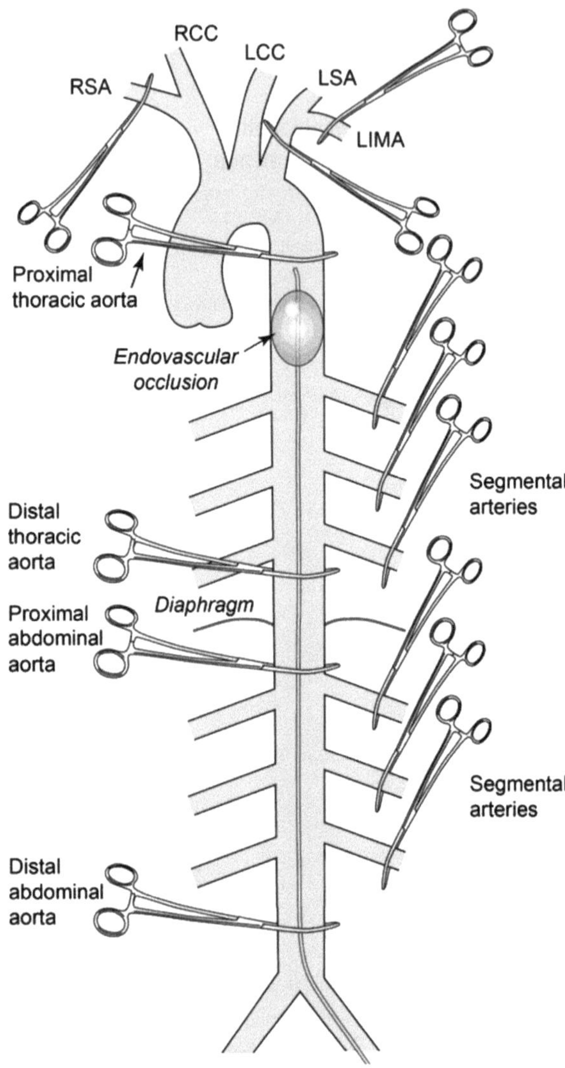

Fig. 2. Anatomical location(s) of aortic occlusion in animal models of spinal cord ischemia. (RSA, right subclavian artery; RCC, right common carotid artery; LCC, left common carotid artery; LSA, left subclavian artery; LIMA, left internal mammary artery). Different approaches used to induce spinal cord ischemia in experimental studies - species (ref): LIMA + LSA + proximal thoracic aorta: *Mouse* (2), Proximal thoracic aorta: *Mouse* (18); *Rat* (19, 20); *Dog* (21, 22); *Sheep* (23), Proximal and distal thoracic aorta: *Rat* (24); *Dog* (25); *Primates* (26, 27), Proximal abdominal aorta: *Rat* (28); *Rabbit* (29); *Dog* (30), Proximal and distal abdominal aorta: *Rat* (31), Segmental arteries: *Rabbit* (32); *Dog* (33); *Pig* (34), Endovascular occlusion: *Rat* (35, 36); *Rabbit* (37); *Dog* (38).

Although current models were developed assuming that rats possess an ARM, further comparative studies are needed to fully characterize rat spinal cord vasculature. Differences in collateral blood flow and/or the location of cross-clamping relative to these poorly defined collateral systems may have influenced the variability that existed in the different reported rat ISCI models (see below and Table 2).

Table 2
Rat models of ischemic spinal cord injury

Study	Anatomical location of induced ischemia	Technique and outcome of induced ischemia
Matsushita and Smith 1970 (19)	Thoracic aorta	• *Approach*: Aorta was exposed through left thoracotomy between the third and fourth ribs and occlusion was done by clip application. • *Ischemia time*: 15–60 min. • *Clinical outcome*: With occlusion time <20 min, $n = 1/5$ rats exhibited extensor rigidity. The remaining $n = 4$ regained normal motor activity within 24 h after the operation. Occlusion of the aorta for >30 min caused death within 24 h in most animals.
Coston et al. 1983 (35)	Descending thoracic aorta	• *Approach*: Aortic occlusion with 2F Fogarty catheter introduced through the femoral artery. • *Ischemia time*: 13–20 min. • *Clinical outcome*: Permanent hind limb motor deficits in 75% of operated animals.
Le May et al. 1987 (20)	Aortic arch between the left common carotid and left subclavian artery (LSA)	• *Approach*: A snare was placed around the aorta through a longitudinal incision in the sternal region. • *Ischemia time*: 10, 15, 20 min. • *Clinical outcome*: 10 min occlusion caused mild neurological deficits with almost complete recovery after 24 h. Occlusion for 15 min caused intermediate but persistent deficits with some improvement by 24 h. The 20 min occlusion caused severe paralysis with minimal improvement over 24 h.
Marsala and Yaksh 1994 (36)	Descending thoracic aorta at the level of LSA	• *Approach*: Aortic occlusion by insertion and inflation of 2F Fogarty catheter inserted through the tail artery. • *Ischemia time*: 10, 15, 20, 30 min. • *Clinical outcome*: 30 min of ischemia caused hindlimb flaccidity at 1 h, spasticity at 4 h, and flaccidity at 8 h. After 20 min of ischemia, initial flaccidity was followed by hind limb spasticity that persisted for 8 h. Ischemia for shorter intervals had minimal effects on motor function.

(continued)

Table 2 (continued)

Study	Anatomical location of induced ischemia	Technique and outcome of induced ischemia
Lopez et al. 1995 (24)	Descending thoracic aorta at the level of the third and the tenth intercostal spaces	• *Approach*: Two left thoracotomies at the third and tenth intercostal spaces. The thoracic aorta was occluded by 2-0 silk snare. • *Ischemia time*: Different time intervals. • *Clinical outcome*: When the aorta was cross-clamped for 15 min or less, all animals recovered completely. Beyond 15 min, the incidence of paraplegia increased proportionately with the duration of aortic occlusion. When the aorta was double cross-clamped for 40 min, none of the rats survived. Aortic occlusion for 30 min was used in follow-up studies in which addition of HCO_3 (20 mM) to the hypothermic (15 °C) solution was found to be neuroprotective.
Usul et al. 2004 (28)	Abdominal aorta just distal to the left renal artery	• *Approach*: An aneurysm clip was applied to the abdominal aorta through a retroperitoneal approach. • *Ischemia time*: 45 min. • *Clinical outcome*: Significant neurological deficits in rats after 45 min of aortic occlusion.
Pashin and Viktorov 2009 (39)	Thoracic aorta	• *Approach*: Photochemical spinal cord ischemia by intravenous (jugular) injection of Bengal pink dye with irradiation of exposed spinal cord. • *Ischemia time*: Duration of radiation. • *Clinical outcome*: Histology revealed the necrosis at location of focal ischemia 1 day after surgery.
Guler et al. 2010 (31)	The abdominal aorta just distal to the left renal artery and proximal to the aortic bifurcation	• *Approach*: Double cross clamping of the aorta via vertical midline incision. • *Ischemia time*: 45 min. • *Clinical outcome*: Significant neurological impairment with histological evidence of ischemic necrosis.

3.2.1. Thoracic Aortic Occlusion in Rat Model of ISCI Using Open or Endovascular Approach

Matsushita and Smith developed a rat model to study the effect of AXC on hind limb spasticity (19). In 1987, Le May et al. developed a rat model of ISCI (20) to replace the rabbit model. At that time, this rabbit model was widely used as a nonspecific representation of neuronal ischemic injury (29, 40–42). Soon after, Le May et al. made additional modifications to the rat model (43). They used the same incision to explore the thoracic aorta, but instead occluded the right and left subclavian arteries and the thoracic aorta for 10 min. Neurological deficits were recorded at different time intervals.

Coston et al. developed another rat model of ISCI using an endovascular approach (35). Specifically, a 2F Fogarty catheter was inserted through the femoral artery into the aorta and was inflated. Fifteen minutes of aortic occlusion caused permanent hind limb motor deficits in 75% of the operated rats. Marsala and Yaksh modified the Coston et al. model by inserting the 2F Fogarty catheter through the tail artery into the thoracic aorta until the catheter tip reached the level of the left subclavian artery (35).

3.2.2. Thoracic Aortic Occlusion in Rat Model of ISCI with Hypovolemia

Kanellopoulos et al. evaluated three different approaches to develop a new rat model, claiming that the previous rat models were not reproducible (44). They concluded that a reproducible model of ISCI was possible in rats by combining aortic arch occlusion with induced hypovolemia (44).

3.2.3. Abdominal AXC in Rat Model of ISCI

Models in which cross clamping of the abdominal aorta has been used to cause spinal cord ischemia have produced variable results. Usul et al. and Guler et al. reported significant neurological deficits in rats after 45 min of aortic occlusion (28, 31). In contrast, Carrillo et al. reported that sub-diaphragmatic AXC is not suitable for causing spinal cord ischemia in rats because this method does not increase blood and tissue concentrations of pyruvate, glucose, or lactate relative to a non-ischemic control population (45). Nouri et al. also concluded that abdominal AXC is not a reliable method to induce spinal cord ischemia in rats (46) because the clamp is placed distal to the origin of the artery of Adamkiewicz (ARM). In rats, the origin of the artery of Adamkiewicz lies slightly more rostral to segments of the thoracic spinal cord (relative to human). Consequently, blood supply to the lumbosacral spinal cord is largely unaffected (46).

3.3. Rabbit ISCI Models

Recently, Mazensky and colleagues found that contrary to other laboratory animals, the blood supply to the spinal cord of rabbits is homo-segmental; each spinal cord segment is supplied with one corresponding radicular artery with minimal or no collateral bloodstream (6). They reported high variability of segmental arteries supplying blood to the ASA inside the spinal canal. Also, when studying the ARM which supplies the caudal half of the spinal cord,

Table 3
Rabbit models of ischemic spinal cord injury

Study	Anatomical location of induced ischemia	Technique of induced ischemia
Zivin and DeGirolami 1980 (29)	The abdominal aorta just caudal to the left renal artery	• *Approach*: Exposure of the aorta through abdominal incision. Occlusion was done by a clip. • *Ischemia time*: 1 h. • *Clinical outcome*: All animals ($n = 40$) were paraplegic, incontinent and unresponsive to noxious stimuli in the hindlimbs.
Cheng et al. 1984 (37)	Abdominal aorta just distal to the left renal artery	• *Approach*: Abdominal aorta was occluded by a balloon catheter that was introduced through the femoral artery. • *Ischemia time*: 15–60 min. • *Clinical outcome*: Frequency of paraplegia varied as a function of ischemia duration: 0% after 15 min, 30% after 17 min, 33% after 20 min, 38% after 25 min, and 100% after 60 min of ischemia.
Maeda et al. 2003 (32)	Five lumbar arteries (L3–L7)	• *Approach*: Selective occlusion of the five lumbar arteries through a flank incision. • *Ischemic time*: Variable. • *Clinical outcome*: Occlusion of the lumbar arteries for 10 and 20 min induced paraplegia within 24 h after ischemia in 2/6 and 4/6 rabbits, respectively. Delayed paraplegia was observed in one rabbit after 10 min ischemia and another one after 20 min ischemia.

a bigger feeding artery arising from the spinal branch of the sixth lumbar artery was observed. Concordantly, there was no collateral blood supply coming from the thoracic spinal cord. These findings suggest the rabbit model is more susceptible to ischemia after abdominal AXC. Table 3 provides a summary of publications detailing rabbit ISCI models.

3.3.1. Abdominal Aortic Occlusion in Rabbit Model of ISCI with Open or Endovascular Approach

Zivin and DeGirolami developed the first reproducible rabbit model of ISCI in 1980 (29). Their intention was to develop a central nervous system ischemia model to avoid variability and clinical complications associated with modeling infarction in the cerebral arteries. They attributed the reproducibility of their model to the uniformity of the spinal cord vasculature in rabbit models, i.e., the main source of blood in the caudal spinal cord is segmental from the abdominal aorta and can be interrupted by a single AXC (29). They later modified their technique of AXC by placing a snare ligature around the abdominal aorta, with the free end of the snare accessible outside the rabbit (40). In related studies, a subcostal incision was performed to expose the infra-renal aorta through a retroperitoneal approach, instead of a midline laparotomy. The use of an externalized snare ligature enabled investigators to cause repeated bouts of ischemia in the conscious animal (47–51).

Cheng et al. used a less invasive approach (52). A balloon catheter was introduced through the femoral artery and advanced into the abdominal aorta until it was just distal to the left renal artery. Occlusion of the aortic lumen was accomplished by inflation of the balloon. This approach has been applied in diverse rabbit models of SCI because it is simple, fast, and less invasive than ligating the aorta through a laparotomy (53–59).

3.3.2. Rabbit Model of ISCI Using Selective Occlusion of Lumbar Arteries

Maeda et al. developed a new rabbit model of SCI (32). Instead of the classical technique of infra-renal AXC, they selectively occluded the five lumbar arteries (L3–L7), utilizing a flank incision. Their goal was to avoid the hemodynamic changes associated with AXC, which include proximal hypertension with subsequent increase in CSF pressure and distal hypotension with subsequent ischemia of the bowel and lower extremities. By selective occlusion of the lumbar arteries, blood flow in the aorta was maintained, but the blood supply to the lumbar spinal cord was still interrupted (Table 3).

The rabbit model with infra-renal AXC is widely used as a model of ISCI. Although the rabbit model is reproducible, owing this quality in part to the simplicity of the surgical approach and the low incidence of adverse reactions associated with the procedure, it does not mimic the consequence of AXC in humans. Indeed, infra-renal AXC in humans undergoing abdominal aortic aneurysm repair rarely causes paraplegia (60, 61).

3.4. Dog ISCI Models

The blood supply to the canine spinal cord is similar to that of humans (Fig. 1). Each spinal segment is supplied by intercostal or lumbar arteries branching from the aorta (7), and like the human spinal cord, rich anastomotic vascular networks exist, especially at the level of the conus medullaris (8). In most dogs, the origin of the ARM lies caudal to the origin in humans; arising from the fifth lumbar artery after following a course similar to the human ARM (8–10). Because dogs have extensive collateral pathways from the aorta to the cord, their motor neurons at any spinal level may survive irreversible ischemic damage when the aorta is cross-clamped, even at high thoracic levels (22).

The idea of using dogs to model ISCI was described long ago. In 1667, Niels Stensen described a procedure he performed on dogs as such: "*. . . . on making a ligature in the descending aorta, without previous cutting, that voluntary movement of all the parts below ceased in proportion to the tightness of the band, while it was restored similarly as the knot was loosened*" (29). Since then, multiple techniques have been used to cause ISCI in dogs (Table 4).

Table 4
Dog models of ischemic spinal cord injury

Study	Anatomical location of induced ischemia	Technique of induced ischemia
Gelfan and Tarlov 1959 (62)	Thoracic aorta	• *Approach*: Adopted from Tureen's technique for cats (1963) (63). • *Ischemia time*: 40–50 min. • *Clinical outcome*: More than 60% of 120 dogs subjected to 40–50 min of aortic occlusion survived the stress. Sixty-two percent of the survivors exhibited permanent rigidity of the hind limbs. The rest exhibited either permanent flaccid paralysis or temporary weakness or were normal.
Nylander et al. 1982 (30)	The abdominal aorta caudal to the renal artery	• *Approach*: Permanent occlusion of the abdominal aorta. • *Clinical outcome*: Paraplegia incidence was 90% in the ischemic group of animals within 24 h post ischemia. High mortality was reported in the ischemic group after 48 h; only 1 animal survived out of 10.
Coles et al. 1982 (25)	The left subclavian artery (LSA), the thoracic aorta just distal to the LSA and the distal thoracic aorta at the level of the diaphragmatic hiatus	• *Approach*: Left thoracotomy through the fourth and the eighth intercostal spaces. • *Ischemia time*: Different time intervals. • *Clinical outcome*: Twelve animals subjected to aortic occlusion demonstrated a characteristic time-related deterioration of the somatosensory evoked potentials (SEPs).
Laschinger 1984 (21)	The thoracic aorta distal to the LSA	• *Approach*: Single aortic cross clamping (AXC) just distal to the origin of the LSA via a small thoracotomy. • *Ischemia time*: Different time intervals guided by loss of SEPs. *Clinical outcome*: Animals with AXC for 5 min after loss of SEPs recovered without neurological deficit. Animals underwent AXC for 10 min after loss of SEPs sustained a 67% incidence of permanent spastic paraplegia.

Kaplan et al. 1987 (38)	The thoracic aorta and positioned 1 cm distal to the origin of LSA	• *Approach*: A balloon catheter was introduced through the carotid artery into the thoracic aorta and ischemia of the spinal cord was induced by inflation of the balloon. • *Clinical outcome*: Acute occlusion of the thoracic aorta (distal pressure 15–25 mmHg) was associated with a change in the SEP. Aortic occlusion for 30 min beyond an evoked potential change resulted in a moderate to severe motor deficit in all cases. Acute occlusion of the thoracic aorta (ten animals, distal pressure 15–25 mmHg) was associated with a change in the SEP.
Kai et al. 1994 (33)	Lumbar arteries between the aortic hiatus and the bifurcation in addition to the middle sacral artery	• *Approach*: Baseline sciatic motor evoked potentials (MEPs) and SEPs were obtained, lumbar arteries were ligated and evoked potentials were recorded continuously. If sciatic MEPs were not lost, intercostal arteries were ligated, and potentials and clinical status were reassessed. • *Clinical outcome*: Following ligation of the lumbar and intercostal arteries, no change in the evoked potentials was recorded from the mid-spine; however, ischemia of the lumbar spine was recognized in all cases.
Sugawara et al. 2005 (22)	Proximal descending aorta	• *Approach*: In addition to AXC, they induced hypovolemia by draining blood from the LSA. A catheter was introduced through the axillary artery to the LSA to drain the blood. • *Ischemia time*: 60 min with AXC only and 40 min in animals with AXC and hypovolemia. • *Clinical outcome*: All animals that underwent the procedure had paraplegia after 40 min of ischemia, while one-third of animals with thoracic aortic occlusion alone demonstrated no spinal cord injury, even with a longer ischemic time.

3.4.1. Abdominal Aortic Occlusion in Dogs

Nylander et al. developed a canine model in which permanent occlusion of the infra-renal aorta was used to cause ISCI (30). The location of aortic ligation was chosen to avoid bowel infarction and kidney damage, i.e., complications that occur with occlusion of the aorta above the celiac artery or the superior mesenteric artery (64, 65).

3.4.2. Thoracic Aortic Occlusion in Dog Model of SCI

Coles et al. described a temporary thoracic aorta occlusion model of dog ISCI (25). By clamping the proximal and distal thoracic aorta, as well as the origin of the left subclavian artery (LSA), it was discovered that collateral blood flow to the spinal cord also was interrupted, thereby increasing the reproducibility of ISCI and paraplegia (63). Coles used this model to evaluate the neuroprotective effects of regional spinal cord cooling and systemic injection of a putative free radical scavenger (63, 66). Subsequent experimental studies by other groups have shown that double aortic cross clamping is also a suitable approach (with minor variations) for causing ISCI in dogs (67–75). Other researchers used single AXC (via a small thoracotomy) just distal to the origin of the LSA, to induce spinal cord ischemia in dogs (21, 76–84). Similarly, Grossi et al. (85), Sugiyama et al. (86) and Shokoku et al. (87) determined the effects of ligating the LSA, right subclavian, and/or double AXC in their models. Kumagai et al. described a different model of canine ISCI (88). This research group was able to simulate TAAA surgery through extensive clamping of the aorta and every other visceral branch. Sugawara et al. (22) added proximal hypovolemia to their model. Specifically, a significant reduction in proximal aortic blood pressure (reduced to ~80 mmHg) was created by withdrawing blood into a reservoir while simultaneously cross clamping the descending aorta. This technique caused substantial ischemia and loss of motor function.

3.4.3. ISCI in Dogs by Selective Occlusion of Segmental Arteries

Because of the extensive collateralization of blood supply to the canine aorta, it is difficult to cause prolonged and consistent ischemia. To overcome this innate property of the canine blood supply, a number of researchers evaluated the effects of ligating lumbar and intercostal segmental arteries. By interrupting the segmental blood supply to 3–5 consecutive segments of the spinal cord (without AXC), Kato et al. (9), Fujimaki et al. (10), and Zhang et al. (89) created reproducible ISCI models. Awad et al. used AXC in addition to permanent interruption of the segmental blood supply to produce irreversible paralysis (90).

3.4.4. Dog Models of ISCI Using an Endovascular Approach

Endovascular occlusion techniques have been used to induce ISCI in dogs. Both Kaplan et al. and Ishizaki et al. used an endo-aortic occlusion approach to produce spinal ischemia in dogs (38, 91). In both cases, a balloon catheter was introduced through the left common carotid artery into the thoracic aorta, then advanced caudally to the lower abdominal aorta. The celiac and superior mesenteric arteries were ligated to eliminate collateral blood flow to the spinal cord. Graded ISCI was caused by stepwise alternation of the level of aortic occlusion via drawing the inflated balloon in the cephalic direction (91).

Overall, because dogs have extensive collateral blood supply to the spinal cord, their spinal cord neurons may escape ischemic injury after single level AXC. Additional ligation of segmental arteries (90), with or without proximal hypovolemia (22), may be required to consistently develop robust spinal cord ischemia in dogs.

3.5. Pig ISCI Models

The availability and cost-effectiveness of pigs compared to other large animals makes the swine model of ISCI a particularly attractive pre-clinical model (92). A comparative anatomical study of pig and human spinal cord vascular supplies revealed some similarities, but also several significant differences. First, pigs have larger internal thoracic and subscapular arteries than humans. This provides extensive collateral flow to the lower body, including the spinal cord. The pig also has a fine caliber vascular plexus that provides blood to the neck area, from which blood flows to reach the spinal cord and the base of the brain. Second, the pig's spinal cord vascular anatomy is plurisegmental, but is paucisegmental (few) in humans. This feature of the pig spinal cord vasculature may render it less susceptible to ischemic injury after ligation of the segmental arteries (34, 93). The segmental thoracic and lumbar arteries are relatively small in pigs, and almost all of them originate as a single branch from the aorta then divide after 3–4 mm. The segmental vessels show a clear diminution after 2–3 cm at the level of the vertebral bodies. The median sacral artery in pigs has an isolated single dorsal branch leading to the spinal cord, and is a large caliber vessel, comparable in size to the common iliac artery (11). See Table 5 for a summary of pig ISCI models.

3.5.1. Pig Model of ISCI Using Open and/or Endovascular Approaches

Wadouh et al. used both endovascular and open approaches to induce spinal cord ischemia in the pig. From their research, the anatomical location of the ARM and the efficacy of the collateral circulation were defined as critical determinants of ischemic spinal cord injury during aortic occlusion (94, 97, 98). They also studied hemodynamic changes in the spinal and aortic collateral circulation after single and double AXC (95, 99, 100).

3.5.2. Pig Model of ISCI Using Thoracic Aortic Occlusion

Like smaller species, the relative simplicity of the single AXC approach has been used by many laboratories to induce spinal cord ischemia in pigs (101–107). The proximal thoracic aorta is usually exposed through a small thoracotomy that is cross clamped distal to the LSA. Other researchers used the double cross clamping approach as a more reliable method to induce ischemia in pig

Table 5
Pig models of ischemic spinal cord injury

Study	Anatomical location of induced ischemia	Technique of induced ischemia
Wadouh et al. 1984 (94)	Thoracic aorta just distal to the left subclavian artery (LSA) Lumbar arteries or arteria radicularis magna (ARM)	• *Two approaches*: Endovascular occlusion by a balloon catheter that was inserted through the left femoral artery. Lumbar arteries or ARM were ligated through left retroperitoneal approach. • *Ischemia time*: Occlusion of the descending aorta for 45 min or permanent ARM ligation. • *Clinical outcome*: The degree of permanent spinal cord damage was 85.7% with thoracic aortic occlusion and 71.4% with permanent ARM ligation.
Wadouh et al. 1986 (95)	The thoracic aorta at the level of T3 to T4 and T13 The abdominal aorta at the level L1 and S1 with ligation of the anterior organ arteries	• *Two approaches*: The thoracic aorta was exposed through two thoracotomies in the bed of fourth and twelfth ribs and the abdominal aorta was exposed through a left retroperitoneal approach. • *Clinical outcome*: Exclusion of the thoracic aorta by double clamping restored the intercostal bed pressure almost to control, whereas exclusion of the abdominal aorta hardly affected the lumbar bed pressure.
De Haan et al. 1999 (34)	Segmental arteries in a cranial to caudal direction, starting just below the level of the LSA	• *Approach*: The thoracoabdominal aorta was mobilized, and the intercostal and lumbar arteries were exposed from the LSA to the bifurcation. Using spinal cord motor evoked potentials (MEPs), a group of segmental arteries not critical for spinal cord blood supply was identified. Before, during, and after clamping of the noncritical segmental arteries, SCI was produced by decreasing spinal cord perfusion pressure by means of increasing cerebrospinal fluid pressure. • *Clinical outcome*: Clamping of originally noncritical segmental arteries significantly reduced the tolerance of the spinal cord to a decrease in spinal cord perfusion pressure.
Hellberg et al. 2000 (96)	The aorta just distal to the LSA and below the level of L1	• *Approach*: The thoracic aorta was exposed through median sternotomy with extended midline incision to expose the abdominal aorta. They used a shunt between the LSA and the left iliac artery to improve the hemodynamics during spinal cord ischemia. • *Ischemia time*: 60 min with the shunt and 30 min without the shunt. • *Clinical outcome*: The shunt between the LSA and the left iliac artery improved the hemodynamics during spinal cord ischemia and obviated the need for pharmacological manipulation.

models, and to mimic the surgical technique used during aortic aneurysm surgery (108–110). For double AXC, the aorta is exposed through two thoracotomies in the fourth and seventh intercostal spaces, then is cross clamped just distal to the LSA above the diaphragm (109, 110).

Hellberg et al. used a modified double AXC technique in which an aorto-aortic shunt with double AXC was used to induce selective ISCI in pigs (96). A modified technique used cross clamping of the right subclavian artery and all lumbar arteries except for L1 (111, 112). By changing the level of distal aortic clamping (caudal to L1 in one group and cranial to T12 in another group), the spinal cord circulation became solely dependent on blood flow through segmental arteries from T12 to L1. This model provides the opportunity to study the alterations in intrathecal oxygen tension during and after 60 min of cross clamping, below the level of L1 and above the level of T12 (111, 112). This latter model has been used to monitor intrathecal Po_2 during various modes of the spinal cord blood supply interruption, including; single and double AXC, clamping of both subclavian arteries, and interruption of intercostal and lumbar segmental arteries (113).

3.5.3. ISCI in Pigs via Selective Occlusion of Segmental Arteries

The spinal cord blood supply in pigs is plurisegmental (i.e., five or more segmental arteries have a radicular branch supplying the spinal cord). Still, ligation of segemental arteries will induce spinal cord ischemia in pigs. This was shown by De Haan et al. (34) and also by the Griepp group (114–116). However, the latter group reported that a given number of segmental vessels could be clamped serially in the pig without causing critical ischemia to the cord (114). Consequently, they suggested that the intercostal and lumbar arteries could be sacrificed in patients undergoing TAAA surgery without causing ischemia and paralysis (117).

3.5.4. Pig Model of ISCI Using Endovascular Approach

Toumpoulis et al. and Papakostas et al. performed an endovascular occlusion technique as a reliable and less invasive method to induce spinal ischemia in pigs (118–121). They used two catheter balloons to occlude the aorta at two levels. The first catheter balloon was introduced through the carotid artery and occluded the aorta at either the LSA or diaphragm level. The second catheter balloon was introduced through the femoral artery, and was then used to occlude the abdominal aorta above the bifurcation. This model of porcine ISCI has been used to evaluate the neuroprotective efficacy of ischemic preconditioning of spinal cord during AXC (118–120).

3.6. Sheep ISCI Models

Sheep may be useful for pre-clinical modeling of ISCI for a number of reasons. Their bodies approximate the human body cavity size. Also, their vessel diameters are suitable for evaluation of endovascular stents (122–125) and their neural axis is similar to that of humans, especially in regard to spinal cord size and CSF volume (126). Sheep have been used in different areas of research, such as spinal cord injury (127, 128), pre-clinical screening of intrathecal drugs (129–131), and cardiovascular research (132). However, few studies exist in which sheep have been used to model ISCI. Consequently, a detailed description of the vascular anatomy of sheep spinal cord blood is not available. Some of the studies that were published are outlined in Table 6. Briefly, Kaplan et al. studied the effects of aortic occlusion on regional spinal cord blood flow and SEPs in a sheep model of ISCI (23). Sheep have also been used to evaluate endovascular therapy, a less invasive alternative to open surgery, for aortic aneurysm disease (122–125). Beygui et al. applied self expanding stent graft in an ovine model of abdominal aortic aneurysm (124). They reported paraplegia in 3 of their first 4 animals subjected to infra-renal aortic and lumbar artery occlusion, which prompted the application of the intra-aortic shunt to prevent paraplegia. Aneurysms were successfully created without paralysis in 35 animals; however, 2 animals developed paraplegia after endo-grafting due to technical failures (124). Lehr et al. also used intra-aortic shunting to prevent paraplegia in another sheep model of ISCI, in which animals underwent double AXC (133). Use of a shunt prevented hind leg paralysis in all 24 animals that survived the procedure.

3.6.1. Endovascular Versus Open Repair in Sheep Model of ISCI

Bockler et al. compared the incidence of ISCI after open or endovascular repair of the descending thoracic aorta in sheep (134). Animals were divided into three groups. In group I, the aorta was exposed through left thoracotomy, and double AXC was applied for 50 min proximal to the origin of the innominate artery (in sheep the left common carotid and subclavian arteries originate from the innominate artery), proximal to the diaphragm. In group II, animals were placed into the supine position and an endo-graft was introduced via the iliac artery under fluoroscopic control. In group III, ischemia was achieved by sequential clipping of the intercostal arteries of the thoracic aorta. Incidence of paraplegia was 100% in group I and 0% in group II. Group III animals showed a 20% infarction rate and no permanent neurological deficits (134).

3.7. Primate ISCI Models

Of all pre-clinical models, the primate spinal cord vascular anatomy is the most similar to that of humans (12). However, relative to all other species that have been used to study ISCI, primates are expensive, and more difficult to obtain and handle. For these reasons, there are relatively few publications using primate ISCI models (Table 7). Still, primates could be critical for testing new cell-based and pharmacological treatments as a prelude to human clinical trials.

Table 6
Sheep models of ischemic spinal cord injury

Study	Anatomical location of induced ischemia	Technique of induced ischemia
Kaplan et al. 1987 (23)	Thoracic aorta 1 cm distal to the origin of the LSA	• *Approach*: The thoracic aorta was exposed through left thoracotomy and then the aortic occlusion was done by cross clamping. • *Ischemia time*: 30 min. • *Clinical outcome*: Aortic occlusion was associated with disappearance of the somatosensory evoked potentials in seven of nine sheep in 7.8 ± 4.1 (SD) min. Spinal cord blood flow at the time of initial cross clamping and 30 min after the onset of ischemia revealed a severe reduction in white and gray matter flow in the thoraco-lumbar cord. Release of the aortic clamp was associated with reactive hyperemia in these ischemic regions.
Lehr 2007 (133)	Proximal and distal thoracic aorta	• *Approach*: The thoracic aorta was exposed through a thoracotomy in the left sixth intercostal space. They occluded the aorta temporarily by two vascular clamps until an intra-aortic shunt was applied. They placed two umbilical tapes around the shunt inside the aorta and removed the clamps. • *Clinical outcome*: A simple intra-aortic shunt was effective in restoring blood pressure and flow in the aorta distal to the operative site, and prevented hind leg paralysis associated with aortic clamping.

Svensson et al. were among the first to study vascular anatomy of the spinal cord in primates (12). They reported that the spinal cord anatomy of baboons is virtually identical to that of Man's. The ASA is continuous in both baboon and Man; however, this is not true in rabbits, dogs, or pigs. Additionally, the origin of ARM is T9–L1 in most baboons (none originated above T9). This is similar to the origin of ARM in humans (13). Radicular arteries were most frequently found in five different locations: mid cervical (C4), lower cervical (C7), upper thoracic (T4), mid thoracic (T7 to T10) and lower thoracic (T12) (12).

3.7.1. ASA Occlusion in Primate Model of ISCI

Sixty years ago, Yoss (135) used rhesus monkeys to induce ischemic spinal cord injury. In that study, the anterior spinal arteries were ligated in two monkeys at either T6 or L2, while the ARM was occluded in the third monkey at the right L2 nerve root. All monkeys became paraplegic after surgery, with palpable softening

Table 7
Primate models of ischemic spinal cord injury

Study	Anatomical location of induced ischemia	Technique of induced ischemia
Yoss 1950 (135)	The anterior spinal arteries and the arteria radicularis magna (ARM)	• *Approach*: The anterior spinal arteries were ligated in two monkeys at T6 and L2 respectively, while the ARM was occluded in the third monkey at the right L2 nerve root. • *Clinical outcome*: The three monkeys were totally paraplegic after surgery. Yoss observed softening of almost the entire transverse diameter of the spinal cord caudal to the site of arterial ligation.
Svensson et al. 1986 (26)	The thoracic aorta just distal to the left subclavian artery (LSA) and just above the diaphragm	• *Approach*: The thoracic aorta was exposed through two left lateral thoracotomies in the fourth and eighth intercostal spaces in five animals. They used an aortic-aortic shunt in addition to double aortic cross clamping (AXC) in the other seven animals. • *Ischemia time*: 60 min. • *Clinical outcome*: Paraplegia occurred in 4 of 5 baboons that underwent double AXC for 60 min; however, only 1 of 7 baboons subjected to the same ischemic technique, but with application of aortic-aortic shunt, had paraplegia.
Rokkas et al. 1993 (27)	The thoracic aorta just distal to the LSA and proximal to the renal arteries	• *Approach*: Thoracic aorta was exposed through thoracotomy and systemic hypothermia was induced by cardiopulmonary bypass in a group of animals. • *Ischemia time*: 60 min. • *Clinical outcome*: Of the 8 animals subjected to AXC under normothermic circumstances, 6 animals were paraplegic and 2 were paraparetic, whereas all 6 animals with hypothermia were neurologically intact ($p = 0.0002$).

of spinal cord caudal to the site of arterial ligation (135). Fried et al. reported that neurological deficits did not occur in monkeys if only the ARM was ligated. Ligation of the ASA above the entering ARM also failed to produce serious motor deficits in $n=9$ monkeys; $n=5/9$ had no motor deficits while the remaining monkeys developed mild hindlimb weakness. Conversely, $n=10/11$ monkeys in which the ASA was ligated below the entering ARM became paraplegic (136).

Histological examination of spinal cords from monkeys subject to this form of ISCI revealed small focal areas of neuronal tissue damage in only $n=2/6$ spinal cords. In contrast, pronounced neuropathology occurred in $n=8/11$ spinal cords in which the anterior spinal artery was occluded below the entrance of the ARM (137).

3.7.2. Thoracic Aorta Occlusion in a Baboon Model of SCI

After characterizing the vascular anatomy of spinal cord in baboons, Svensson et al. used this species to investigate ISCI (12, 26, 138). They also studied the changes of spinal cord blood flow during AXC in baboons, and observed that by using an intra-aortic shunt, the spinal cord blood flow to the lumbar cord increased, but blood flow to the thoracic cord remained unchanged (26). They also used the baboon model in their subsequent study to evaluate the neuroprotective effects of multiple adjuncts against spinal cord ischemic injury (138).

Rokkas et al. also used a baboon model of SCI to test the efficacy of systemic hypothermia as a neuroprotective intervention against SCI injury (27). Double AXC was performed distal to the LSA and proximal to the renal arteries for 60 min, to induced ISCI. Profound hypothermia (15 °C rectal temperature) was established through hypothermic cardiopulmonary bypass. Of the 8 animals subjected to AXC under normothermic circumstances, 6 animals became paraplegic and 2 became paraparetic, whereas all 6 animals with hypothermia stayed neurologically intact ($p=0.0002$) (Table 7) (27).

4. Discussion

ISCI injury that complicates thoracic and TAAA surgery is a major cause of morbidity even after a successful surgery. Development of reliable and clinically-relevant animal models is necessary to better understand the mechanisms underlying ISCI neuropathology. These models will likely lead to the development of more effective treatments. Although few will question the importance of preclinical modeling and the utility of these systems for mechanistic studies, there is no consensus as to which model best mimics the pathophysiological sequelae of ISCI in humans (Fig. 1).

Various species of animals have been used as experimental models of ISCI, including mice, rats, rabbits, dogs, pigs, sheep and primates. Each of these animal models have unique differences that must be considered to appropriately interpret the expected outcomes, either in the context of the species' unique characteristics, or by the method in which ischemia is induced. Between species, the spinal cord vasculature varies somewhat making location of the AXC critical; otherwise, the reproducibility of the ischemic insult will vary (Fig. 2). Also, ischemia duration and the surgical approach used (whether open surgery or endovascular approach), can impact the outcomes measured in various ways, some of which are obvious while others remain obscure. Considering these variables concurrently may help to define the clinical relevance of an individual model, especially if the goal is to mimic what occurs in patients undergoing thoracic and TAAA surgery (Fig. 2).

As an example, the open surgery approach (e.g., via thoracotomy) is a surgical trauma that stimulates a systemic immune response. This response occurs before AXC and subsequent spinal cord ischemia. Whether surgery and systemic inflammation affect post-ischemic pathology in the spinal cord is unknown; however, these variables cannot be easily dismissed and animal models that incorporate a thoracotomy may more accurately simulate clinical aortic aneurysm surgery than models that use a laparotomy or endovascular approach.

The location of the AXC is also important to consider (Fig. 2). AXC distal to the LSA mimics the surgical technique used in TAAA repair. On the other hand, abdominal AXC, similar to what has been used extensively in rabbit models, does not produce significant spinal cord dysfunction in human patients.

The duration of ischemia, and its correlation to intraspinal pathology and loss of function, seems to be related to animal size. For example, in mice, aortic occlusion for 9–11 min causes paralysis, while 45–60 min of AXC is needed to produce paralysis in dogs or pigs. In small animal models (e.g., mouse, rat), long periods of AXC are associated with short post-surgical survival intervals and high mortality rates.

There are practical and logistical issues that must be considered when performing pre-clinical studies. Because body sizes and vascular diameters are comparable to humans, large animal models (e.g., pig, dog, sheep, primates) are ideal for testing new surgical techniques and other potential clinical adjunct therapies that cannot be easily tested in smaller species. The downside of using large animals is cost, availability, institutional regulatory concerns and comparatively smaller group sizes. In contrast, small animal models (e.g., mouse, rat, rabbit) are easier to handle, less expensive, readily available, and require relatively simple surgical procedures to induce ischemia. The mouse model of ISCI also offers an additional advantage. With the mouse genome defined, transgenic/knockout mice can be used to understand specific molecular and genetic mechanisms that cause or contribute to ischemic neuronal injury.

5. Conclusions and Future Directions

With the marked advances in surgical technique and neuroprotective adjuncts that have been clinically introduced to aortic aneurysm surgery protocol for spinal cord protection, the incidence rates of paraplegia/paraparesis have decreased to 3% in descending thoracic aneurysm, and 5–10% in thoracocabdominal aneurysm repair surgery (117). More research relevant to spinal cord ischemic injury is still needed to develop more effective interventions that can eliminate this devastating complication. Experimental studies with diverse animal models of ISCI can be used for developing and evaluating different treatments before clinical application.

Small animal models are suitable for basic neuroscience research, which focuses on exploring the neuronal ischemic injury mechanisms at cellular and molecular levels. We believe that interruption of the ischemic injury cascade at the molecular level could provide optimum spinal cord protection during ischemic events. Mice provide the best model for this research track as their genome is already known.

In the meanwhile, clinical relevant research is required as a second parallel research track. Large animal models are suitable for experimental studies that mimic the clinical aspects of ISCI during AXC. Multiple adjuncts have been evaluated in large animal models before clinical application, including hypothermia, CSF drainage, neuroprotective drugs, distal aortic perfusion and motor and sensory evoked potentials (27, 110, 114, 133, 138–140).

Large and small animal model research should be integrated to prevent spinal cord damage that complicates unavoidable ischemic insults, hopefully eliminating this serious complication.

Acknowledgements

We wish to acknowledge Dennis Mathias, publications editor with The Ohio State University Department of Surgery, for creating the figures, and Keri Hudec, technical editor with the Department of Anesthesiology.

References

1. Svensson LG, Crawford ES, Hess KR, Coselli JS, Safi HJ (1993) Experience with 1509 patients undergoing thoracoabdominal aortic operations. J Vasc Surg 17(2):357–368
2. Lang-Lazdunski L, Matsushita K, Hirt L, Waeber C, Vonsattel JP, Moskowitz MA et al (2000) Spinal cord ischemia. Development of a model in the mouse. Stroke 31(1):208–213
3. Tveten L (1976) Spinal cord vascularity. IV. The spinal cord arteries in the rat. Acta Radiol Diagn (Stockh) 17(4):385–398
4. Tveten L (1976) Spinal cord vascularity. II. Extraspinal sources of spinal cord arteries in the rat. Acta Radiol Diagn (Stockh) 17(2):167–179
5. Schievink WI, Luyendijk W, Los JA (1988) Does the artery of Adamkiewicz exist in the albino rat? J Anat 161:95–101
6. Mazensky D, Radonak J, Danko J, Petrovova E, Frankovicova M (2011) Anatomical study of blood supply to the spinal cord in the rabbit. Spinal Cord 49(4):525–528
7. Fletcher TF, Kitchell RL (1966) Anatomical studies on the spinal cord segments of the dog. Am J Vet Res 27(121):1759–1767
8. Sindou M, Chignier E, Mazoyer JF, Pialat J, Fischer G, Descotes J (1979) Revascularization of the spinal cord by micro-anastomoses in dogs. Surg Neurol 12(6):492–495
9. Kato S, Kawahara N, Tomita K, Murakami H, Demura S, Fujimaki Y (2008) Effects on spinal cord blood flow and neurologic function secondary to interruption of bilateral segmental arteries which supply the artery of Adamkiewicz: an experimental study using a dog model. Spine (Phila Pa 1976) 33(14):1533–1541
10. Fujimaki Y, Kawahara N, Tomita K, Murakami H, Ueda Y (2006) How many ligations of bilateral segmental arteries cause ischemic spinal cord dysfunction? An experimental study using a dog model. Spine (Phila Pa 1976) 31(21):E781–E789
11. Strauch JT, Lauten A, Zhang N, Wahlers T, Griepp RB (2007) Anatomy of spinal cord blood supply in the pig. Ann Thorac Surg 83(6):2130–2134
12. Svensson LG, Klepp P, Hinder RA (1986) Spinal cord anatomy of the baboon–comparison with man and implications for spinal cord blood flow during thoracic aortic cross-clamping. S Afr J Surg 24(1):32–34
13. Di Chiro G, Doppman JL (1969) Paraplegia after resection of aneurysm. N Engl J Med 281(14):799
14. Brockstein B, Johns L, Gewertz BL (1994) Blood supply to the spinal cord: anatomic and physiologic correlations. Ann Vasc Surg 8(4):394–399
15. Nijenhuis RJ, Jacobs MJ, Jaspers K, Reinders M, van Engelshoven JM, Leiner T et al (2007) Comparison of magnetic resonance with computed tomography angiography for preoperative localization of the Adamkiewicz artery in thoracoabdominal aortic aneurysm patients. J Vasc Surg 45(4):677–685
16. Gaviria M, Bonny JM, Haton H, Jean B, Teigell M, Renou JP et al (2006) Time course of acute phase in mouse spinal cord injury monitored by ex vivo quantitative MRI. Neurobiol Dis 22(3):694–701
17. Wang Z, Yang W, Britz GW, Lombard FW, Warner DS, Sheng H (2010) Development of a simplified spinal cord ischemia model in mice. J Neurosci Methods 189(2):246–251
18. Awad H, Ankeny DP, Guan Z, Wei P, McTigue DM, Popovich PG (2010) A mouse model of ischemic spinal cord injury with delayed paralysis caused by aortic cross-clamping. Anesthesiology 113(4):880–891
19. Matsushita A, Smith CM (1970) Spinal cord function in postischemic rigidity in the rat. Brain Res 19(3):395–410
20. LeMay DR, Neal S, Neal S, Zelenock GB, D'Alecy LG (1987) Paraplegia in the rat induced by aortic cross-clamping: model characterization and glucose exacerbation of neurologic deficit. J Vasc Surg 6(4):383–390
21. Laschinger JC, Cunningham JN Jr, Cooper MM, Krieger K, Nathan IM, Spencer FC (1984) Prevention of ischemic spinal cord injury following aortic cross-clamping: use of corticosteroids. Ann Thorac Surg 38(5):500–507
22. Sugawara Y, Kumagai H, Sueda T (2005) A novel canine model of spinal cord ischemia with reproducible neurologic outcomes. Surg Today 35(8):649–652
23. Kaplan BJ, Friedman WA, Gravenstein N, Richards R, Davis RF (1987) Effects of aortic occlusion on regional spinal cord blood flow and somatosensory evoked potentials in sheep. Neurosurgery 21(5):668–675
24. Lopez S, Manahan E, Evans JR, Kao RL, Browder W (1995) Third place winner of the Conrad Jobst Award in the gold medal paper competition. Prevention of spinal cord dysfunction in a new model of spinal cord ischemia. Am Surg 61(1):16–20
25. Coles JG, Wilson GJ, Sima AF, Klement P, Tait GA (1982) Intraoperative detection of

spinal cord ischemia using somatosensory cortical evoked potentials during thoracic aortic occlusion. Ann Thorac Surg 34(3):299–306

26. Svensson LG, Rickards E, Coull A, Rogers G, Fimmel CJ, Hinder RA (1986) Relationship of spinal cord blood flow to vascular anatomy during thoracic aortic cross-clamping and shunting. J Thorac Cardiovasc Surg 91(1):71–78
27. Rokkas CK, Sundaresan S, Shuman TA, Palazzo RS, Nitta T, Despotis GJ, Burns TC, Wareing TH, Kouchoukos NT (1993) Profound systemic hypothermia protects the spinal cord in a primate model of spinal cord ischemia. J Thorac Cardiovasc Surg 106(6):1024–1035
28. Usul H, Cakir E, Cobanoglu U, Alver A, Peksoylu B, Topbas M et al (2004) The effects of tyrphostine Ag 556 on experimental spinal cord ischemia reperfusion injury. Surg Neurol 61(1):45–54
29. Zivin JA, De Girolami U (1980) Spinal cord infarction: a highly reproducible stroke model. Stroke 11(2):200–202
30. Nylander WA Jr, Plunkett RJ, Hammon JW Jr, Oldfield EH, Meacham WF (1982) Thiopental modification of ischemic spinal cord injury in the dog. Ann Thorac Surg 33(1):64–68
31. Guler A, Sahin MA, Ucak A, Onan B, Inan K, Oztas E et al (2010) Protective effects of angiotensin II type-1 receptor blockade with olmesartan on spinal cord ischemia-reperfusion injury: an experimental study on rats. Ann Vasc Surg 24(6):801–808
32. Maeda T, Mori K, Shiraishi Y, Tatebayashi K, Kawai Y (2003) Selective occlusion of lumbar arteries as a spinal cord ischemia model in rabbits. Jpn J Physiol 53(1):9–15
33. Kai Y, Owen JH, Allen BT, Dobras M, Davis C (1994) Relationship between evoked potentials and clinical status in spinal cord ischemia. Spine (Phila Pa 1976) 19(10):1162–1167
34. de Haan P, Kalkman CJ, Meylaerts SA, Lips J, Jacobs MJ (1999) Development of spinal cord ischemia after clamping of noncritical segmental arteries in the pig. Ann Thorac Surg 68(4):1278–1284
35. Coston A, Laville M, Baud P, Bussel B, Jalfre M (1983) Aortic occlusion by a balloon catheter: a method to induce hind limb rigidity in rats. Physiol Behav 30(6):967–969
36. Marsala M, Yaksh TL (1994) Transient spinal ischemia in the rat: characterization of behavioral and histopathological consequences as a function of the duration of aortic occlusion. J Cereb Blood Flow Metab 14(3):526–535
37. Cheng MK, Robertson C, Grossman RG, Foltz R, Williams V (1984) Neurological outcome correlated with spinal evoked potentials in a spinal cord ischemia model. J Neurosurg 60(4):786–795
38. Kaplan BJ, Gravenstein N, Friedman WA, Blackmore J, Curran M (1987) Thoracic aortic occlusion: somatosensory evoked potential monitoring and neurologic outcome in a canine model. Surg Neurol 28(5):351–360
39. Pashin SS, Viktorov IV (2009) Morphofunctional changes in the rat spinal cord after focal photothrombosis. Neurosci Behav Physiol 39(2):203–206
40. DeGirolami U, Zivin JA (1982) Neuropathology of experimental spinal cord ischemia in the rabbit. J Neuropathol Exp Neurol 41(2):129–149
41. Ball TD, Lundy EF, Zelenock GB, D'Alecy LG (1987) Effects of lodoxamide tromethamine on paraplegia that occurs after infrarenal aortic occlusion in the rabbit. J Vasc Surg 6(6):572–577
42. Oldfield EH, Plunkett RJ, Nylander WA Jr, Meacham WF (1982) Barbiturate protection in acute experimental spinal cord ischemia. J Neurosurg 56(4):511–516
43. LeMay DR, Lu AC, Zelenock GB, D'Alecy LG (1988) Insulin administration protects from paraplegia in the rat aortic occlusion model. J Surg Res 44(4):352–358
44. Kanellopoulos GK, Kato H, Hsu CY, Kouchoukos NT (1997) Spinal cord ischemic injury. Development of a new model in the rat. Stroke 28(12):2532–2538
45. Carrillo SE, Guimaraes SB, Vasconcelos PR, Vasconcelos PR (2006) Is subdiaphragmatic aortic cross-clamping a suitable model for spinal cord ischemia/reperfusion injury study in rats? Acta Cir Bras 21(4):219–222
46. Nouri M, Rasouli M, Shafiei S, Tavasoly A, Dehpour AR, Rahimi-Movaghar V (2006) Does abdominal aorta clamping, as a method of spinal ischemia in rats, really work? Surg Neurol 66(3):332–333
47. Vanicky I, Marsala M, Galik J, Marsala J (1993) Epidural perfusion cooling protection against protracted spinal cord ischemia in rabbits. J Neurosurg 79(5):736–741
48. Malatova Z, Vanicky I, Galik J, Marsala M (1995) Epidural perfusion cooling protects against spinal cord ischemia in rabbits. An evaluation of cholinergic function. Mol Chem Neuropathol 25(2–3):81–96
49. de Haan P, Vanicky I, Jacobs MJ, Bakker O, Lips J, Meylaerts SA et al (2000) Effect of ischemic pretreatment on heat shock protein

72, neurologic outcome, and histopathologic outcome in a rabbit model of spinal cord ischemia. J Thorac Cardiovasc Surg 120(3): 513–519
50. Kiyoshima T, Fukuda S, Matsumoto M, Iida Y, Oka S, Nakakimura K et al (2003) Lack of evidence for apoptosis as a cause of delayed onset paraplegia after spinal cord ischemia in rabbits. Anesth Analg 96(3):839–846
51. Hamaishi M, Orihashi K, Isaka M, Kumagai H, Takahashi S, Okada K et al (2009) Low-dose edaravone injection into the clamped aorta prevents ischemic spinal cord injury. Ann Vasc Surg 23(1):128–135
52. Chen WF, Sung CS, Jean YH, Su TM, Wang HC, Ho JT et al (2010) Suppressive effects of intrathecal granulocyte colony-stimulating factor on excessive release of excitatory amino acids in the spinal cerebrospinal fluid of rats with cord ischemia: role of glutamate transporters. Neuroscience 165(4):1217–1232
53. Robertson CS, Grossman RG (1987) Protection against spinal cord ischemia with insulin-induced hypoglycemia. J Neurosurg 67(5):739–744
54. Cuevas P, Carceller-Benito F, Reimers D (1989) Administration of bovine superoxide dismutase prevents sequelae of spinal cord ischemia in the rabbit. Anat Embryol (Berl) 179(3):251–255
55. Lindsberg PJ, Siren AL, Feuerstein GZ, Hallenbeck JM (1995) Antagonism of neutrophil adherence in the deteriorating stroke model in rabbits. J Neurosurg 82(2):269–277
56. Sakurai M, Hayashi T, Abe K, Sadahiro M, Tabayashi K (1998) Delayed selective motor neuron death and fas antigen induction after spinal cord ischemia in rabbits. Brain Res 797(1):23–28
57. Sakurai M, Hayashi T, Abe K, Sadahiro M, Tabayashi K (1998) Delayed and selective motor neuron death after transient spinal cord ischemia: a role of apoptosis? J Thorac Cardiovasc Surg 115(6):1310–1315
58. Lips J, de Haan P, Bodewits P, Vanicky I, Dzoljic M, Jacobs MJ et al (2000) Neuroprotective effects of riluzole and ketamine during transient spinal cord ischemia in the rabbit. Anesthesiology 93(5):1303–1311
59. Yamauchi T, Sakurai M, Abe K, Matsumiya G, Sawa Y (2007) Impact of the endoplasmic reticulum stress response in spinal cord after transient ischemia. Brain Res 1169:24–33
60. Rosenthal D (1999) Spinal cord ischemia after abdominal aortic operation: is it preventable? J Vasc Surg 30(3):391–397
61. Mallick IH, Kumar S, Samy A (2003) Paraplegia after elective repair of an infrarenal aortic aneurysm. J R Soc Med 96(10): 501–503
62. Gelfan S, Tarlov IM (1959) Interneurones and rigidity of spinal origin. J Physiol 146(3):594–617
63. Coles JC, Ahmed SN, Mehta HU, Kaufmann JC (1986) Role of free radical scavenger in protection of spinal cord during ischemia. Ann Thorac Surg 41(5):551–556
64. Giuseffi J, Greene LW Jr, Vetto M (1957) Experiments on occlusion of the abdominal aorta in normothermic dogs. Surg Gynecol Obstet 105(4):427–434
65. Burch BH, Traphagen DW, Folkman MJ, Rosenbaum DA, Mueller EC (1954) Temporary aortic occlusion in abdominal surgery. Surgery 35(5):684–689
66. Coles JG, Wilson GJ, Sima AF, Klement P, Tait GA, Williams WG et al (1983) Intraoperative management of thoracic aortic aneurysm. Experimental evaluation of perfusion cooling of the spinal cord. J Thorac Cardiovasc Surg 85(2):292–299
67. Berguer R, Porto J, Fedoronko B, Dragovic L (1992) Selective deep hypothermia of the spinal cord prevents paraplegia after aortic cross-clamping in the dog model. J Vasc Surg 15(1):62–71
68. Rose WW 3rd, Reddy DJ, Ernst CB, Reickert CA, Patterson JS (1997) Protective effect of hypothermia and left heart bypass on spinal ischemia in the dog. Arch Surg 132(6): 633–639
69. Elmore JR, Gloviczki P, Harper CM Jr, Murray MJ, Wu QH, Bower TC et al (1992) Spinal cord injury in experimental thoracic aortic occlusion: investigation of combined methods of protection. J Vasc Surg 15(5): 789–798
70. Marsala M, Vanicky I, Galik J, Radonak J, Kundrat I, Marsala J et al (1993) Panmyelic epidural cooling protects against ischemic spinal cord damage. J Surg Res 55(1):21–31
71. Wisselink W, Becker MO, Nguyen JH, Money SR, Hollier LH (1994) Protecting the ischemic spinal cord during aortic clamping: the influence of selective hypothermia and spinal cord perfusion pressure. J Vasc Surg 19(5):788–795
72. Ohtake H, Urayama H, Katada S, Tamura M, Kawasuji M, Watanabe Y (1998) Prevention of spinal cord ischemia by selective intercostal arterial infusion of prostaglandin E1. J Vasc Surg 28(2):301–307

73. Ishizaki M, Sugiyama S, Uchida H, Nawa S, Shimizu N (1999) Identification and selective perfusion of the spinal cord-feeding arteries by intrathecal pO2 monitoring for spinal cord protection. Eur J Vasc Endovasc Surg 18(1):17–24
74. Nagy G, Dzsinich C, Selmeci L, Sepa G, Dzsinich M, Kekesi V et al (2002) Biochemical alterations in cerebrospinal fluid during thoracoabdominal aortic cross-clamping in dogs. Ann Vasc Surg 16(4):436–441
75. Contreras IS, Moreira LF, Ballester G, de Monaco BA, Lancellotti CL, Dias AR et al (2005) Immediate ischemic preconditioning based on somatosensory evoked potentials seems to prevent spinal cord injury following descending thoracic aorta cross-clamping. Eur J Cardiothorac Surg 28(2):274–279
76. Lim KH, Connolly M, Rose D, Siegman F, Jacobowitz I, Acinapura A et al (1986) Prevention of reperfusion injury of the ischemic spinal cord: use of recombinant superoxide dismutase. Ann Thorac Surg 42(3): 282–286
77. Laschinger JC, Cunningham JN Jr, Cooper MM, Baumann FG, Spencer FC (1987) Monitoring of somatosensory evoked potentials during surgical procedures on the thoracoabdominal aorta. I. Relationship of aortic cross-clamp duration, changes in somatosensory evoked potentials, and incidence of neurologic dysfunction. J Thorac Cardiovasc Surg 94(2):260–265
78. Simpson JI, Eide TR, Schiff GA, Clagnaz JF, Hossain I, Tverskoy A et al (1994) Intrathecal magnesium sulfate protects the spinal cord from ischemic injury during thoracic aortic cross-clamping. Anesthesiology 81(6): 1493–1499
79. DelRossi AJ, Cernaianu AC, Cilley JH Jr, Spence RK, Camishion RC, Yu Y et al (1990) Preventive effect of Fluosol-DA for paraplegia encountered after surgical treatment of the thoracic aorta. Preliminary results in a dog model. J Thorac Cardiovasc Surg 99(4):665–669
80. Simpson JI, Eide TR, Schiff GA, Newman SB, Clagnaz JF, Hossain I et al (1996) Effect of nitroglycerin on spinal cord ischemia after thoracic aortic cross-clamping. Ann Thorac Surg 61(1):113–117
81. Matsuyama K, Chiba Y, Ihaya A, Kimura T, Tanigawa N, Muraoka R (1997) Effect of spinal cord preconditioning on paraplegia during cross-clamping of the thoracic aorta. Ann Thorac Surg 63(5):1315–1320
82. Katircioglu SF, Ulus AT, Gokce P, Surucu S (2000) Iloprost protects the spinal cord during aortic cross-clamping in a canine model. J Cardiovasc Surg (Torino) 41(1):89–93
83. Winnerkvist A, Bartoli S, Iliopoulos DC, Hess KR, Miller CC, Safi HJ (2002) Spinal cord protection during aortic cross clamping: retrograde venous spinal cord perfusion, distal aortic perfusion, and cerebrospinal fluid drainage. Scand Cardiovasc J 36(1):6–10
84. Benicio A, Moreira LF, Monaco BA, Castelli JB, Mingrone LE, Stolf NA (2007) Comparative study between ischemic preconditioning and cerebrospinal fluid drainage as methods of spinal cord protection in dogs. Rev Bras Cir Cardiovasc 22(1):15–23
85. Grossi EA, Laschinger JC, Krieger KH, Nathan IM, Colvin SB, Weiss MR et al (1988) Epidural-evoked potentials: a more specific indicator of spinal cord ischemia. J Surg Res 44(3):224–228
86. Sugiyama S, Ishizaki M, Uchida H (1993) Spinal epidural oxygen partial pressure and evoked spinal cord potential in relation to the severity of spinal ischemia during cross-clamping of the thoracic aorta. Acta Med Okayama 47(6):369–376
87. Shokoku S, Uchida H, Teramoto S (1993) An experimental study on spinal cord ischemia during cross-clamping of the thoracic aorta: the monitoring of spinal cord ischemia with motor evoked potential by transcranial stimulation of the cerebral cortex in dogs. Surg Today 23(12):1068–1073
88. Kumagai H, Isaka M, Sugawara Y, Okada K, Imai K, Orihashi K et al (2006) Intra-aortic injection of propofol prevents spinal cord injury during aortic surgery. Eur J Cardiothorac Surg 29(5):714–719
89. Zhang JS, Huan Y, Sun LJ, Ge YL, Zhang XX, Chang YJ (2007) Temporal evolution of spinal cord infarction in an in vivo experimental study of canine models characterized by diffusion-weighted imaging. J Magn Reson Imaging 26(4):848–854
90. Awad H, Suntres Z, Heijmans J, Smeak D, Bergdall-Costell V, Christofi FL et al (2008) Intracellular and extracellular expression of the major inducible 70 kDa heat shock protein in experimental ischemia-reperfusion injury of the spinal cord. Exp Neurol 212(2):275–284
91. Ishizaki M, Sugiyama S, Uchida H, Nawa S, Shimizu N (1997) Intrathecal oxygen concentration as a new indicator of spinal cord ischemia. Acta Med Okayama 51(2):71–77
92. Qayumi AK, Janusz MT, Lyster DM, Gillespie KD (1997) Animal model for investigation of spinal cord injury caused by aortic cross-clamping. J Invest Surg 10(1–2):47–52

93. Maharajh GS, Pascoe EA, Halliday WC, Grocott HP, Thiessen DB, Girling LG et al (1996) Neurological outcome in a porcine model of descending thoracic aortic surgery. Left atrial-femoral artery bypass versus clamp/repair. Stroke 27(11):2095–2100
94. Wadouh F, Lindemann EM, Arndt CF, Hetzer R, Borst HG (1984) The arteria radicularis magna anterior as a decisive factor influencing spinal cord damage during aortic occlusion. J Thorac Cardiovasc Surg 88(1):1–10
95. Wadouh F, Arndt CF, Oppermann E, Borst HG, Wadouh R (1986) The mechanism of spinal cord injury after simple and double aortic cross-clamping. J Thorac Cardiovasc Surg 92(1):121–127
96. Hellberg A, Christiansson L, Tulga Ulus A, Bergqvist D, Wiklund L, Karacagil S et al (2000) A prolonged spinal cord ischaemia model in pigs. Passive shunting offers stable central haemodynamics during aortic occlusion. Eur J Vasc Endovasc Surg 19(3):318–323
97. Wadouh F, Hartmann M, Arndt CF, Metzger H, Borst HG (1985) Lumbar and thoracic spinal cord oxygen tension during aortic cross-clamping. Adv Exp Med Biol 191:173–180
98. Wadouh F, Arndt CF, Metzger H, Hartmann M, Wadouh R, Borst HG (1985) Direct measurements of oxygen tension on the spinal cord surface of pigs after occlusion of the descending aorta. J Thorac Cardiovasc Surg 89(5):787–794
99. Wadouh F, Wadouh R, Hartmann M, Crisp-Lindgren N (1990) Prevention of paraplegia during aortic operations. Ann Thorac Surg 50(4):543–552
100. Wadouh F (1992) New surgical techniques for the prevention of paraplegia during aortic surgery. Thorac Cardiovasc Surg 40(1):26–32
101. Qayumi AK, Janusz MT, Jamieson WR, Lyster DM (1992) Pharmacologic interventions for prevention of spinal cord injury caused by aortic crossclamping. J Thorac Cardiovasc Surg 104(2):256–261
102. Follis F, Dragan R, Blisard KS, Hartshorne M, Temes T, Pett SB et al (1999) Retrograde perfusion of the spinal cord during aortic crossclamping: initial observations in the swine model. J Thorac Cardiovasc Surg 118(4):597–602, discussion 603
103. Ross SD, Kern JA, Gangemi JJ, St Laurent CR, Shockey KS, Kron IL et al (2000) Hypothermic retrograde venous perfusion with adenosine cools the spinal cord and reduces the risk of paraplegia after thoracic aortic clamping. J Thorac Cardiovasc Surg 119(3):588–595
104. Christiansson L, Hellberg A, Koga I, Thelin S, Bergqvist D, Wiklund L et al (2000) A new method of intrathecal PO2, PCO2, and pH measurements for continuous monitoring of spinal cord ischemia during thoracic aortic clamping in pigs. Surgery 127(5):571–576
105. Mori A, Ueda T, Hachiya T, Kabei N, Okano H, Yozu R et al (2005) An epidural cooling catheter protects the spinal cord against ischemic injury in pigs. Ann Thorac Surg 80(5):1829–1833
106. Pocar M, Rossi V, Addis A, Monaco A, Sichel S, Addis F et al (2007) Spinal cord retrograde perfusion: review of the literature and experimental observations. J Card Surg 22(2):124–128
107. Lee JS, Hong JM, Kim YJ (2008) Ischemic preconditioning to prevent lethal ischemic spinal cord injury in a swine model. J Invest Surg 21(4):209–214
108. Rokkas CK, Cronin CS, Nitta T, Helfrich LR Jr, Lobner DC, Choi DW et al (1995) Profound systemic hypothermia inhibits the release of neurotransmitter amino acids in spinal cord ischemia. J Thorac Cardiovasc Surg 110(1):27–35
109. Yoshitake A, Mori A, Shimizu H, Ueda T, Kabei N, Hachiya T et al (2007) Use of an epidural cooling catheter with a closed countercurrent lumen to protect against ischemic spinal cord injury in pigs. J Thorac Cardiovasc Surg 134(5):1220–1226
110. Ishikawa A, Mori A, Kabei N, Yoshitake A, Suzuki T, Katori N et al (2009) Epidural cooling minimizes spinal cord injury after aortic cross-clamping through induction of nitric oxide synthase. Anesthesiology 111(4):818–825
111. Hellberg A, Koga I, Christiansson L, Stiernstrom H, Wiklund L, Bergqvist D et al (2000) Influence of segmental spinal cord perfusion on intrathecal oxygen tension d uring experimental thoracic aortic crossclamping. J Vasc Surg 31(1 Pt 1):164–170
112. Hellberg A, Ulus AT, Christiansson L, Westman J, Leppanen O, Bergqvist D et al (2001) Monitoring of intrathecal oxygen tension during experimental aortic occlusion predicts ultrastructural changes in the spinal cord. J Thorac Cardiovasc Surg 121(2):316–323
113. Christiansson L, Ulus AT, Hellberg A, Bergqvist D, Wiklund L, Karacagil S (2001) Aspects of the spinal cord circulation as assessed by intrathecal oxygen tension monitoring during various arterial interruptions in

the pig. J Thorac Cardiovasc Surg 121(4): 762–772

114. Strauch JT, Spielvogel D, Lauten A, Zhang N, Shiang H, Weisz D et al (2003) Importance of extrasegmental vessels for spinal cord blood supply in a chronic porcine model. Eur J Cardiothorac Surg 24(5):817–824

115. Etz CD, Homann TM, Plestis KA, Zhang N, Luehr M, Weisz DJ et al (2007) Spinal cord perfusion after extensive segmental artery sacrifice: can paraplegia be prevented? Eur J Cardiothorac Surg 31(4):643–648

116. Etz CD, Homann TM, Luehr M, Kari FA, Weisz DJ, Kleinman G et al (2008) Spinal cord blood flow and ischemic injury after experimental sacrifice of thoracic and abdominal segmental arteries. Eur J Cardiothorac Surg 33(6):1030–1038

117. Griepp RB, Griepp EB (2007) Spinal cord perfusion and protection during descending thoracic and thoracoabdominal aortic surgery: the collateral network concept. Ann Thorac Surg 83(2):S865–S869

118. Toumpoulis IK, Anagnostopoulos CE, Drossos GE, Malamou-Mitsi VD, Pappa LS, Katritsis DG (2003) Does ischemic preconditioning reduce spinal cord injury because of descending thoracic aortic occlusion? J Vasc Surg 37(2):426–432

119. Toumpoulis IK, Anagnostopoulos CE, Drossos GE, Malamou-Mitsi VD, Pappa LS, Katritsis DG (2003) Early ischemic preconditioning without hypotension prevents spinal cord injury caused by descending thoracic aortic occlusion. J Thorac Cardiovasc Surg 125(5):1030–1036

120. Toumpoulis IK, Papakostas JC, Matsagas MI, Malamou-Mitsi VD, Pappa LS, Drossos GE et al (2004) Superiority of early relative to late ischemic preconditioning in spinal cord protection after descending thoracic aortic occlusion. J Thorac Cardiovasc Surg 128(5):724–730

121. Papakostas JC, Matsagas MI, Toumpoulis IK, Malamou-Mitsi VD, Pappa LS, Gkrepi C et al (2006) Evolution of spinal cord injury in a porcine model of prolonged aortic occlusion. J Surg Res 133(2):159–166

122. Orend KH, Heymer B, Pamler RS, Hahn J, Sunder-Plassmann L (1996) Histomorphological examination of endoluminal stent grafting in the descending aorta in a sheep model. Thorac Cardiovasc Surg 44(3):132–135

123. Boudghene FP, Sapoval MR, Bonneau M, LeBlanche AF, Lavaste FC, Michel JB (1998) Abdominal aortic aneurysms in sheep: prevention of rupture with endoluminal stent-grafts. Radiology 206(2):447–454

124. Beygui RE, Kinney EV, Pelc LR, Krievins D, Whittemore J, Fogarty TJ et al (1999) Prevention of spinal cord ischemia in an ovine model of abdominal aortic aneurysm treated with a self-expanding stent-graft. J Endovasc Surg 6(3):278–284

125. Yamada K, Pavcnik D, Uchida BT, Timmermans HA, Corless CL, Yin Q et al (2001) Endoluminal treatment of ruptured abdominal aortic aneurysm with small intestinal submucosa sandwich endografts: a pilot study in sheep. Cardiovasc Intervent Radiol 24(2):99–105

126. Detweiler DJ, Eisenach JC, Tong C, Jackson C (1993) A cholinergic interaction in alpha 2 adrenoceptor-mediated antinociception in sheep. J Pharmacol Exp Ther 265(2):536–542

127. Yeo JD, Payne W, Hinwood B, Kidman AD (1975) The experimental contusion injury of the spinal cord in sheep. Paraplegia 12(4):279–298

128. Cain CC, Fraser RD (1995) Bony and vascular anatomy of the normal cervical spine in the sheep. Spine (Phila Pa 1976) 20(7):759–765

129. Cousins MJ, Mather LE (1984) Intrathecal and epidural administration of opioids. Anesthesiology 61(3):276–310

130. Payne R, Inturrisi CE (1985) CSF distribution of morphine, methadone and sucrose after intrathecal injection. Life Sci 37(12):1137–1144

131. Kyles AE, Waterman AE, Livingston A (1993) The spinal antinociceptive activity of the alpha 2-adrenoceptor agonist, xylazine in sheep. Br J Pharmacol 108(4):907–913

132. Herijgers P, Ozaki S, Verbeken E, Van Lommel A, Racz R, Zietkiewicz M et al (1999) Calcification characteristics of porcine stentless valves in juvenile sheep. Eur J Cardiothorac Surg 15(2):134–142

133. Lehr EJ, Coe JY, Ross DB (2007) An intra-aortic shunt prevents paralysis during aortic surgery in sheep. J Surg Res 141(1):78–82

134. Bockler D, Kotelis D, Kohlhof P, von Tengg-Kobligk H, Mansmann U, Zink W et al (2007) Spinal cord ischemia after endovascular repair of the descending thoracic aorta in a sheep model. Eur J Vasc Endovasc Surg 34(4):461–469

135. Yoss RE (1950) Vascular supply of the spinal cord: the production of vascular syndromes. Med Bull (Ann Arbor) 16(11):333–345

136. Fried LC, Di Chiro G, Doppman JL (1969) Ligation of major thoraco-lumbar spinal cord

arteries in monkeys. J Neurosurg 31(6): 608–614

137. Fried LC, Aparicio O (1973) Experimental ischemia of the spinal cord. Histologic studies after anterior spinal artery occlusion. Neurology 23(3):289–293
138. Svensson LG, Von Ritter CM, Groeneveld HT, Rickards ES, Hunter SJ, Robinson MF, Hinder RA (1986) Cross-clamping of the thoracic aorta. Influence of aortic shunts, laminectomy, papaverine, calcium channel blocker, allopurinol, and superoxide dismutase on spinal cord blood flow and paraplegia in baboons. Ann Surg 204(1):38–47
139. McCullough JL, Hollier LH, Nugent M (1988) Paraplegia after thoracic aortic occlusion: influence of cerebrospinal fluid drainage. Experimental and early clinical results. J Vasc Surg 7(1):153–160
140. Monaco BA, Benicio A, Contreras IS, Mingrone LE, Ballester G, Moreira LF (2007) Ischemic preconditioning and spinal cord function monitoring in the descending thoracic aorta approach. Arq Bras Cardiol 88(3):291–296

Chapter 12

Magnetic Resonance Imaging of Experimental Spinal Cord Injury

Laura E. Gonzalez-Lara, Freda Jawan, and Paula J. Foster

Abstract

Magnetic resonance imaging (MRI) is a valuable in vivo imaging method for investigating experimental spinal cord injury (SCI). In this chapter a variety of MRI techniques will be described including anatomical, functional, and cellular MRI. The benefits and challenges of each technique are addressed and examples of the application of these imaging tools are presented for mouse and rat SCI.

Key words: MRI, Rodent, Spinal cord injury, Mice, Rat, Imaging

1. Introduction

In order to better understand the mechanisms of spinal cord injury (SCI) and to explore potential treatments, animal models of SCI have been developed. Experimental models of SCI are often performed in rodents using methods to mimic the initial mechanical injury that occurs during SCI in humans, such as contusion (1), compression (2), and transection (3). Rats have been used most widely (4) due in part to size and model development (the first model described was in dogs and initially adapted to rats), preference, and availability (5) and, most importantly, pathology similarities to SCI in humans (6). The lesion in rats is characterized by extensive degeneration of gray matter (GM) and white matter (WM), apparent at 1 week post-injury, which increases as rats develop cysts filled with substantial debris (7). Mouse models are being increasingly utilized due to the recent availability of many interesting and valuable genetically modified mice (8). In mice the relative lesion (with respect to cord size) is initially larger, but mice show wound contraction as connective tissue develops at the lesion site, similar to the process observed in skin wounds, and while they

Håkan Aldskogius (ed.), *Animal Models of Spinal Cord Repair*, Neuromethods, vol. 76,
DOI 10.1007/978-1-62703-197-4_12, © Springer Science+Business Media, LLC 2013

do present small cavities initially, these disappear over time (7). Currently histopathology is the most commonly used method for evaluating experimental SCI. This traditional method is invasive, cross-sectional, labor intensive, and provides only a “snapshot” of the overall pathology, especially for the spinal cord since it is practically impossible to section and analyze the cord in its entirety.

Magnetic resonance imaging (MRI) can provide high-resolution images with exquisite soft tissue contrast in a noninvasive, nondestructive, and three-dimensional (3D) manner, which allows a dynamic view of changing pathology, function, and cellular events in vivo in entire organs or animals. Furthermore, cellular MRI allows for monitoring cells in vivo in SCI models, which can provide valuable information regarding the spatial and temporal infiltration of immune cells, changes in pathology in response to cell therapy, anti-inflammatory treatment, or other treatment protocols.

2. Anatomical MRI

Current micro-MRI technology can achieve three-dimensional spatial resolutions on the order of tens of microns (9). MRI uses no ionizing radiation and is considered safe and noninvasive. Anatomical MRI is focused on generating optimal image contrast to allow the differentiation of various tissues. Image contrast is determined by the choice of imaging pulse sequence and the choice of various imaging parameters. Conventional MRI pulse sequences include the spin echo (SE) and gradient echo (GE). SE images are typically acquired as proton density (PD)-, T1-, or T2-weighted images by varying the timing parameters: repetition time (TR) and echo time (TE). For example, PD-weighted images would be generated using a long TR and a short TE, T1-weighted images with a short TR and short TE, and T2-weighted images with a long TR and long TE. GE images are often used to generate what is known as T2-star (T2*) weighted images. All tissues have distinct T1 and T2 relaxation times and proton densities and therefore optimal contrast between tissues can be achieved by manipulation of these timing parameters.

MRI of SCI models is usually performed using high magnetic field strengths and dedicated animal scanners. Mouse MRI systems have magnetic field strengths of 4 T and higher. At higher field strengths the image signal-to-noise ratio (SNR) is greater. Imaging the rodent spinal cord requires the acquisition of very small voxels (volume elements), or high spatial resolution. The smaller the voxel, the lesser the signal; therefore, signal gained by operating at high magnetic field strength is an advantage and improves image quality. Animal MRI systems tend to have higher gradient strengths as well. The gradient strength determines the maximum spatial resolution that can be achieved. There are some limitations of imaging at the higher magnetic fields. Image artifacts due to magnetic susceptibility and dielectric effects are much more pronounced at high field and may worsen image quality (image distortions, signal loss, and nonuniformity).

In vivo anatomical imaging of experimental disease models in the rodent spinal cord is challenging due to the very small size of the cord itself (cross-sectional area of 1.5 mm^2 for mice and 6.26 mm^2 for rats (7)), motion artifacts in the region of interest due to breathing and cerebrospinal fluid (CSF) pulsation, and health issues (diminished thermal regulation, decreased breathing capacity, and deficits in cardiovascular function) which make repetitive in vivo imaging under anesthesia more difficult (10). Ohta et al. performed the first in vivo MRI of SCI, in a cervical compression model in the rat at 4 weeks after the injury. They described areas of signal hypointensity due to hemorrhage and areas of signal hyperintensity that reflected edema, gliosis, and cavity formation (11). The severity of SCI in a rat contusion model was assessed by Mihai et al. in images acquired on a 4.7 T MRI system by measuring edema and lesion length (12). Longitudinal MRI used to study syringomyelia at 9.4 T (13) showed spontaneous recovery between 2 and 8 weeks post-injury in rats using an implantable RF coil (14). Changes in SCI pathology such as cyst formation, hemosiderin deposition, edema, and fibroglial scar have been monitored in a rat contusion model of SCI at 17.6 T (15).

There have not been nearly as many MRI studies in mouse SCI models as compared to rats. This is in part because the very small size of the mouse cord requires ultrahigh spatial resolution to obtain informative image data and this involves longer imaging times. In addition, the lesion in SCI mice is not as pronounced as in other species, which makes visualization by MRI more difficult. Most reports in the current literature suggest that while rats and other mammals develop fluid-filled cysts after SCI, mice do not (7, 16, 17). Microcysts have been reported in mice, but not central cavities (1, 2). Bilgen et al. reported regions of signal loss in images of mice with contusion SCI. Histology showed that these could be attributed to either fibrous connective tissue, necrosis, or hemorrhage (18).

Contrast agents are often used along with anatomical MRI to enhance the detection of pathology in contrast to normal tissues. The most conventionally used contrast agent is a gadolinium chelate (Gd-DTPA). Normally, T1-weighted images are acquired pre- and post-intravenous (i.v.) injection of Gd-DTPA and if the blood–brain or blood–spinal cord barrier is compromised then the contrast agent leaks in and causes signal enhancement in the affected area. Anatomical and contrast enhanced MRI were used together in one study to compare the MRI appearance of the cord in mouse and rat SCI models at 7 T (Fig. 1) (19).

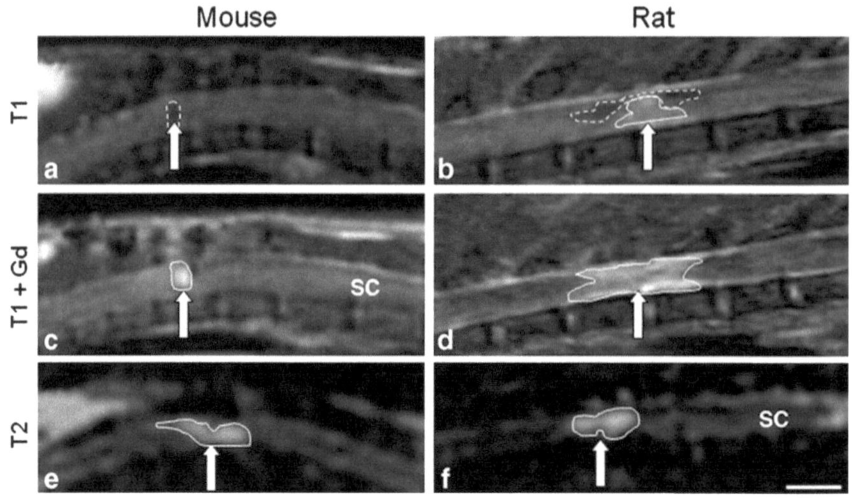

Fig. 1. Mouse (*left*) and rat (*right*) MRI at 28 days following a moderate SCI. Images were taken using a T1-weighted protocol (**a**, **b**), a T1-weighted protocol with Gd contrast enhancement (**c**, **d**), or a T2-weighted protocol (**e**, **f**). SCI lesion (*arrows*) appears as a hypointense region in mice with T1 imaging, a combination of hypo- and hyperintense in rats, and a hyperintense region with Gd contrast in both species. With T2 imaging, the hyperintensity is less contained and demonstrates a larger lesion area in both rats and mice. Hyperintense regions are outlined with a *solid line*; hypointense regions are outlined with a *dashed line*. SC = Spinal cord. Scale bar = 0.45 cm for (**a**–**d**); 0.74 cm for (**e**); 0.4 cm for (**f**). © 2010 Wiley. Used with permission from Byrnes et al., Neuropathological differences between rats and mice after spinal cord injury, J Magn Reson Imaging, John Wiley & Sons.

Gonzalez-Lara et al. have also studied mouse SCI with MRI (20). Their approach was to use a lower magnetic field strength, clinical 3 T scanner, equipped with a high strength insertable gradient and an SNR efficient 3D imaging pulse sequence. Balanced steady state free precession (bSSFP) images showed a heterogeneous signal at the lesion epicenter with areas of both signal hyperintensity and signal hypointensity, representing the mixed pathology present in the injured cord. The areas of hyperintensity represented edema and acute hemorrhage and the regions of signal loss could be related to the accumulation of hemosiderin-laden macrophages, associated with hemorrhage breakdown (20). The 3D whole mouse body images presented in this paper also revealed the presence of large cystic lesions rostral to the epicenter of the SCI, at a distance much farther than would typically be interrogated by histology (20). These cysts resembled what has been described in the literature as post-traumatic syringomyelia (21), a relatively rare complication secondary to SCI which is not fully understood.

Another study has shown that high-resolution MRI of rat SCI can be performed on a clinical system. Sandner et al. demonstrated that hemorrhage, edema, and cyst formation could be visualized at 3 T in rats with contusion SCI (22). The authors discuss the advantages of using a clinical scanner, including: bore size, which allowed for imaging of older, larger rats; decreased mortality, since it allows for better ventilation; and scanner availability. This study was followed up by a comprehensive description of optimization methods to study CNS pathologies, including SCI, in rat models at 3 T, where the authors emphasized that lack of access to a dedicated high-field scanner should not be a limitation to perform small-animal MRI (23).

3. Magnetization Transfer Imaging

Magnetization transfer imaging (MTI) is based on the magnetization interaction between bulk water protons and macromolecular protons, such as proteins (24). MT contrast is generated by applying off-resonance RF irradiation designed to preferentially saturate the immobile, or bound, protons of macromolecules. This saturates the energy level of the bound protons, inducing energy transfer to the mobile, or free water, protons and consequently reducing their signal intensity (25). Thus, MTI indirectly measures the interaction between these two proton pools. The relative difference in signal intensity (SI) with and without magnetization transfer can be measured as the magnetization transfer ratio (MTR). The MTR has been very useful for assessing tissue pathology and most MTI has focused on intrinsic markers like myelin and collagen (26, 27). Much of the work evaluating the utility of the MTR as a marker of disease has focused on multiple sclerosis (MS) and the associated demyelination (28, 29). Demyelination results in a reduction in the MTR, as protons associated with macromolecules are lost. Several papers show a strong correlation between the MTR and myelin content (28–30). Schmierer et al. proved that the MTR is affected by myelin content in MS WM by measuring the MTR in fresh postmortem brains from MS patients (29).

The MTR has not been widely applied to study SCI. Franconi et al. measured the MTR of the cervical cord in a rat SCI model at 7 T and found the MTR to be most useful in defining the thin CSF surrounding the cord but less effective for distinguishing between WM and GM (31). McGowan et al. used MTI at 1.7 T for histogram analysis in a rat weight-drop injury model and showed that the WM MTR was predictive of myelination status and neurofilament damage (32). Gareau et al. used MTI to compare severe and mild SCI due to contusion in rats and showed that the WM MTR was lower for severe SCI on the first day postsurgery, but not by 1 week, and that the WM MTR was insensitive to differences between mild SCI and the control uninjured spinal cord (33). When using the MTR to evaluate SCI, or other CNS damage, it is important to consider the extent of other pathology in the tissue since the MTR can also be influenced by edema; the accumulation of fluid (and so more free water protons) in tissue will also lower the MTR (33).

4. Functional MRI

4.1. fMRI

Functional MRI (fMRI) is a specialized MRI technique used to measure the hemodynamic response related to neural activity in the brain or spinal cord (34, 35). When a neuron fires, it takes up large amounts of oxygen from hemoglobin molecules. The local response to energy utilization, which occurs with neural activity, is to increase blood flow to the region. Changes in the local concentration of oxyhemoglobin are detectable by MRI through paramagnetic effects. Blood-oxygen-level-dependent (BOLD) imaging is the fMRI method used to visualize where in the brain or spinal cord oxygen is being consumed, thereby mapping neural activation (34, 36). T2*-weighted images are acquired which take advantage of the fact that deoxygenated hemoglobin is magnetic whereas oxygenated hemoglobin is not. BOLD fMRI is a relative technique; images taken during one state are compared to another. Paradigms therefore have periods of rest alternating with periods of activation or task. An area is more "activated" when it has more oxygenated blood compared to another point in time.

One limitation of fMRI is that it is extremely sensitive to motion. In rodents this is dealt with through the use of anesthesia; however, the anesthetic agent and dosage can have a direct effect on the baseline BOLD signal (37, 38). Some cause vasodilatation in the cerebral vasculature, which results in uncoupling between flow and neuronal activation (39). Proper choice of anesthetic and dosage is crucial for the success of BOLD fMRI small-animal experiments. A number of papers describe the proper use of anesthesia for fMRI (40–42). BOLD fMRI is also limited in that artifacts are often present in regions that are close to air (i.e., sinuses). Another problem is that areas of activation may be observed in large draining veins rather than directly at a capillary bed near the site of neuronal activation.

BOLD fMRI works best at ultrahigh magnetic field strengths. This is because the volume of the smallest voxel affects the likelihood of distinguishing activity from distinct functional units, and the ability to segregate microvasculature signals in the gray matter from nonspecific contributions originating from large veins. Furthermore, high spatial resolution (small voxels) is constrained by the SNR and contrast-to-noise ratio (CNR) and ultrahigh fields provide an advantage because of field-dependent gains in SNR and CNR.

fMRI was initially developed to study the human brain and there have been significant challenges associated with implementing fMRI to study the spinal cord in both humans and small animals (43). Porszasz et al. were the first to perform fMRI in the rat spinal cord (44). Malisza and Stroman further refined the technique acquiring images of the normal rat cord at 9.4 T in response to painful stimulation applied to the dorsal forepaw. Activation was observed in regions of the spinal cord associated with motor and pain responses demonstrating the utility of fMRI as a tool to study animal models of pain and cord injury (45).

BOLD fMRI has been proposed as a method to monitor functional reorganization in the CNS after injury. Ramu et al. used fMRI to study brain reorganization in SCI rats at 4 and 8 weeks post-injury (46). To do this, electrical stimulation of the rat paws was performed while imaging the brain. In normal rats this resulted in brain activation confined to the contralateral somatosensory cortex. In rat SCI activation was widespread indicating extensive and ongoing neural brain reorganization. To study plastic changes in the spinal cord in response to injury, Endo et al. applied fMRI in rats subjected to complete transection of the mid-thoracic cord along with electrical hind limb stimulation (47). Stimulation, at levels that did not show activation in the normal rat spinal cord, resulted in BOLD contrast in the ipsilateral dorsal horn of the injured spinal cord at multiple time points, from 1 day to 6 months after SCI. These results suggest exaggerated responsiveness of spinal neurons after SCI, a condition that might be associated with mechanisms causing neuropathic pain and autonomic dysreflexia.

4.2. Perfusion Imaging

Perfusion MRI measures regional blood flow. Two perfusion MRI methods have been developed, initially for measuring cerebral blood flow. The first method, called bolus tracking, involves an i.v. injection of a gadolinium-containing contrast agent and measurement of the T2*-weighted signal as it perfuses through the tissue brain over a short period of time (48). Changes in blood flow over time after a single injection can be measured to generate a perfusion map. The second method is known as arterial spin labeling (ASL). In ASL blood flowing into the image slices of interest is magnetically "labeled" or "tagged" using radiofrequency pulses (49). A perfusion-weighted image can be generated by the subtraction of an image in which inflowing spins have been "labeled" from an image in which spin labeling has not been performed. Quantitative perfusion maps can be calculated if the tissue T1 and the efficiency of spin labeling are also measured. This method takes advantage of arterial water as a freely diffusible tracer; contrast agents are not required. The drawbacks of this technique relate mainly to its weak SNR and the sensitivity to motion as a source of error. ASL methods benefit from high magnetic field strengths because not only is the sensitivity higher, but T1 also lengthens, allowing more label to accumulate.

Perfusion refers to the delivery of oxygen and nutrients to tissues by means of blood flow, and is one of the most fundamental physiological parameters. In SCI perfusion can be severely compromised. Duhamel et al. have explored the use of ASL for measuring spinal cord blood flow (SCBF) in mice (50, 51). Absolute SCBF measurements obtained at the cervical (C3) and lumbar (L1) levels of the normal mouse cord showed significantly lower GM perfusion at L1 compared to C3. The technique was also successfully applied to a mouse model of hemisection performed at L1 to characterize the pathology.

4.3. Diffusion Imaging

Diffusion-weighted imaging (DWI) is very sensitive to the random movement of protons in water molecules (Brownian movement) (52). In DWI each image voxel has a signal that reflects the rate of water diffusion at that location. The amount of water diffusion for a given voxel can be calculated and is called the apparent diffusion coefficient (ADC) (53). Areas with low ADC values (i.e., low diffusion) appear more intense. ADC values are direction sensitive. For instance, if images are taken perpendicular to myelin fiber tracts (i.e., corpus callosum), ADC values will be lower than if the images are taken along the length of these fibers. Diffusion tensor imaging (DTI) is a quantitative imaging technique used to measure the amount of diffusion as a function of direction. Typically a full diffusion tensor analysis requires the acquisition of diffusion-weighted images along at least six independent directions of diffusion-sensitizing gradients. Therefore, DTI measurements of the rodent CNS are time consuming. From the diffusion tensor, diffusion anisotropy measures such as the fractional anisotropy and the mean diffusivity can be computed. Because DTI can provide information about the location, orientation, and anisotropy of WM tracts, it has been applied widely to study the injured spinal cord in animal models (54–59).

Ellingson et al. used DTI to explore lesion growth and degeneration patterns from the acute through the chronic stages of rat SCI (54). Significant growth of the traumatic lesion was detected up to 15 weeks post-injury. The mean diffusivity was sensitive to overall spinal cord integrity, whereas fractional anisotropy showed specificity to sites of cavity formation. DTI has also been used to assess preservation of disrupted rubrospinal tract axons and changes in glial orientation in the GM (55) and microstructural WM changes (56). In mouse SCI models, DTI has been used to study the evolution of ventrolateral WM changes following contusive SCI at 1, 3, 7, and 14 days post-injury. Measurements of directional diffusivity derived from DTI clearly showed rostral–caudal asymmetry of the injury and a trend of acute primary axonal injury followed by delayed, subacute myelin damage at the impact site (Fig. 2) (57). Sundberg et al. used high-resolution anatomical MRI and DTI at 7 T to study WM integrity following mouse SCI for 2 months post-injury (58). WM in regions up to 1 cm away from the injury epicenter appeared normal on anatomical MRI but exhibited significant changes by DTI. Histology confirmed a significant decrease in myelin and oligodendrocytes, suggesting that DTI is a more sensitive measure of the evolving injury in mouse SCI.

Tu et al. have reported that two-directional DWI provides similar diffusion measurements as full tensor DTI in normal and injured mouse spinal cord (59). Diffusion indices derived using

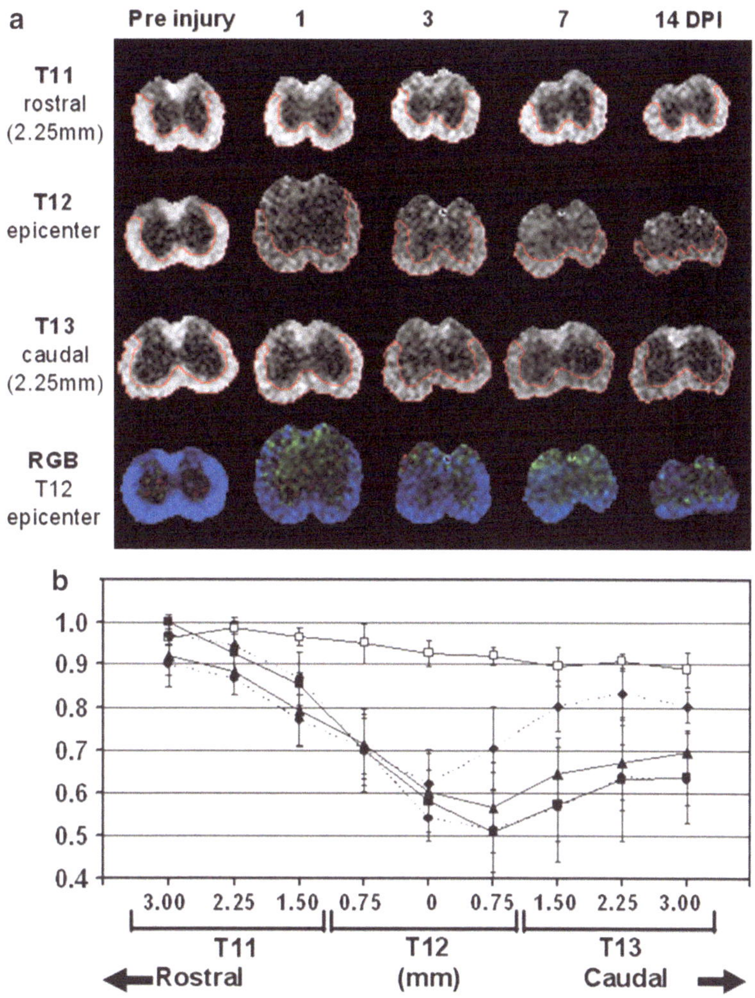

Fig. 2. Representative pre- and post-injury RA maps (scale = 0–1.414) at the impact site (T12), 2.25 mm rostral (T11), and 2.25 mm caudal (T13) to the impact site demonstrate the temporal and spatial evolution of acute SCI (**a**). At the impact site (T12), a red–green–blue (RGB) color-coded anisotropy map (red = *x*, green = *y*, and blue = *z*) is shown in (**a**), with homogeneous *blue* color suggesting the consistent orientation of the ventrolateral WM tracts. The time course of RA derived using quantitative ROI analysis from ventrolateral WM at T11, T12, and T13 is shown in (**b**). GM/WM contrast attenuated over time but remained discernible within the impact epicenter at all measured time points. Rostrally, RA declined linearly as measurements approached the epicenter. At T11, ventrolateral WM appeared essentially normal at 14 DPI. In contrast, progressively diminishing RA was observed in ventrolateral WM at all measured levels caudal to the epicenter, primarily at 1–3 DPI. Ventrolateral WM ROIs are outlined in *red*. *Open rectangle*: pre-injury control; *solid diamond*: 1 DPI; *solid triangle*: 3 DPI; *solid circle*: 7 DPI; *solid square*: 14 DPI. © 2007 Wiley. Used with permission from Kim et al., Noninvasive diffusion tensor imaging of evolving white matter pathology in a mouse model of acute spinal cord injury, Magn Reson Med, John Wiley & Sons.

two-direction DWI (parallel and perpendicular to axonal tracts) were compared with those obtained using six-direction DTI in a mouse contusion model of SCI and the measurements of the extent of spared WM after injury, estimated by either method, were in close agreement and correlated well with histology and behavioral tests.

5. Cellular MRI

Cellular MRI is a relatively young field of MRI research that combines the ability to obtain high-resolution image data with the use of magnetic contrast agents for labeling specific cells, thereby enhancing their detection (60). Superparamagnetic iron oxide (SPIO) nanoparticles represent a class of magnetic contrast agents used for cellular MRI that exhibit extremely high relaxivity (61). A variety of iron oxide-based labels are now available (62). Iron-labeled cells are usually imaged using either T2*-weighted GE or T2-weighted SE sequences (62). The presence of the magnetic label causes a distortion in the magnetic field and leads to abnormal signal hypointensities. Areas containing iron-labeled cells appear as regions of low signal intensity, creating negative contrast. The large magnetic susceptibility of these particles affects an area much larger than the actual particle size. This effect is known as a "blooming artifact," and leads to an exaggeration of the region occupied by iron oxide (63).

Cellular MRI studies use one of two ways to label cells with iron particles, in situ or in vitro. In situ labeling is performed by injecting iron particles intravenously with the goal of labeling circulating immune cells or immune cells at the site of inflammatory lesions. This approach has been used to detect and monitor macrophages in many different preclinical disease models, including experimental autoimmune encephalomyelitis (EAE) (64, 65), ischemic stroke (66), atherosclerosis (67, 68), peripheral nervous system injury (69), arthritis (70), and rejection in a variety of transplantation models (71–73). To determine the contribution of blood-derived macrophages to the signal loss observed in MR images of inflammatory lesions after i.v. SPIO, Oweida et al. performed in situ cell labeling in a mouse model of MS using transgenic mice that expressed enhanced green fluorescent protein (EGFP) specifically in hematopoietic cells of the myelomonocytic lineage (64). MRI revealed the presence of discrete regions of signal loss throughout the brains of animals that were administered Feridex and histology showed that regions of signal loss on corresponded anatomically with regions of PPB- and GFP-positive cells. This experiment provided the first direct evidence that macrophages of hematogenous origin are labeled with SPIO after i.v. administration of SPIO, and contribute to the regions of signal loss detected

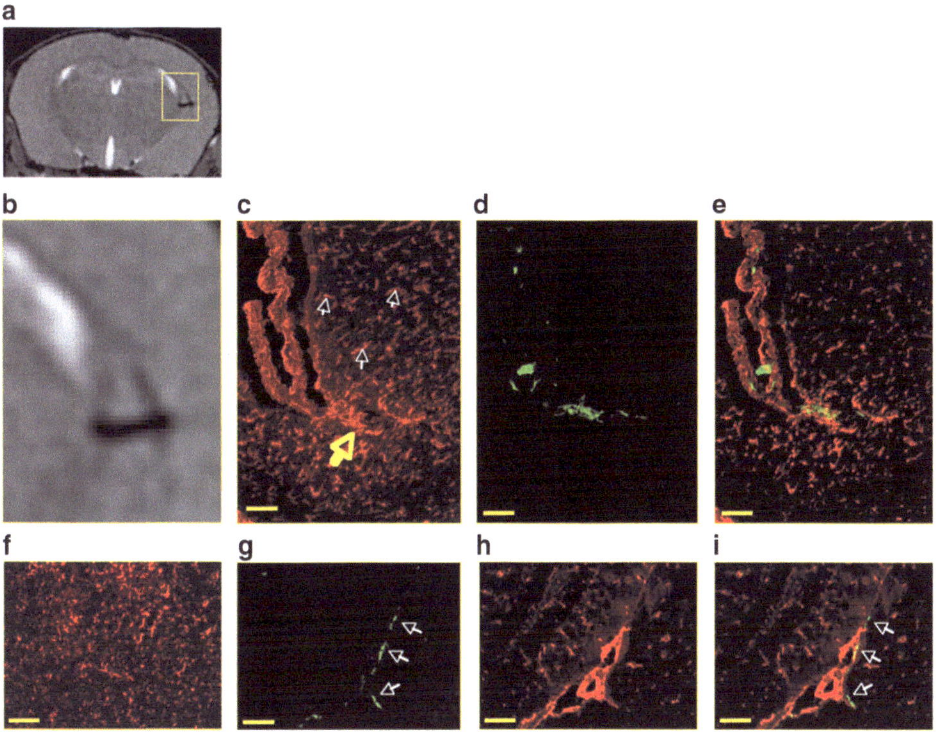

Fig. 3. Correlation of signal loss on MR with macrophage- and GFP-positive histology. (**a**) An MR image shows a periventricular region of signal loss. An enlarged view of the region of signal loss is shown in (**b**). Corresponding histology sections show macrophage accumulation with F4/80 staining and (**c**) GFP expression with anti-GFP staining (**d**). *White arrows* point to microglia, and *yellow arrow* points to macrophage accumulation. An overlay image of (**c**) and (**d**) is shown in (**e**) demonstrating colocalization of stains. Ventricle region is indicated with letter V. Control sections from non-EAE animals injected with Feridex show microglia-positive staining (**f**) that does not correspond to GFP-positive regions. When GFP-positive regions are detected (**g**) they are within blood vessels, as evidenced by tomato lectin staining (**h**, **i**). Scale bar = 6 μm. © 2007 Wiley. Used with permission from Oweida et al., Iron-oxide labeling of hematogenous macrophages in a model of experimental autoimmune encephalomyelitis and the contribution to signal loss in fast imaging employing steady state acquisition (FIESTA) images, J Magn Reson Imaging, John Wiley & Sons.

in MR images (Fig. 3). Dunn et al. used cellular MRI to detect macrophages in a clip compression model of SCI in the rat ex vivo MRI at 1.5 T (74).

In vitro cell labeling involves prelabeling cells prior to their injection or transplantation and following their fate with MRI. Many types of cells will readily take up iron particles by simple coincubation with iron oxide particles. For example, macrophages, cancer cells, and stem cells, are relatively easy to label this way. Other cells, in particular nonadherent cells, are more difficult to label and require specialized techniques. Various methods have been developed to enhance the amount of iron taken up by cells such as the use of transfection agents (75), functionalized iron particles (76), magnetoelectroporation (77–79), magnetic fields, and cell penetrating peptides (80). SPIO has been used to label and track a wide variety of cell types including T lymphoctyes (81), macrophages (82), pancreatic islets (83), cancer cells (63), and

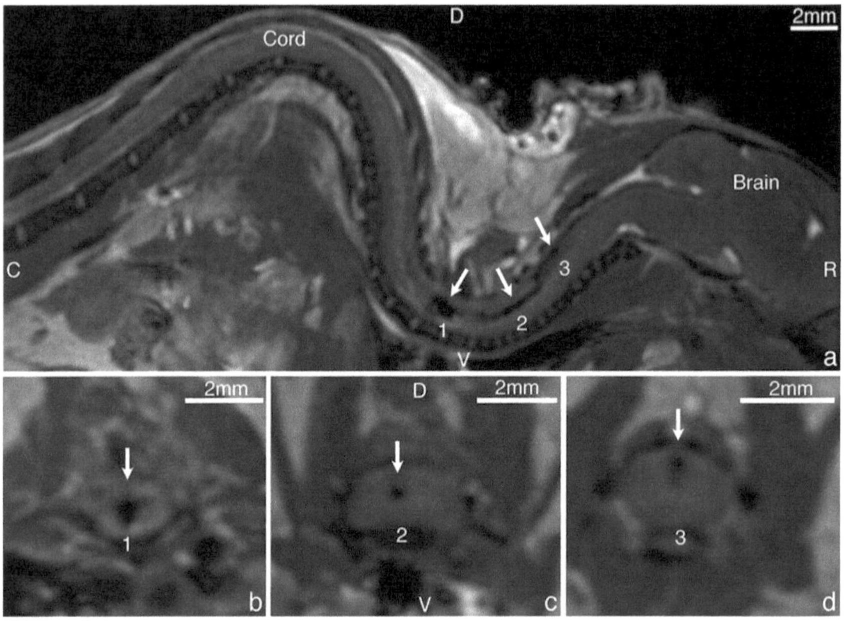

Fig. 4. Representative sagittal image ((**a**), *top*) of a mouse that received a successful transplant of iron-labeled MSC and corresponding cord areas in an axial plane (*bottom*). Clear areas of signal loss were observed within the cord: 1: close to the central canal (**b**), 2: by the white dorsal column (**c**) and 3: closer to the CSF (**d**) *D-dorsal, V-ventral, R-rostral, C-caudal.* © Gonzalez-Lara, Laura E., "In Vivo Cellular MRI In Experimental Traumatic Spinal Cord Injury" (2011). Electronic Thesis and Dissertation Repository. Paper 238.

stem cells (84) with minimal impact on cell function over a period of several weeks and over multiple cell divisions.

Sykova et al. showed that iron-labeled mesenchymal stem cells (MSC) administered i.v. in SCI rats can be detected as regions of signal loss in ex vivo images of the cord at 4 weeks after the cell injection (84). They also showed that iron-positive cells in the cord did not colocalize with macrophage staining, suggesting that iron-labeled MSC persist and can be tracked with MRI. Dunning et al. have used SPIO to label and track Schwann cells (SC) and olfactory ensheathing cells (OEC) in vivo for up to 4 weeks in SCI rats. They showed by histology transplanted cells remyelinating the demyelinated axons, demonstrating that the iron-labeled cells retained their biological function and that the majority of the label had remained within the transplanted cells (85). Lee et al. tracked SPIO labeled OEC in rat SCI for 2 months and found that MRI was able to show that the OEC were unable to bridge the transection lesion (86). Gonzalez-Lara et al. were the first group to use cellular MRI to track transplanted stem cells in the injured mouse cord (20, 87). The presence of transplanted cells in the cord was easily visualized as a large localized region of signal loss and could be monitored over time in vivo (Fig. 4). MRI was also useful for evaluating the accuracy of the transplantation technique since iron-labeled cells inadvertently placed outside of the spinal cord tissue could be readily identified and these mice were removed from the longitudinal study. This short-term cell tracking is valuable because endpoint histological analyses of these transplantation studies normally report on the presence or absence of stem cells in the tissue, and knowledge of the success/failure of the original transplantation is crucial for the interpretation of these data.

The biggest limitation of using iron nanoparticles and MRI for cell tracking in SCI models is that when iron-labeled cells die (or free iron is released) the iron may be taken up by bystander cells (macrophages), which will confuse the image interpretation (88–90). In any direct transplantation model a percentage of transplanted cells are expected to die due to the injection procedure itself (91). Cells labeled peripherally after i.v. SPIO administration, or the viable transplanted cells, will also eventually die at the SCI lesion site due to the harsh inflammatory conditions. This is compounded in SCI models by disrupted blood flow at the lesion site, which may result in an accumulation of iron particles due to cell death. Gonzalez-Lara et al. recently examined this issue in a longitudinal imaging study that followed SCI mice for 6 weeks after the direct transplantation of viable iron-labeled MSC, dead iron-labeled MSC, or free iron particles (87). The MRI appearance of the region of signal loss at the transplant site was similar for all three groups. In all transplanted SCI mice the volume of signal loss decreased over time. This decrease was most rapid in SCI mice that received live iron-labeled cells. The persistence of signal loss in the images of cords from mice that received dead MSC or free iron was attributed to iron particles lodged in the damaged cord tissue at 6 weeks postinjection, as shown by iron stained histology. The method of transplantation, host tissue, cell type, cell lifespan, and the extent of inflammation in the model system no doubt play important roles in the cell survival and clearance. The direct transplantation models for the injured spinal cord may represent the worst-case scenario for cell tracking with MRI. Long-term tracking of stem cells by MRI in models of SCI should be carefully validated through histology.

6. Summary

MRI is a powerful tool for investigating SCI. The noninvasive nature of MRI allows for studies of the evolution of the SCI lesion over time and for monitoring changes in the pathology that occur in response to treatments. High magnetic field strengths and advanced MRI hardware and software permit the acquisition of detailed, high-resolution images of the rat and mouse cord in vivo. Compared to histological analysis, MRI provides a more rapid and comprehensive view of the spinal cord; 3D MRI produces images in all three orientations (axial, sagittal, and coronal) without gaps and permits repeated sampling of individual animals over long periods of time. Anatomical MRI can be combined with other functional or cellular MRI techniques to obtain multilayered information on the structure and function of the injured spinal cord.

References

1. Ma M, Basso DM, Walters P, Stokes BT, Jakeman LB (2001) Behavioral and histological outcomes following graded spinal cord contusion injury in the C57Bl/6 mouse. Exp Neurol 169(2):239–254
2. Joshi M, Fehlings MG (2002) Development and characterization of a novel, graded model of clip compressive spinal cord injury in the mouse: Part 1. Clip design, behavioral outcomes, and histopathology. J Neurotrauma 19(2):175–190
3. Siegenthaler MM, Tu MK, Keirstead HS (2007) The extent of myelin pathology differs following contusion and transection spinal cord injury. J Neurotrauma 24(10):1631–1646
4. Jacob JE, Pniak A, Weaver LC, Brown A (2001) Autonomic dysreflexia in a mouse model of spinal cord injury. Neuroscience 108(4):687–693
5. Onifer SM, Rabchevsky AG, Scheff SW (2007) Rat models of traumatic spinal cord injury to assess motor recovery. ILAR J 48(4):385–395
6. Fleming JC, Norenberg MD, Ramsay DA, Dekaban GA, Marcillo AE, Saenz AD et al (2006) The cellular inflammatory response in human spinal cords after injury. Brain 129(Pt 12):3249–3269
7. Inman DM, Steward O (2003) Physical size does not determine the unique histopathological response seen in the injured mouse spinal cord. J Neurotrauma 20(1):33–42
8. Guertin PA (2005) Paraplegic mice are leading to new advances in spinal cord injury research. Spinal Cord 43(8):459–461
9. Johnson GA, Cofer GP, Gewalt SL, Hedlund LW (2002) Morphologic phenotyping with MR microscopy: the visible mouse. Radiology 222(3):789–793
10. Blight AR (2002) Miracles and molecules–progress in spinal cord repair. Nat Neurosci 5(Suppl):1051–1054
11. Ohta K, Fujimura Y, Nakamura M, Watanabe M, Yato Y (1999) Experimental study on MRI evaluation of the course of cervical spinal cord injury. Spinal Cord 37(8):580–584
12. Mihai G, Nout YS, Tovar CA, Miller BA, Schmalbrock P, Bresnahan JC et al (2008) Longitudinal comparison of two severities of unilateral cervical spinal cord injury using magnetic resonance imaging in rats. J Neurotrauma 25(1):1–18
13. Bilgen M, Rumboldt Z (2008) Neuronal and vascular biomarkers in syringomyelia: investigations using longitudinal MRI. Biomark Med 2(2):113–124
14. Narayana PA, Grill RJ, Chacko T, Vang R (2004) Endogenous recovery of injured spinal cord: longitudinal in vivo magnetic resonance imaging. J Neurosci Res 78(5):749–759
15. Weber T, Vroemen M, Behr V, Neuberger T, Jakob P, Haase A et al (2006) In vivo high-resolution MR imaging of neuropathologic changes in the injured rat spinal cord. AJNR Am J Neuroradiol 27(3):598–604
16. Brown A, Jacob JE (2006) Genetic approaches to autonomic dysreflexia. Prog Brain Res 152: 299–313
17. Sroga JM, Jones TB, Kigerl KA, McGaughy VM, Popovich PG (2003) Rats and mice exhibit distinct inflammatory reactions after spinal cord injury. J Comp Neurol 462(2): 223–240
18. Bilgen M, Al-Hafez B, Alrefae T, He YY, Smirnova IV, Aldur MM et al (2007) Longitudinal magnetic resonance imaging of spinal cord injury in mouse: changes in signal patterns associated with the inflammatory response. Magn Reson Imaging 25(5): 657–664
19. Byrnes KR, Fricke ST, Faden AI (2010) Neuropathological differences between rats and mice after spinal cord injury. J Magn Reson Imaging 32(4):836–846
20. Gonzalez-Lara LE, Xu X, Hofstetrova K, Pniak A, Brown A, Foster PJ (2009) In vivo magnetic resonance imaging of spinal cord injury in the mouse. J Neurotrauma 26(5):753–762
21. Brodbelt AR, Stoodley MA (2003) Post-traumatic syringomyelia: a review. J Clin Neurosci 10(4):401–408
22. Sandner B, Pillai DR, Heidemann RM, Schuierer G, Mueller MF, Bogdahn U et al (2009) In vivo high-resolution imaging of the injured rat spinal cord using a 3.0 T clinical MR scanner. J Magn Reson Imaging 29(3):725–730
23. Pillai DR, Heidemann RM, Kumar P, Shanbhag N, Lanz T, Dittmar MS et al (2011)

Comprehensive small animal imaging strategies on a clinical 3 T dedicated head MR-scanner; adapted methods and sequence protocols in CNS pathologies. PLoS One 6(2):e16091
24. Henkelman RM, Stanisz GJ, Graham SJ (2001) Magnetization transfer in MRI: a review. NMR Biomed 14(2):57–64
25. Wolff SD, Balaban RS (1989) Magnetization transfer contrast (MTC) and tissue water proton relaxation in vivo. Magn Reson Med 10(1): 135–144
26. Levesque IR, Giacomini PS, Narayanan S, Ribeiro LT, Sled JG, Arnold DL et al (2010) Quantitative magnetization transfer and myelin water imaging of the evolution of acute multiple sclerosis lesions. Magn Reson Med 63(3):633–640
27. Harel A, Eliav U, Akselrod S, Navon G (2008) Magnetization transfer based contrast for imaging denatured collagen. J Magn Reson Imaging 27(5):1155–1163
28. Rausch M, Hiestand P, Baumann D, Cannet C, Rudin M (2003) MRI-based monitoring of inflammation and tissue damage in acute and chronic relapsing EAE. Magn Reson Med 50(2):309–314
29. Schmierer K, Scaravilli F, Altmann DR, Barker GJ, Miller DH (2004) Magnetization transfer ratio and myelin in postmortem multiple sclerosis brain. Ann Neurol 56(3):407–415
30. Chen JT, Kuhlmann T, Jansen GH, Collins DL, Atkins HL, Freedman MS et al (2007) Voxel-based analysis of the evolution of magnetization transfer ratio to quantify remyelination and demyelination with histopathological validation in a multiple sclerosis lesion. Neuroimage 36(4):1152–1158
31. Franconi F, Lemaire L, Marescaux L, Jallet P, Le Jeune JJ (2000) In vivo quantitative microimaging of rat spinal cord at 7 T. Magn Reson Med 44(6):893–898
32. McGowan JC, Berman JI, Ford JC, Lavi E, Hackney DB (2000) Characterization of experimental spinal cord injury with magnetization transfer ratio histograms. J Magn Reson Imaging 12(2):247–254
33. Gareau PJ, Weaver LC, Dekaban GA (2001) In vivo magnetization transfer measurements of experimental spinal cord injury in the rat. Magn Reson Med 45(1):159–163
34. Ogawa S, Menon RS, Kim SG, Ugurbil K (1998) On the characteristics of functional magnetic resonance imaging of the brain. Annu Rev Biophys Biomol Struct 27:447–474
35. Yoshizawa T, Nose T, Moore GJ, Sillerud LO (1996) Functional magnetic resonance imaging of motor activation in the human cervical spinal cord. Neuroimage 4(3 Pt 1):174–182
36. Madi S, Flanders AE, Vinitski S, Herbison GJ, Nissanov J (2001) Functional MR imaging of the human cervical spinal cord. AJNR Am J Neuroradiol 22(9):1768–1774
37. Hyder F, Rothman DL, Shulman RG (2002) Total neuroenergetics support localized brain activity: implications for the interpretation of fMRI. Proc Natl Acad Sci U S A 99(16): 10771–10776
38. Maandag NJ, Coman D, Sanganahalli BG, Herman P, Smith AJ, Blumenfeld H et al (2007) Energetics of neuronal signaling and fMRI activity. Proc Natl Acad Sci U S A 104(51):20546–20551
39. Lee JG, Smith JJ, Hudetz AG, Hillard CJ, Bosnjak ZJ, Kampine JP (1995) Laser-Doppler measurement of the effects of halothane and isoflurane on the cerebrovascular CO2 response in the rat. Anesth Analg 80(4):696–702
40. Pawela CP, Biswal BB, Hudetz AG, Schulte ML, Li R, Jones SR et al (2009) A protocol for use of medetomidine anesthesia in rats for extended studies using task-induced BOLD contrast and resting-state functional connectivity. Neuroimage 46(4):1137–1147
41. Weber R, Ramos-Cabrer P, Wiedermann D, van Camp N, Hoehn M (2006) A fully noninvasive and robust experimental protocol for longitudinal fMRI studies in the rat. Neuroimage 29(4):1303–1310
42. Hutchison RM, Mirsattari SM, Jones CK, Gati JS, Leung LS (2010) Functional networks in the anesthetized rat brain revealed by independent component analysis of resting-state FMRI. J Neurophysiol 103(6):3398–3406
43. Stroman PW (2005) Magnetic resonance imaging of neuronal function in the spinal cord: spinal FMRI. Clin Med Res 3(3):146–156
44. Porszasz R, Beckmann N, Bruttel K, Urban L, Rudin M (1997) Signal changes in the spinal cord of the rat after injection of formalin into the hindpaw: characterization using functional magnetic resonance imaging. Proc Natl Acad Sci U S A 94(10):5034–5039
45. Malisza KL, Stroman PW (2002) Functional imaging of the rat cervical spinal cord. J Magn Reson Imaging 16(5):553–558
46. Ramu J, Bockhorst KH, Mogatadakala KV, Narayana PA (2006) Functional magnetic resonance imaging in rodents: Methodology and application to spinal cord injury. J Neurosci Res 84(6):1235–1244
47. Endo T, Spenger C, Westman E, Tominaga T, Olson L (2008) Reorganization of sensory processing below the level of spinal cord injury as revealed by fMRI. Exp Neurol 209(1):155–160
48. Ostergaard L (2004) Cerebral perfusion imaging by bolus tracking. Top Magn Reson Imaging 15(1):3–9
49. Paiva FF, Tannus A, Silva AC (2007) Measurement of cerebral perfusion territories

using arterial spin labelling. NMR Biomed 20(7):633–642

50. Duhamel G, Callot V, Cozzone PJ, Kober F (2008) Spinal cord blood flow measurement by arterial spin labeling. Magn Reson Med 59(4):846–854
51. Duhamel G, Callot V, Decherchi P, Le Fur Y, Marqueste T, Cozzone PJ et al (2009) Mouse lumbar and cervical spinal cord blood flow measurements by arterial spin labeling: sensitivity optimization and first application. Magn Reson Med 62(2):430–439
52. Koh DM, Collins DJ (2007) Diffusion-weighted MRI in the body: applications and challenges in oncology. AJR Am J Roentgenol 188(6):1622–1635
53. Hagmann P, Jonasson L, Maeder P, Thiran JP, Wedeen VJ, Meuli R (2006) Understanding diffusion MR imaging techniques: from scalar diffusion-weighted imaging to diffusion tensor imaging and beyond. Radiographics 26(Suppl 1):S205–S223
54. Ellingson BM, Schmit BD, Kurpad SN (2010) Lesion growth and degeneration patterns measured using diffusion tensor 9.4-T magnetic resonance imaging in rat spinal cord injury. J Neurosurg Spine 13(2):181–192
55. Schwartz ED, Duda J, Shumsky JS, Cooper ET, Gee J (2005) Spinal cord diffusion tensor imaging and fiber tracking can identify white matter tract disruption and glial scar orientation following lateral funiculotomy. J Neurotrauma 22(12):1388–1398
56. Cheung MM, Li DT, Hui ES, Fan S, Ding AY, Hu Y et al (2009) In vivo diffusion tensor imaging of chronic spinal cord compression in rat model. Conf Proc IEEE Eng Med Biol Soc 2009:2715–2718
57. Kim JH, Loy DN, Liang HF, Trinkaus K, Schmidt RE, Song SK (2007) Noninvasive diffusion tensor imaging of evolving white matter pathology in a mouse model of acute spinal cord injury. Magn Reson Med 58(2):253–260
58. Sundberg LM, Herrera JJ, Narayana PA (2010) In vivo longitudinal MRI and behavioral studies in experimental spinal cord injury. J Neurotrauma 27(10):1753–1767
59. Tu TW, Kim JH, Wang J, Song SK (2010) Full tensor diffusion imaging is not required to assess the white-matter integrity in mouse contusion spinal cord injury. J Neurotrauma 27(1): 253–262
60. Modo M, Hoehn M, Bulte JW (2005) Cellular MR imaging. Mol Imaging 4(3):143–164
61. Wang YX, Hussain SM, Krestin GP (2001) Superparamagnetic iron oxide contrast agents: physicochemical characteristics and applications in MR imaging. Eur Radiol 11(11):2319–2331
62. Kim D, Hong KS, Song J (2007) The present status of cell tracking methods in animal models using magnetic resonance imaging technology. Mol Cells 23(2):132–137
63. Heyn C, Ronald JA, Ramadan SS, Snir JA, Barry AM, MacKenzie LT et al (2006) In vivo MRI of cancer cell fate at the single-cell level in a mouse model of breast cancer metastasis to the brain. Magn Reson Med 56(5):1001–1010
64. Oweida AJ, Dunn EA, Karlik SJ, Dekaban GA, Foster PJ (2007) Iron-oxide labeling of hematogenous macrophages in a model of experimental autoimmune encephalomyelitis and the contribution to signal loss in fast imaging employing steady state acquisition (FIESTA) images. J Magn Reson Imaging 26(1): 144–151
65. Dousset V, Ballarino L, Delalande C, Coussemacq M, Canioni P, Petry KG et al (1999) Comparison of ultrasmall particles of iron oxide (USPIO)-enhanced T2-weighted, conventional T2-weighted, and gadolinium-enhanced T1-weighted MR images in rats with experimental autoimmune encephalomyelitis. AJNR Am J Neuroradiol 20(2):223–227
66. Kim J, Kim DI, Lee SK, Kim DJ, Lee JE, Ahn SK (2008) Imaging of the inflammatory response in reperfusion injury after transient cerebral ischemia in rats: correlation of superparamagnetic iron oxide-enhanced magnetic resonance imaging with histopathology. Acta Radiol 49(5):580–588
67. Schmitz SA, Coupland SE, Gust R, Winterhalter S, Wagner S, Kresse M et al (2000) Superparamagnetic iron oxide-enhanced MRI of atherosclerotic plaques in Watanabe heritable hyperlipidemic rabbits. Invest Radiol 35(8):460–471
68. Klug K, Gert G, Thomas K, Christan Z, Marco P, Elisabeth B et al (2009) Murine atherosclerotic plaque imaging with the USPIO Ferumoxtran-10. Front Biosci 14:2546–2552
69. Bendszus M, Stoll G (2003) Caught in the act: in vivo mapping of macrophage infiltration in nerve injury by magnetic resonance imaging. J Neurosci 23(34):10892–10896
70. Beckmann N, Falk R, Zurbrugg S, Dawson J, Engelhardt P (2003) Macrophage infiltration into the rat knee detected by MRI in a model of antigen-induced arthritis. Magn Reson Med 49(6):1047–1055
71. Beckmann N, Cannet C, Zurbruegg S, Haberthur R, Li J, Pally C et al (2006) Macrophage infiltration detected at MR imaging in rat kidney allografts: early marker of chronic rejection? Radiology 240(3):717–724
72. Ye Q, Wu YL, Foley LM, Hitchens TK, Eytan DF, Shirwan H et al (2008) Longitudinal tracking of recipient macrophages in a rat

chronic cardiac allograft rejection model with noninvasive magnetic resonance imaging using micrometer-sized paramagnetic iron oxide particles. Circulation 118(2):149–156

73. Ho C, Hitchens TK (2004) A non-invasive approach to detecting organ rejection by MRI: monitoring the accumulation of immune cells at the transplanted organ. Curr Pharm Biotechnol 5(6):551–566
74. Dunn EA, Weaver LC, Dekaban GA, Foster PJ (2005) Cellular imaging of inflammation after experimental spinal cord injury. Mol Imaging 4(1):53–62
75. Cromer Berman SM, Walczak P, Bulte JW (2011) Tracking stem cells using magnetic nanoparticles. Wiley Interdiscip Rev Nanomed Nanobiotechnol 3(4):343–355
76. Tang KS, Shapiro EM (2011) Enhanced magnetic cell labeling efficiency using -NH(2) coated MPIOs. Magn Reson Med 65(6): 1564–1569
77. Walczak P, Kedziorek DA, Gilad AA, Lin S, Bulte JW (2005) Instant MR labeling of stem cells using magnetoelectroporation. Magn Reson Med 54(4):769–774
78. Walczak P, Ruiz-Cabello J, Kedziorek DA, Gilad AA, Lin S, Barnett B et al (2006) Magnetoelectroporation: improved labeling of neural stem cells and leukocytes for cellular magnetic resonance imaging using a single FDA-approved agent. Nanomedicine 2(2): 89–94
79. Tai JH, Foster P, Rosales A, Feng B, Hasilo C, Martinez V et al (2006) Imaging islets labeled with magnetic nanoparticles at 1.5 Tesla. Diabetes 55(11):2931–2938
80. Smith CA, de la Fuente J, Pelaz B, Furlani EP, Mullin M, Berry CC (2010) The effect of static magnetic fields and tat peptides on cellular and nuclear uptake of magnetic nanoparticles. Biomaterials 31(15):4392–4400
81. Baeten K, Adriaensens P, Hendriks J, Theunissen E, Gelan J, Hellings N et al (2010) Tracking of myelin-reactive T cells in experimental autoimmune encephalomyelitis (EAE) animals using small particles of iron oxide and MRI. NMR Biomed 23(6):601–609
82. Heyn C, Ronald JA, Mackenzie LT, MacDonald IC, Chambers AF, Rutt BK et al (2006) In vivo magnetic resonance imaging of single cells in mouse brain with optical validation. Magn Reson Med 55(1):23–29
83. Jirak D, Kriz J, Strzelecki M, Yang J, Hasilo C, White DJ et al (2009) Monitoring the survival of islet transplants by MRI using a novel technique for their automated detection and quantification. MAGMA 22(4):257–265
84. Sykova E, Jendelova P (2007) In vivo tracking of stem cells in brain and spinal cord injury. Prog Brain Res 161:367–383
85. Dunning MD, Lakatos A, Loizou L, Kettunen M, ffrench-Constant C, Brindle KM et al (2004) Superparamagnetic iron oxide-labeled Schwann cells and olfactory ensheathing cells can be traced in vivo by magnetic resonance imaging and retain functional properties after transplantation into the CNS. J Neurosci 24(44):9799–9810
86. Lee IH, Bulte JW, Schweinhardt P, Douglas T, Trifunovski A, Hofstetter C et al (2004) In vivo magnetic resonance tracking of olfactory ensheathing glia grafted into the rat spinal cord. Exp Neurol 187(2):509–516
87. Gonzalez-Lara LE, Xu X, Hofstetrova K, Pniak A, Chen Y, McFadden CD et al (2011) The use of cellular magnetic resonance imaging to track the fate of iron-labeled multipotent stromal cells after direct transplantation in a mouse model of spinal cord injury. Mol Imaging Biol 13(4):702–711
88. Pawelczyk E, Jordan EK, Balakumaran A, Chaudhry A, Gormley N, Smith M et al (2009) In vivo transfer of intracellular labels from locally implanted bone marrow stromal cells to resident tissue macrophages. PLoS One 4(8):e6712
89. Terrovitis J, Stuber M, Youssef A, Preece S, Leppo M, Kizana E et al (2008) Magnetic resonance imaging overestimates ferumoxide-labeled stem cell survival after transplantation in the heart. Circulation 117(12):1555–1562
90. Amsalem Y, Mardor Y, Feinberg MS, Landa N, Miller L, Daniels D et al (2007) Iron-oxide labeling and outcome of transplanted mesenchymal stem cells in the infarcted myocardium. Circulation 116(11 Suppl):I38–I145
91. Coyne TM, Marcus AJ, Woodbury D, Black IB (2006) Marrow stromal cells transplanted to the adult brain are rejected by an inflammatory response and transfer donor labels to host neurons and glia. Stem Cells 24(11):2483–2492

Chapter 13

Spinal Cord Injuries: Principles and Methods for Outcome Assessment

Elisabet Åkesson, Cinzia Calzarossa, and Erik Sundström

Abstract

Animal models are indispensable for research on central nervous system disorders and particularly relevant and useful in spinal cord injury (SCI) research. In contrast to neurodegenerative diseases, trauma inflicted to animals is inherently similar (albeit not identical) to the situation after trauma in human beings. There are still many factors to consider when the choice of animal model is to be made. How should the injury be induced? At what segmental level? Which parameters can be evaluated? Ultimately, the purpose of the study should be defined before an animal model is chosen. Assessing acute and chronic structural changes, studying motor and sensory functions after injury, and evaluating treatments such as protective drugs or substances enhancing regeneration all place different demands on the animal model and how results from the model are interpreted. With a focus on small animal (rodent) models of SCI, we give a background on pathology and pathophysiology, review the most common rodent injury models and functional tests used, present examples of experimental studies using various methods, and discuss difficulties in translating results from animal models to the clinical situation.

Key words: Spinal cord injury, Secondary injury, Oxidative damage, Glutamate, Inflammation, Animal models, Rodent, Functional assessment

1. Introduction

In this chapter we will review the most commonly used rodent models of traumatic spinal cord injury (SCI), summarize pathophysiological mechanisms, discuss the purpose of studying models, and how tests of motor function and sensorimotor integration can be used. In addition and most importantly, we will discuss what to consider when animal models are used to develop therapies for SCI patients.

The functional analyses we will discuss are those used to assess motor performance, locomotion, and to some extent sensorimotor integration, the focus of most experimental studies on SCI. There are also methods to analyze the various modalities of sensory and autonomic functions. The analysis of sensory functions has inherent

Håkan Aldskogius (ed.), *Animal Models of Spinal Cord Repair*, Neuromethods, vol. 76,
DOI 10.1007/978-1-62703-197-4_13, © Springer Science+Business Media, LLC 2013

difficulties, one of which is that almost all of them are assessed by analyzing motor activity such as paw retraction or vocalization in response to sensory stimuli. It is still crucial to evaluate perception of pain in experimental treatment strategies involving regeneration or plasticity. Allodynia, the perception of pain in response to non-painful stimuli, is such a potentially severe side effect that it has to be controlled for.

Autonomic functions can be more accurately quantified than sensory functions, and as pointed out in a study by Anderson (1), they are also perceived by many patients as the most important symptoms to target with future treatment strategies. Experimental studies of autonomic functions are beyond the scope of this chapter, and interested readers are referred to the comprehensive and useful review on methods to study autonomic functions in SCI animal models by Inskip et al. (2).

As an additional introductory note, we want to emphasize that symptoms of SCI depend on two anatomical consequences of the injury. The first and most commonly addressed is the degeneration of ascending and descending tracts in the spinal cord white matter. The loss of these tracts will affect motor and sensory functions mediated by spinal segments at and below the injury site. The commonly used experimental injury models of thoracic spinal cord typically focus on this, and these animals are functionally evaluated by assessing hindlimb function, for which there are a number of tests. The second, much less considered effect in experimental studies is the degeneration of gray matter, having consequences for segmental spinal circuitry at the level of injury. This is often much more difficult to study, and more or less impossible in low thoracic injuries, which affect muscles of the abdomen. Although segmental gray matter degeneration will contribute to the functional consequences of lumbar injuries, cervical injuries are preferable for studies of this issue. As we will discuss below, some aspects of neuronal degeneration is presumably different in gray and white matter, and the regeneration/replacement strategies including cell therapy will have to be developed to match the type of degeneration targeted.

2. Pathophysiology of Traumatic SCI

Similar to other traumatic and ischemic injuries in the central nervous system, the pathophysiology of SCI is typically divided into two phases: primary and secondary injury. The most common SCIs are contusion and compression injuries, as a result of an external trauma to the vertebral column. Penetrating trauma such as gunshot or knife stab wounds results in direct transection of ascending and descending axons. In most countries this is however not a very common type of injury.

2.1. Primary Versus Secondary Injury

The primary injury after contusive, compressive, and stretch injuries is dominated by stretch and shear forces as a consequence of the dislocation of the vertebral column (3, 4). In addition to these primary mechanical events, blood flow to the injured tissue is rapidly compromised, due to a combination of local hemorrhage, edema, and vasospasm (5–7). Shortly after the primary SCI the spinal cord tissue becomes edematous as a result of two processes: appearance of vasogenic edema with increased perineuronal space and cytotoxic edema resulting in swelling of neurons and glia (8). When edema takes place in the closed compartment of the vertebral column, the intraspinal pressure increases, which contributes to the ischemia in the injured tissue. Systemic factors such as reduced cardiac output and low arterial blood pressure add to the hypoperfusion (9, 10). It is a matter of definition if ischemia should be considered a part of the primary injury. The vasogenic edema may be considered a part of the primary injury. Since the cytotoxic edema is caused by potentially reversible biochemical changes in the tissue, it could rather be seen as a part of secondary injury processes. However, both types of edema mechanisms will aggravate the ischemia.

The primary and secondary injury processes were recognized already by Allen, almost 100 years ago (11, 12). During the last decades, a number of mechanisms of secondary injury have been studied in animal models of SCI (see, e.g., (13)). Two main degenerative mechanisms can be identified: oxidative damage and excitotoxicity (or more correctly glutamate toxicity), with inflammation as a third process contributing to extending the degeneration. Common to the first two mechanisms of secondary injury is increased intracellular Ca^{2+}, a central event in the degenerative process, as was shown 40 years ago by Schlaepfer and Bunge (14). Primary injury will typically lead to necrotic degeneration of affected cells, but in some neurons apoptosis will be initiated, for example, when axons are severed far distal to the cell body. The processes of secondary injury will trigger apoptotic processes, and if the injury is severe, these processes may also lead to necrosis. Both types of cell death are present in SCI, with a central necrotic cavity surrounded by more intact nervous tissue containing apoptotic cells, in animal models (15, 16) as well as in patients (17).

2.2. Secondary Injury: Oxidative Damage

Oxidative damage—free radical-induced lipid peroxidation and protein nitration—stands out as the only degenerative mechanism for which there is a neuroprotective therapy used clinically in SCI (albeit with a very modest effect), high-dose methylprednisolone (18). A comprehensive review of the molecular mechanisms of oxidative damage in the injured spinal cord was recently published (19). Acidosis, which rapidly appears in the injured spinal cord tissue, is an important contributing factor to oxidative damage. An equally important factor is the hemorrhage in the traumatized spinal cord, and these two factors probably interact. Iron is a major catalyzing agent for free radical production. At pH around and below 6, which typically presents locally after injury, iron is released from hemoglobin and the transport protein transferrin in plasma. Intracellularly, iron is released from ferritin at acidic conditions. The presence of free iron at and around the site of injury thereby catalyzes oxidative damage in the traumatized parenchyma.

Targets of free radicals and oxidative damage are mitochondria and mechanisms maintaining ion homeostasis such as Na^{+}/K^{+}-ATPase and Ca^{2+}-APTase (20). The large increase in intracellular Na^{+} concentrations in combination with energy depletion will reverse the Na^{+}/Ca^{2+} antiporter, mediating transport of Ca^{2+} into the cell instead of its extrusion. Increased cellular Ca^{2+} entrance, concomitant with impaired clearance of cytoplasmic Ca^{2+} to intracellular stores or over the plasma membrane will rapidly raise the intracellular Ca^{2+} concentration. Once it has reached critical levels, a number of Ca^{2+}-dependent mechanisms will be initiated, some of which further boost pathological Ca^{2+} increase, activating pro-apoptotic signaling and/or mechanisms leading to necrosis. In addition, oxidative damage has a direct effect on plasma membrane integrity, with leakage of ions through ruptures in the membrane.

2.3. Secondary Injury: Glutamate Toxicity

Another key mechanism for secondary injury is elevated extracellular glutamate concentrations, triggered by tissue ischemia and energy depletion. It is the consequence of compromised, possibly reversed, glutamate uptake and leakage of intracellular glutamate through damaged cell membranes. Presynaptic depolarization can also, prior to severe energy depletion, evoke increased synaptic glutamate release adding to the extracellular glutamate levels. The result is excessive activation of post-synaptic *N*-methyl-D-aspartate (NMDA) and amino-3-hydroxy-5-methyl-4-isoxazolepropionic acid (AMPA) receptors. With the majority of AMPA receptors having a subunit composition that renders them impermeable to Ca^{2+} (but see, e.g., (21) regarding spinal motor neurons), glutamate activation of AMPA receptors will primarily lead to influx of Na^{+} and membrane depolarization. As described above, in situations of energy failure the Na^{+}/K^{+} ATPase is not functional and may even operate in reverse mode, exaggerating the intracellular Na^{+} accumulation and depolarization further. This results in further increase of intracellular Ca^{2+} due to reversal of the Na^{+}/Ca^{2+} antiporter. These effects are sufficient to cause neurodegeneration, as shown by the fact that AMPA receptor antagonists are neuroprotective in SCI (22).

In contrast to most AMPA receptors, NMDA receptors are permeable to Ca^{2+}, but only when the Mg^{2+}-block of the ion channel is removed by depolarization of the plasma membrane. As discussed, several mechanisms contribute to the depolarization of the membrane after trauma, allowing massive Ca^{2+} entry through

NMDA receptors excessively stimulated by glutamate (in concert with glycine). Accordingly, NMDA receptor antagonists can also be neuroprotective in the spinal cord (23). Similar to the consequence of oxidative damage, elevated intracellular Ca^{2+}-concentration caused by glutamate toxicity triggers a number of intracellular mechanisms ultimately leading to cell death, either through apoptosis or necrosis.

In vitro activation of synaptic NMDA receptors and L-type calcium channels evokes Ca^{2+}-dependent intracellular responses such as CREB activation, but not neuronal degeneration, which instead is mediated by extra-synaptic NMDA receptors (24, 25). The subunits of glutamate receptors are physically linked to a large web of intracellular proteins in the post-synaptic density, participating in the control of intracellular signaling triggered by receptor activation. The constituents of this Web will vary with the location of the receptor. Consequently, activation of NMDA receptors at extra-synaptic sites will result in a different intracellular response than activation of synaptic NMDA receptors. In this context, Sun et al. (26) showed that blocking the interaction between the intracellular C-termini of NMDA receptor subunits and one of the main scaffold proteins, PSD95, reduced degeneration after focal brain ischemia in rats. Differences in interaction between membrane glutamate receptors and intracellular pro-apoptotic pathways may explain the different role of synaptic and extra-synaptic NMDA receptors in neuronal degeneration.

Glutamate toxicity is not restricted to neurons, as was initially believed. For some time the glutamate receptors expressed by oligodendrocytes were believed to be AMPA receptors (27). It was later shown that these cells also express NMDA receptors (28), allowing both receptor types to contribute to excitotoxic injury of oligodendrocytes in gray matter. However, the recent demonstration of receptors in the optic nerve, which are assemblies of the NMDA receptor subunits NR1 and NR3 (29), complicates the situation, since these receptors are not glutamate receptors (or NMDA receptors) but Ca^{2+}-permeable atypical excitatory glycine receptors (30). It is not clear if these receptors participate in excitotoxicity, and what the role of glycine is in excitotoxicity in a more general sense.

Reactive oxygen species are produced in situations of glutamate toxicity (see (31) for review). Ruptures in plasma membranes resulting from lipid peroxidation will allow efflux of intracellular glutamate, and oxidative stress will enhance Na^{+}-mediated Ca^{2+}-independent release of glutamate (32). Thus, there is a close interaction between glutamate toxicity and oxidative damage. For more details on excitotoxicity/glutamate toxicity and neurodegeneration, there are reviews on SCI (33) and stroke and other insults (34), for which most of the processes are believed to be similar.

2.4. Secondary Injury: White Matter Degeneration

White matter will obviously be lost after SCI as a consequence of the necrosis caused by the primary injury, as well as through degeneration of axons from neurons undergoing apoptosis during secondary injury. But there are also secondary injury processes specific for white matter. Apoptosis of oligodendrocytes has been described in experimental (18) and clinical (17) SCI resulting in demyelination, a pathological condition, which may persist for years (35). In addition, degeneration of axons may occur, not just as a consequence of the loss of the neuronal cell body or the demyelination, but also by a direct fatal disturbance of ion homeostasis in the axonal cylinder. In a series of studies, Stys and colleagues (36) reported a combined effect of failing Na-K-ATPase and depolarization resulting in an intracellular rise in Na^+ reversing Na^+–Ca^{2+} exchange. The rapid increase of intracellular calcium activates Ca-dependent enzymes with deleterious consequences for the structural integrity of the axon.

Thus, when animal models of SCI are used to evaluate potential neuroprotective treatments, the relative contribution of gray matter circuitry versus far projecting white matter axons to outcome must be considered.

2.5. Secondary Injury: Inflammation

Shortly after SCI, inflammatory mechanisms come into play, with innate immune cells (neutrophils, monocytes, and dendritic cells) being recruited within hours to days. The monocytes differentiate into macrophages and regulate additional leukocyte recruitment, including T and B lymphocytes (as reviewed in (37)). The peak of T cell influx into the lesion epicenter occurs approximately 3–7 days after SCI. Lymphocyte expansion also takes place in the spleen and lymph nodes and auto-antibodies targeting gangliosides, myelin-associated glycoprotein, glutamate receptors, β-III tubulin, and nuclear antigens have been described.

Microglia are the dominant monocyte occupying the injury site during the first 3 days post-injury, preceding an extensive monocytes/macrophage invasion of the lesion (38). Microglial activation at the lesion epicenter peaks 3–7 days after SCI in rodent models (39). Microglial reaction rostral and caudal to the lesion is mainly observed in white matter regions, and plateaus in intensity after 2–4 weeks post-injury. Activated microglia persist for months with inflammation-related genes chronically upregulated (40). As a consequence of inflammatory processes in SCI, secondary degenerative mechanisms will result in necrosis as well as apoptosis in gray and white matter of experimental animals and human patients (41, 42).

However, activated macrophages exhibit at least two phenotypes, an M1 type, which produces bactericidal substances and neurotoxins, and an M2 type acting more immunoregulatory supporting wound healing. In SCI the M2 response occurs early and is transitory, whereas the M1 response persists indefinitely (43). In the epicenter of injury there is a dominance of toxic and destructive inflammatory processes with the presence of reactive oxygen and nitrogen species, interleukin-1, and tumor necrosis factor-α. On the other hand, trophic factor release supporting regeneration is mainly seen at some distance from the injury.

Inflammation after SCI will further trigger neurodegenerative processes as described above, but may also support neuroprotection and regeneration enhancing trophic factor effects in SCI (44). Treatment with in vitro-activated macrophages may improve function by neuroprotective immune actions in transection and contusion SCI animal models (45). A Phase I–IIa clinical study was conducted where incubated in vitro-activated autologous macrophages were administered to patients with acute complete SCI, without negative effects (46).

3. Experimental Traumatic SCI Models

Compared to many other conditions in the central nervous system, animal models of traumatic SCI have much less problems with relevance to the clinical situation. Neurodegenerative disorders such as Parkinson's disease, Alzheimer's disease, and amyotrophic lateral sclerosis (ALS) progress over many years from a subclinical condition to a clinically overt degeneration of specific neuronal populations, and eventually to a more widespread nonspecific neuronal degeneration. Modeling the various aspects of these diseases in animals is commonly performed using acute lesions or transgenic animals, both with apparent drawbacks regarding translation to the human disease. For traumatic injuries on the other hand, the primary insult in the human patients can be much more closely mimicked in the experimental animal. Still, the choice of trauma inflicted on the animal needs careful consideration in relation to the experimental objectives.

3.1. Contusion and Compression Injury Models

The most commonly used SCI model is probably the contusion injury. In the pioneering studies by Allen (11), a weight was dropped from a defined height onto the exposed dorsal spinal cord surface of dogs. This basic concept was later improved and adapted for rats and mice, and further improved by monitoring the biophysical parameters such as force of impact and displacement, combined with feedback control of the impactor (47–54). Alternatively the cord can be subject to a compressive injury, for example, using an aneurysm clip with a defined closure force as described by Tator and collaborators (55). Instead of a clip, a pair of forceps modified to prevent complete closure of the legs can be used. With the aneurysm clip the applied force is fixed, while the forceps typically induce a defined displacement. Compression injury can also be accomplished with a fairly intact vertebral column by introducing an epidural balloon, which is inflated once positioned within the vertebral column (56–58). In contrast to the millisecond duration of the impact in contusion injury models, compression of the cord is typically applied for half a minute up to several minutes. Consequently, in compression injury models the shearing forces exerted by the compression will be accompanied by local ischemia due to compression of superficial blood vessels, which will contribute to the primary injury. Reduced local blood flow is also seen in the acute stage of impact injuries although it is likely that significant ischemia only appears after severe impact (5). Other pathological mechanisms are probably also different in contusion and compression injuries, but these aspects have rarely been studied systematically.

3.2. Transection Injury Models

In addition to contusion and compression animal models, complete (59, 60) or partial transections (61, 63–65) may be used, however, for different purposes. To study mechanisms specifically related to axotomy (in the acute stage) or to determine if regeneration of severed axons occur after a treatment (at later stages), transection injuries have clear advantages. In order to unequivocally demonstrate axonal regeneration and/or sprouting, it is necessary to use anatomically complete injuries for proof-of-concept studies, since incomplete injuries present inherent problems with regard to separating regeneration/sprouting from compensatory plasticity of spared neuronal circuitry. In addition, functional improvement due to axon regeneration can be proven if a second transection of the regenerating/sprouting axons eliminates the improvement that took place following the first injury. Elegant examples of this strategy was applied in the studies by Fradiakis et al. (60) and Houle et al. (61), both on transplantation of peripheral nerves as conduits for regeneration.

Transection injuries are also quite common in models of cervical injury. Cervical contusion or compression injuries, which typically lead to bilateral injury, will at least transiently cause tetraparesis with inherent difficulties to manage the animals. These injury models are nevertheless used (53, 62) with the main rationale that about 50% of the SCI patients have suffered a bilateral cervical injury. However, to achieve unilateral injuries or other types of more limited and anatomically defined lesions, scissors or scalpel is used to produce transection of the dorsal columns, hemisection, or over-hemisection of the cord or transections of specific funiculi (see, e.g., (59–61, 63–65)).

3.3. Choosing SCI Experimental Model

In addition to the contusion, compression, and (hemi)transection models described above, there are models of ischemic SCI, including the photochemical lesion (66), irradiation injury (67), and vessel ligation models (68) to mention some of the most common.

In a subpopulation of SCI patients a progressive posttraumatic cystic myelopathy will develop with further neurological deterioration.

Tethering of the spinal cord to the vertebral column due to arachnoiditis disturbs cerebrospinal fluid flow and in addition to the parenchymal injury this is believed to cause the progressing syringomyelia. To develop treatment strategies for syringomyelia, animal models including kaolin injection into the subarachnoid space have been developed by Yang et al. (69) and Fehlings and Seki (70). All SCI models mentioned here are summarized in Table 1.

The various injury models all have advantages and disadvantages. It is not possible to conclude that any of them is more relevant to clinical situations in general, since clinical SCI is often the result of a sudden impact of the bone structures transmitted to the spinal cord (contusion), followed by a sustained compression resulting from the dislocation of vertebrae, and sometimes also transection of axons by penetrating bone fragments. In a clinical cohort of patients, the contribution of these factors will vary considerably, and their relative impact is rarely possible to determine in the individual patient. This heterogeneity is a problem in clinical trials, particularly if a novel intervention is to be studied acutely. Rather than arguing for the choice of a particular SCI animal model, referring to supreme clinical relevance, a candidate treatment should be studied in several injury models. In this context, it is also questionable if any added value is achieved by applying a multi-site-one-model design to experimental SCI studies, as has been suggested (71). In our view, positive effects in two or more animal models would give much better support for possible clinical utility of the treatment evaluated.

Choosing a relevant animal model for proof-of-concept depends on the therapeutic mechanism studied; neuroprotection, regeneration, plasticity, or replacement. When potentially neuroprotective drugs are studied, contusion or compression injury models are preferred over transection injuries since they are characterized by extensive secondary injury that can be modulated (i.e., reduced or increased). Since degenerative mechanisms at different spinal levels are likely to be similar, the level of injury is probably of minor importance when assessing the efficacy of a neuroprotective treatment. In contrast, for an elaborate design of conduits guiding regenerative growth, the anatomy of the axonal tracts is crucial, and consequently also the choice of spinal level studied. The neuronal connections that need to be reestablished for improving respiration after a cervical injury are different from those required to restore urogenital functions after a thoracic injury. An anatomically complete transection is the best choice for achieving unequivocal proof of a functional regenerative effect. In this situation, the choice of spinal level is however limited by the fact that animals with complete transection at cervical or high thoracic levels are very difficult to manage and care for, and therefore most often avoided. The choice of model for replacement therapy will depend on the cell type to be replaced. If, for example, cells with the potential to

Table 1
Summary of spinal cord injury models

SCI model	Uni/bilateral	Region	Species: Strain	References
Contusion				
Weight-drop	Bilateral	Cervical/ thoracic	Dog Rat: Sprague-Dawley Mouse: C57BLC	(11, 12) (47) (50) (62)
Weight-drop (NYU)		Thoracic	Rat: Long-Evans	(48)
Weight-drop, electro-magnetic spinal cord injury device (ESCID)	Bilateral	Thoracic	Rat: Wistar Mouse: Balbc-B10/PL	(49) (51)
Infinite horizons (IH) impactor		Cervical or thoracic	Rat: Sprague-Dawley	(52) (53)
Louisville Injury System Apparatus (LISA)	Bilateral	Thoracic	Rat: Sprague-Dawley	(54)
Compression				
Aneurysm clip	Bilateral (unilateral)	Thoracic	Rat: Sprague-Dawley	(55)
Extra/subdural balloon	Bilateral	Thoracic	Rat: Sprague-Dawley Rat: Wistar	(56) (57) (58)
Transection				
Scissors/microscissors	Bilateral	Cervical Thoracic	Rat: Wistar Rat: Sprague-Dawley	(59) (60)
Hemisection/over-hemisection/partial transection				
Lateral hemisection lesion cavity by aspiration	Unilateral	Cervical	Rat: Sprague-Dawley	(61)
Lateral hemisection using microknife	Unilateral	Cervical	Mouse: C57B1/6	(63)
Dorsal hemisection by microscissors	Bilateral dorsal hemisection	Cervical	Mouse: Glast-CreER	(65)
Lateral over-hemisection by iridectomy scissors	Bilateral dorsal columns/ unilateral	Cervical/ thoracic	Rat: Sprague-Dawley	(64)
Ischemia				
Photochemical lesion		Thoracic	Rat: Wistar	(66)
Irradiation lesion		Thoracic/ lumbar	Cat	(67)
Vessel ligation		Thoracic/ lumbar	Rat: Sprague-Dawley	(68)
Syringomyelia				
Quisqualic acid and kaolin injections	Bilateral	Cervical	Rat: Sprague-Dawley	(69)
Clip compression and kaolin injections	Bilateral	Thoracic	Rat: Wistar	(70)

remyelinate axons are evaluated, demyelinated axons must of course be present, surrounding the central lesion of an incomplete injury. If neurons are to be replaced, any type of model can be used.

4. Tools for Functional Analyses

Morphological and stereological methods can be used to quantify a neuroprotective effect of a drug or cell therapy, anatomical tract tracing, and other techniques can help us assess the extent of regenerative growth or anatomical plasticity. Still, every treatment intended for SCI patients has to be proven effective with respect to functional outcome. Functional analyses in SCI animal models are therefore absolutely necessary. The requirements of the tests used will however be different depending on the type of injury as well as the mechanism of the treatment studied.

It is important to recognize that all test situations except those that concern spontaneous locomotion in the animal's home cage are to some extent stressful. To minimize variation between animals it is crucial to give the animal sufficient time to get accustomed to the test equipment and test situation. If reward is necessary in the test situation, sweets (cereal, syrup, etc.) can be used. Starving the animals to appreciate food pellets is not recommended, since it is important that injured animals have optimal conditions to recover.

4.1. Tests of Open Field Locomotion

A number of functional assessment methods to evaluate hindlimb movement during open field locomotion have been developed during the last decades. The Combined Behavioral Score (CBS) (72) and the Motor Performance Score (MPS) (22) are examples of compound variables based on several tests, either with the results just summed (CBS) or applied in a hierarchical design (MPS). The most widely used test of locomotion in rodents with SCI during the last 15 years is the Basso–Beattie–Bresnahan locomotor score (BBB-score) (73, 74) and variants thereof such as the BBB-subscore (75) and the Basso Mouse Scale (76).

In contrast to CBS and MPS, the BBB-score does not include any other tests; it solely includes a refined qualitative analysis of hindlimb movements, coordination, and trunk stability. Tests of open field locomotion have the important advantage that motor functions ranging from complete paralysis to normal locomotion can be assessed. Although used very extensively, the BBB-score do suffer from some significant limitations. The scoring is by nature subjective and the data is ordinal, nonlinear, but may nevertheless be statistically analyzed with parametric statistical methods according to Scheff et al. (77). In addition, the sensitivity is not sufficient at all parts of the assessment scale. Importantly, the BBB-score is not suitable for cervical injuries, with lack of sensitivity as a particular problem (78). Finally, all parameters of normal locomotion may not be reproduced in different strains of rats. As an example, we have found that paw placement in normal Harlan Sprague-Dawley rnu/rnu immune-deficient rats is different from standard Harlan Sprague-Dawley rats (unpublished observation).

To assess open field locomotion the animal is required to be spontaneously active. To increase the motivation to move, injured and normal rats can be evaluated simultaneously. This helps the observer to identify mild pathological changes in locomotion. Importantly, to keep the observer blinded, the normal rat must not be a rat from a control group, but rather taken from a reference group which is not a part of the study. A "house" or similar hiding place from the animal's home cage placed on the other side of the open field is a strong incentive for the animal to move. Still, the overall health status of the animal can have effect on the performance, and certain animals also seem to be more affected by the test situation, showing significantly worse performance in the open field test situation compared to the home cage.

4.2. Beam and Grid Walk Tests

There are a number of functional tests that require that the animals can perform stepping. These tests are commonly used for mild-to-moderate thoracic contusion or compression injuries, but can be applied in animals with unilateral cervical injuries, in which the fore- and hindlimbs on one side of the animal are affected, provided some degree of stepping still is present. In the beam walk (or narrow beam walk) test for rats (79), the runway is 1–2 m long beams, width ranging from a few centimeters to less than 1 cm, with a flat surface and a soft mattress or elastic net below. In our lab, the test is performed by having the animal traverse a number of beams with successively decreasing width, each beam three times, until it falls off the beam, makes more foot slips than predetermined as exclusion criteria or refuses to walk the runway (22). The smallest width passed is recorded as the test result. Alternatively, a single beam, which gets successively narrower, can be used, again recording the smallest width passed.

In the grid walk test, the animal has to cross a meshwork (80) or a horizontal ladder (81), the test track typically about 10–30 cm wide and 1–2 m long with holes a few centimeters wide. Slightly raising the end of the track (meshwork or ladder) may increase the motivation for the animals to cross it. The number of paw misplacements and slips (i.e., when a paw slips through a hole in the mesh or between two rungs) is usually recorded. Since a mildly injured (or normal) rat can retract a misplaced paw very fast to correct the mistake, video recording is necessary for the evaluation. A major problem is that severely injured rats drag their hindlimbs without them slipping through the holes. To eliminate this problem, the number of steps when the animal takes a grip of the grid or rung can instead be used as outcome parameter. However, if animals with mild injuries are used, it will be necessary to quantify a large number of correct grips, making it a labor-intensive test.

A variant of the test, which has been used for mice, is to use a shorter ladder in a close to vertical position (82). The position of the ladder ascertains that paralyzed hindlimbs slip between the rungs of the ladder, thus counting of misplaced hindlimbs (or grip of rungs) correlates with motor function. Our experience is that the grid walk test is very sensitive to small functional differences in mildly injured rats, which was also noted by Lankhorst et al. (75) when they compared the grid walk test to BBB. However, since the beam walk and grid walk tests require stepping to have any discriminative power, they can only be used for mild-to-moderate SCI.

4.3. Gait Analysis

As an alternative to subjective scoring of locomotion, gait analysis of SCI animals can be used. The CatWalk (Noldus Information Technology, The Netherlands) or the TreadScan (CleverSys, USA) systems are commercially available alternatives with digital acquisition of foot-prints and automated analysis of a number of parameters such as swing and stance time, stride length, toe spread, track width, etc. Although as a general rule, it is advantageous to use objective quantitative analyses, the rather expensive equipment adds surprisingly little to the analytical power of common locomotion scoring (see, e.g., (83–85)). Gait analysis also requires weight-bearing stepping and thus shares this limitation with the beam- and grid walk tests. Consequently only mild-to-moderate injuries can be studied. Moderately injured animals that for some reason temporarily lose hindlimb weight support have to be assigned a non-quantitative (or "pseudo-quantitative") result or be excluded from the analysis. Either way, it will lead to biased or skewed data. Thus, these tests have limited use, but certainly add very important objective information in selected SCI models.

4.4. Swimming Tests

Locomotion in paraparetic rodents can also be analyzed using tests of swimming performance, and similar to open field locomotion, scoring systems have been suggested (86, 87), which assess limb movements, coordination, trunk position and stability, tail movement, etc. One of the potential benefits of analyzing SCI animals during swimming is that trunk and hindlimb movement parameters in severely injured animals are more easily detected. In the absence of weight-bearing capacity, weak movements of the hindlimbs may be more easily evaluated in a freely floating animal, while such movements can be difficult to detect during open field locomotion. However, none of the scoring systems have been validated, and according to our experience they are less reliable than BBB and other assessments of locomotion. Importantly, while normal rats consistently get the maximum score in locomotion tests, swimming scores of normal rats vary (see, e.g., (87)).

In addition to scoring movements, kinematic analysis of high speed video recordings of swimming can be used to objectively quantify components of movements such as angular velocity, rather than being limited to a subjective scoring (88, 89). So far, the validity and reliability of these methods for SCI animals have not been studied.

Water at room temperature is a mild stressor particularly to mice. With a platform at the end of the swim tank, most rats will immediately swim to the platform, and climb it to get out of the water. It is our experience that during sessions of repeated swimming, it is important to leave the rat on the platform for some time (10–15 s) and then take it out of the tank to reinforce the reward of reaching the platform. If a rat is immediately brought back from the platform to the other end of the tank for a new trial, some rats will stop swimming after 3–4 trials, something we interpret as learned helplessness.

4.5. Tests of Forelimb Function

As mentioned, from a translational aspect one may argue that cervical injuries, representing about 50% of all SCI, should be specifically modeled. Functional analysis of unilateral cervical injuries requires test methods largely different from those used for lower thoracic and lumbar injuries. The asymmetrical nature of the injury places certain demands on the test, but most importantly, skilled forepaw movements, manual dexterity involving the functions of digital muscles can be studied. Methods to analyze these functions include testing skilled reaching for food pellets (62), removing pieces of adhesive tape applied to the forepaw or forehead (90), and forepaw use during vertical exploration-rearing (91) sometimes called forelimb preference test. Gait analysis or analysis of forelimb use during swimming as well as grid walk test for analysis of ipsilateral forelimb–hindlimb coordination have also been used to evaluate cervical spinal cord injuries. Analysis of these latter tests is complicated by the fact that the cervical injury will affect hindlimb motor function, even if gross motor functions are preserved.

If the involvement of sensorimotor integration is to be excluded from the test, a defined forepaw motor function can be studied using a test of grip strength (92). An important advantage using this type of test is its quantitative nature. As we have seen, most quantitative objective tests are much more complicated than grip tests, while simpler tests with few exceptions rely on subjective nonlinear scoring. In this respect, forepaw grip strength test is exceptionally useful. For a more extensive review on functional effects of cervical SCI in rats, see (93).

4.6. Choice of Tests

The importance of using multiple analysis methods has been nicely illustrated in another disease model. When analyzing locomotor deficits in a mouse model of Huntington's disease, Pallier and colleagues (94) found that gait abnormalities present in the home cage, did not appear in an automated treadmill task. This illustrates that there can be major differences how non-motor factors affect performance in locomotion tests.

The anatomical substrates for different motor tests are of course different, and this can have surprisingly large consequences. In an interesting study on unilateral cervical injuries in rats, Anderson and colleagues (92) showed that recovery of forepaw motor function after medial or lateral hemisections depended on the test used. While rats with dorsal-medial transection often permanently lost gripping ability but improved in the food retrieval test, rats with lateral transections recovered in grip strength but suffered permanent deficits in food retrieval. Although it is not known if recovery in these tests is due to plasticity or regeneration, one must assume that a truly regenerative treatment tested in a cervical injury model would show very different effectiveness depending on the functional tests applied. Again, this emphasizes the need to use a wide range of functional tests when neuroprotective or regenerative treatments are evaluated after SCI.

While the relevance of SCI models can be inferred from the comparison of the biophysics of the injuries as well the resulting progressive pathological changes in rodents and human patients, the situation is different when it comes to functional evaluation of these injury models. There are neuroanatomical and neurophysiological differences between rodents and human beings that make functional comparison complicated. But it is an even more complex task to determine how a given improvement in rodent hindlimb motor performance can be translated to a functional improvement in human patients.

For validation of tools for functional analyses, the result of the functional assessment is usually correlated with a morphological parameter describing the extent of the lesion. Most commonly this parameter reflects the amount of spared white matter at the epicenter of the injury or the volume of the remaining spinal cord tissue. The rationale for measuring spared white matter is that hindlimb motor function mainly depends on descending (and ascending) axonal tracts in the white matter. If instead spared total spinal cord tissue is used, gray matter including neuronal tissue participating in local reflex circuitry is included. For most experimental injuries there is a close correlation between loss of gray and white matter (see, e.g., (57)). Thus, correlating spared white matter or spared total tissue will often give similar results for thoracic injuries, while gray matter will be relatively more important for loss of function after cervical injuries.

However, correlation of functional outcome with a morphological parameter is not evidence that the test is suitable for all purposes. It tells us how functions are lost with increasing injury severity. For studies on recovery, neuroprotection, or regeneration, some aspects of locomotion may be more affected than others. An example of this led to the introduction of the so-called BBB-subscore (75). These researchers observed that toe clearance and paw position are parameters lost after mild injuries while forelimb–hindlimb coordination is only affected after moderate injuries. However, limb coordination most often does not recover in spite of recovery of other functions (see (95) for discussion). Likewise, certain pathways may be more amenable to drug-enhanced regeneration, which could lead to a recovery of functions that do not match the expected sequence. To avoid the risk of missing important effects, a variety of functional tests should be included in the outcome assessment.

Functional and behavioral tests are not just used as outcome measures for neuroprotective and regenerative treatments, but also to detect side effects of treatments. Substances such as chondroitinase ABC (degrading inhibitory chondroitin sulfate proteoglycans) and Nogo antisera promote regeneration and plasticity, and enhance motor recovery after experimental SCI (96, 97). However, such treatments as well as neurotrophic factor or cell delivery may increase the risk of adverse effects. Among side effects, the greatest concern is probably increased pain, hyperalgesia, or allodynia, i.e., pain perception in response to non-painful stimuli (98, 99). Hence, analysis of pain perception should be added to the functional tests performed. The modalities typically studied are mechanical and thermal, assessed by applying (1) pressure, using calibrated filaments of different dimensions ("von Frey hairs") which are pressed onto the skin until the filament becomes bent (100, 101); (2) cold, by applying ethyl chloride or acetone on the paw or the shaved skin (102); (3) heat, by placing the rat on a hot plate, or focusing light on the paw or shaved skin (103). Rats react to painful stimuli in several ways, but vocalization has been shown to be the most reliable response. The functional assessment methods here described are summarized in Table 2.

5. Timing of Treatment and Clinical Translation

A major factor concerning clinical translation is the timing of treatment. If the therapeutic window according to animal data is 24 h, what does this correspond to in human patients? When models of SCI are used to study new potential therapies, three endogenous mechanisms can conceivably be affected by the treatment: degeneration, regeneration, and plasticity. In addition, replacement may be the result of cell therapy or bioengineering.

5.1. Protection

With regard to the first mechanism, protective effects are only possible for treatments initiated acutely, before the reversible degenerative processes are complete. Most scientists would probably agree that the major part of the reversible changes takes place hours

Table 2
Summary of functional assessment methods

Functional test	Function assessed	Uni-/bilateral	SCI type	SCI severity	Species/strain	References
Open field locomotion						
Combined Behavioral Score (CBS)	Sensory–motor function (hindlimb)	Bilateral	Contusion—weight-drop	Mild/moderate/severe	Rat: Sprague-Dawley	(72)
Motor performance score (MPS)	Sensory–motor function (hindlimb)	Bilateral	Photochemical lesion/transection	Mild/moderate/severe	Rat: Sprague-Dawley	(22)
Basso–Beattie–Bresnahan (BBB)	Open field locomotion (hindlimb)	Bilateral	Contusion—weight-drop(NYU, OSU)/transection	Mild/moderate (contusion) Severe (transection)	Rat: Sprague-Dawley/Long-Evans	(73, 74)
Basso Mouse Score (BMS)	Open field locomotion (hindlimb)	Bilateral	Contusion—weight-drop (ESCID, OSU, IH)/transection	Mild/moderate/severe	Mouse: C57BL/6, C57BL/10, B10.PL, BALB/c, C57BL/6 × 129S6	(76)
Grid walk/ladder walk	Sensory–motor function (hind- and fore limb)	Bilateral/unilateral	Dorsal hemisection Compression injury	Mild/moderate	Rat: Sprague-Dawley Mouse: C57BL/6J	(80) (81) (82)
Beam walk	Sensory–motor function	Bilateral/unilateral	Cortical lesion Photochemical lesion/transection	Mild/moderate	Rat: Charles River Rat: Sprague-Dawley	(79) (22)
Gait analysis						
CatWalk/TreadScan	Sensory–motor function (hind- and forelimb)	Bilateral/unilateral	Contusion—weight-drop (NYU)/dorsal hemisection	Mild/moderate	Rat: Wistar	(83)
			Contusion-LISA		Rat: Wistar Mouse: C57Bl/6	(84) (85)

(continued)

Table 2 (continued)

Functional test	Function assessed	Uni-/bilateral	SCI type	SCI severity	Species/strain	References
Swimming tests						
LSS	Sensory–motor function (hind- and forelimb)	Bilateral/ unilateral	Contusion—weight-drop (NYU)	Mild/moderate/ severe	Rat: Sprague-Dawley	(86)
Kinematics			Dorsal or lateral hemisection	Moderate/severe	Rat: Sprague-Dawley	(88)
					Rat: Lewis Mouse: C57BL/6	(89)
Other tests						
Forelimb reaching test	Sensory–motor function (forelimb)	Bilateral/ unilateral	Contusion—weight-drop	Moderate	Rat: Sprague-Dawley	(62)
Forelimb preference test	Sensory–motor function (forelimb)	Unilateral/ bilateral	Hemisection-cervical	Mild/moderate	Rat: Sprague-Dawley	(91)
Grip Strength Meter test	Motor function (forelimb)	Unilateral/ bilateral	Lateral hemisection/ midline transection	Moderate	Rat: Sprague-Dawley	(92)
Sensory test	Perception of					
von Frey hair test	Pressure	Unilateral/ bilateral	Ligation of L5-6 spinal nerve	–	Rat: Sprague-Dawley	(100)
Acetone/ethyl chloride spray	Cold		Photochemical lesion		Rat: Sprague-Dawley	(102)
Tail flick test	Heat			Moderate	Rat: Sprague-Dawley	(101)

to days after injury, but the definition of a reversible change depends on the existence of treatments that prevent those changes.

With regard to small molecule drugs, protective substances such as free radical scavengers and glutamate receptor antagonists have to be administered within hours after experimental injuries. These substances can prevent secondary degeneration of different constituents of the parenchyma, cell bodies and axons, oligodendrocytes, and myelin sheaths. The neuroprotective effect of growth factors is different. They may rescue axotomized neurons by providing trophic support through sustained release of growth factors. Loss of axotomized neurons can be a slow process. In a study on degeneration of spinal motor neurons after root avulsion, brain-derived neurotrophic factors delivered 14 days after injury could still reduce neuronal loss (albeit much less effectively than when administered immediately after injury) (104). It is a reasonable assumption that the delayed neuroprotection that has often been demonstrated for transplanted stem and progenitor cells is due to a sustained release of growth factors.

5.2 .Timing of Cell Therapy

A crucial question is if timing also is important for treatments aimed to enhance regeneration. Transplantation of oligodendrocyte progenitor cells derived from human embryonic stem cells is an example of a treatment proven effective only when performed shortly after injury (105). This would be expected if the main effect of the transplanted cells is neuroprotection, but these cells are believed to improve outcome by remyelinating axons in and around the injury. In many other studies on cell therapy aiming at enhancing regeneration, improved functional recovery has been demonstrated after transplantation in the acute stage of a SCI. Consequently, it is not possible to exclude that the effect is due to neuroprotection (106, 107), which could explain the need for early treatment.

In a recent study we found that animals transplanted with human neural precursor cells within 1 week after injury had an improved functional recovery associated with sparing of host spinal cord neurons, which was not seen in animals transplanted at 7 weeks after injury (Emgård et al., submitted). With regard to clinical translation, it is unfortunate that so many treatments with the aim to enhance regeneration have only been studied in acute–subacute experimental designs (see, e.g., (108)) without analysis of the mechanism(s) involved. Consequently, there are few treatment strategies today for which there are unequivocal evidence that they stimulate regeneration, and therefore very limited data that can be used to rationalize a therapeutic window for regenerative treatments in human patients.

5.3. Timing of Regenerative Treatment

It is not known how late after injury a treatment enhancing regeneration can be given. Spontaneous regeneration (109) and the option to enhance endogenous regeneration (110) decrease with time after injury, partly because death of axotomized neurons progresses, reducing the number of potentially regenerating axons (111). With respect to the local environment, the expression of a number of endogenous neurite-inhibiting factors changes during the first few weeks after injury (112–114), which makes the terrain less permissive for regenerative axon growth.

A consequence of these changes on regeneration was recently shown. Antibodies to the neurite growth-inhibitory factor Nogo-A increase regeneration and functional recovery after experimental SCI (87). Gonzenbach and colleagues in the Schwab research group recently found that this treatment only could be delayed by 1, but not 2 weeks (115). This difference correlated with much less regenerative growth after late treatment and probably reflects the rapid change of the injured cord towards a milieu less permissive to regenerative neurite growth even if Nogo-A is inhibited. In concordance, chondroitinase ABC treatment after SCI has a time-limited beneficial effect, for example, effects on motor parameters decrease already with a delay of a few days (116).

5.4. Translating Experimental Data to Clinical Practice

Thus, the vast majority of SCI treatments are most effective when given in the acute–subacute stage, i.e., up to 14 days, apparently irrespective of the underlying mechanism. Unfortunately, since it is difficult to translate the time scale of various biochemical and morphological processes from rodents to human beings, the more narrow the therapeutic window is in experimental animals, the more difficult it will be to correctly determine the corresponding window in human patients. If the mechanism(s) of the treatment is clarified, the limits of the therapeutic window can be estimated. By investigating the relevant conditions in human tissue, it will be possible to translate this window to the clinical situation. However, if the treatment mechanisms are not fully understood, or if the corresponding conditions in the human spinal cord are not known, a prediction about the therapeutic window will at best be an educated guess.

Whatever the mechanism and the target, protective drugs will be effective only as long as the injury has not reached its final extent. A study by Andrade and colleagues (117) is important with respect to the therapeutic window of such treatments. They performed a careful stereological analysis of the morphological development of SCI 7 and 30 days after moderate and severe contusion. The lesion volume in moderately injured rats increased considerably between the two time points, while the lesion volume in severely injured rats reached its final size already after 7 days. Similarly, the loss of neurons caudal to the injury declined between 7 and 30 days in moderately injured rats, but in severely injured rats reached its final extent already at 7 days.

With regard to clinical application of experimental data, these results suggest that to determine the therapeutic window for treatment, it will not just be necessary to translate the temporal changes per se from animal models to the patient situation. We must also consider the severity of the experimental injury and if possible compare it to the representative injury of patients recruited for a clinical trial. If severe injuries are selected for a clinical trial (a reasonable strategy to minimize the risk of doing harm), and the experimental data are derived from mild-to-moderate injuries, the set-up of the clinical study may be inappropriate. The translation of experimental studies to clinical practice, and estimating the limits of therapeutic windows, is complex at several levels.

6. Summary

Compared to many other disorders of the central nervous system, research in SCI offers the advantage to use several highly relevant animal models, both with respect to the injuries per se and the methods to evaluate the outcome of treatments tested. However, the choice of model has to be carefully considered, as their usefulness depends on the processes studied and the specific questions addressed.

When experimental studies on the relevant animal models eventually result in the identification of an effective treatment, the therapeutic window established in the animal model must be translated to the human situation. If this is to be more than guesswork, detailed knowledge of the treatment mechanism(s) is necessary, with sufficient knowledge of the corresponding conditions in the injured human spinal cord. With that information available, it is possible to answer the crucial question if the therapeutic window is clinically realistic, i.e., if there will be enough time to treat a patient?

Experimental data suggest that combining several treatments is more efficient than a single treatment. Likewise, for human SCI patients it is not likely that we will find a single treatment with superior effect (see (118) for a discussion). But with several treatments combined, the more important it is to understand the mechanism of action of each of these interventions, to make a sensible design of treatment regimes for human SCI patients, maximizing the chances of success.

References

1. Anderson KD (2004) Targeting recovery: priorities of the spinal cord-injured population. J Neurotrauma 21:1371–1383
2. Inskip JA, Ramer LM, Ramer MS, Krassioukov AV (2008) Autonomic assessment of animals with SCI: tools, techniques and translation. Spinal Cord 47:2–35
3. Norenberg MD, Smith J, Marcillo A (2004) The pathology of human SCI: defining the problem. J Neurotrauma 21:429–440
4. Onose G, Anghelescu A, Muresanu DF, Padure L, Haras MA, Chendreanu CO et al (2009) A review of published reports on neuroprotection in spinal cord injury. Spinal Cord 47:716–726
5. Hall ED, Wolf DL (1987) Post-traumatic spinal cord ischemia: relationship to injury severity and physiological parameters. Cent Nerv Syst Trauma 4:15–25
6. Anthes DL, Theriault E, Tator CH (1996) Ultrastructural evidence for arteriolar vasospasm after spinal cord trauma. Neurosurgery 39:804–814
7. Tei R, Kaido T, Nakase H, Sakaki T (2005) Secondary spinal cord hypoperfusion of

circumscribed areas after injury in rats. Neurol Res 27:403–408
8. Balentine JD (1978) Pathology of experimental spinal cord trauma I. The necrotic lesion as a function of vascular injury. Lab Invest 39: 236–253
9. Guha A, Tator CH, Rochon J (1989) Spinal cord blood flow and systemic blood pressure after experimental SCI in rats. Stroke 20: 372–377
10. Levi L, Wolf A, Belzberg H (1993) Hemodynamic parameters in patients with acute cervical cord trauma: description, intervention, and prediction of outcome. Neurosurgery 33:1007–1016
11. Allen AR (1911) Surgery of the experimental lesion of spinal cord equivalent to crush injury of fracture dislocation of spinal column. J Am Med Assoc 57:878–890
12. Allen AR (1914) Remarks on histopathological changes in the spinal cord due to impact. An experimental study. J Nerv Ment Dis 41: 141–147
13. Bramlett HM, Dietrich WD (2009) Progressive damage after brain and SCI: pathomechanisms and treatment strategies. Prog Brain Res 161:125–141
14. Schlaepfer WW, Bunge RP (1973) Effects of calcium ion concentration on the degeneration of amputated axons in tissue culture. J Cell Biol 59:456–470
15. Crowe MJ, Bresnahan JC, Shuman SL, Masters JN, Beattie MS (1997) Apoptosis and delayed degeneration after SCI in rats and monkeys. Nat Med 3:73–76
16. Springer JE, Azbill RD, Knapp PE (1999) Activation of the caspase-3 apoptotic cascade in traumatic SCI. Nat Med 5:943–946
17. Emery E, Aldana P, Bunge MB, Pucket W, Srinivaan A, Keane RW et al (1998) Apoptosis after traumatic human SCI. J Neurosurg 89: 911–920
18. Bracken MB, Shepard MJ, Holford TR, Leo-Summers L, Aldrich EF, Fazi M et al (1997) Administration of methylprednisolone for 24 or 48 hours or tirilazad mesylate for 48 hours in the treatment of acute SCI. Results of the third national acute SCI randomized controlled trial. National Acute SCI Study. JAMA 277:1597–1604
19. Hall ED (2011) Antioxidant therapies for acute SCI. Neurotherapeutics 8:152–167
20. Rohn TT, Hinds TR, Vincenzi FF (1993) Ion transport ATPases as targets for free radical damage. Protection by an aminosteroid of the Ca^{2+} pump ATPase and Na^+/K^+ pump ATPase of human red blood cell membranes. Biochem Pharmacol 46:525–534
21. van den Bosch L, Vandenberghe W, Klaassen H, van Houtte E, Robberecht W (2000) Ca(2+)-permeable AMPA receptors and selective vulnerability of motor neurons. J Neurol Sci 180:29–34
22. von Euler M, Akesson E, Samuelsson EB, Seiger A, Sundstrom E (1996) Motor performance score: a new algorithm for accurate behavioral testing of SCI in rats. Exp Neurol 137:242–254
23. Faden AI, Lemke M, Simon RP, Noble LJ (1988) N-methyl-D-aspartate antagonist MK801 improves outcome following traumatic SCI in rats: behavioral, anatomic, and neurochemical studies. J Neurotrauma 5:33–45
24. Hardinghan GE, Fukunaga Y, Bading H (2002) Extrasynaptic NMDARa oppose synaptic NMDARs by triggering CREB shut-off and cell death pathways. Nat Neurosci 5:405–414
25. Stanika RI, Pivovarova NB, Brantner CA, Watts CA, Winters CA, Andrews SB (2009) Coupling diverse routes of calcium entry to mitochondrial dysfunction and glutamate excitotoxicity. Proc Natl Acad Sci USA 106: 9854–9859
26. Sun HS, Doucette TA, Liu Y, Fang Y, Teves L, Aarts M et al (2008) Effectiveness of PSD95 inhibitors in permanent and transient focal ischemia in the rat. Stroke 39:2544–2553
27. Li S, Stys PK (2000) Mechanisms of ionotropic glutamate receptor-mediated excitotoxicity in isolated spinal cord white matter. J Neurosci 20:1190–1198
28. Kárasdóttir R, Cavelier P, Bergersen LH, Attwell D (2005) NMDA receptors are expressed in oligodendrocytes and activated in ischemia. Nature 438:1162–1166
29. Piña-Crespo JC, Talantova M, Micu I, States B, Chen HS, Tu S et al (2010) Excitatory glycine responses of CNS myelin mediated by NR1/NR3 "NMDA" receptor subunits. J Neurosci 30:11501–11505
30. Chatterton JE, Awobuluyi M, Premkumar LS, Takahashi H, Talantova M, Shin Y et al (2002) Excitatory glycine receptors containing the NR3 family of NMDA receptor subunits. Nature 415:793–798
31. Atlante A, Calissano P, Bobba A, Giannattasio S, Marra E, Passarella S (2001) Glutamate neurotoxicity, oxidative stress and mitochondria. FEBS Lett 497:1–5
32. Tretter L, Adam-Vizi V (2002) Glutamate release by an Na+ load and oxidative stress in nerve terminals: relevance to ischemia/reperfusion. J Neurochem 83:855–862
33. Park E, Velumian AA, Fehlings MG (2004) The role of excitotoxicity in secondary mechanisms of spinal cord injury: a review with an

emphasis on the implications for white matter degeneration. J Neurotrauma 21:754–774
34. Mattson MP (2003) Excitotoxic and excitoprotective mechanisms: abundant targets for the prevention and treatment of neurodegenerative disorders. Neuromolecular Med 3: 65–94
35. Guest JD, Hiester ED, Bunge RP (2005) Demyelination and Schwann cell responses adjacent to injury epicenter cavities following chronic human SCI. Exp Neurol 192: 384–393
36. Stys P (2004) White matter injury mechanisms. Curr Mol Med 4:113–130
37. Benowitz LI, Popovich PG (2011) Inflammation and axon regeneration. Curr Opin Neurol 24:577–583
38. Popovich PG, Hickey WF (2001) Bone marrow chimeric rats reveal the unique distribution of resident and recruited macrophages in the contused rat spinal cord. J Neuropathol Exp Neurol 60:676–685
39. Popovich PG, Wie P, Stokes BT (1997) Cellular inflammatory response after SCI in Sprague-Dawley and Lewis rats. J Comp Neurol 377:443–464
40. Byrnes KR, Washington PM, Knoblach SM, Hoffman E, Faden AI (2011) Delayed inflammatory mRNA and protein expression after spinal cord injury. J Neuroinflammation 8:130
41. Grossman SD, Rosenberg LJ, Wrathall JR (2001) Temporal-spatial pattern of acute neuronal and glial loss after spinal cord contusion. Exp Neurol 168:273–282
42. Yu WR, Fehlings MG (2011) Fas/FasL-mediated apoptosis and inflammation are key features of acute human spinal cord injury: implications for translational, clinical application. Acta Neuropathol 122(6):747–761
43. Kigerl KA, Gensel JC, Ankeny DP, Alexander JK, Donnelly DJ, Popovich PG (2009) Identification of two distinct macrophage subsets with divergent effects causing either neurotoxicity or regeneration in the injured mouse spinal cord. J Neurosci 29:13435–13444
44. Chen Q, Smith GM, Shine HD (2008) Immune activation is required for NT-3-induced axonal plasticity in chronic spinal cord injury. Exp Neurol 209:497–509
45. Bomstein Y, Marder JB, Vitner K, Smirnov I, Lisaey G, Butovsky O et al (2003) Features of skin-coincubated macrophages that promote recovery from spinal cord injury. J Neuroimmunol 142:10–16
46. Knoller N, Auerbach G, Fulga V, Zelig G, Attias J, Bakimer R et al (2005) Clinical experience using incubated autologous macrophages as a treatment for complete spinal cord injury: phase I study results. J Neurosurg Spine 3:173–181
47. Wrathall JR, Pettegrew RK, Harvey F (1985) Spinal cord contusion in the rat: production of graded, reproducible, injury groups. Exp Neurol 88:108–122
48. Gruner JA (1992) A monitored contusion model of SCI in the rat. J Neurotrauma 9:123–128
49. Stokes BT, Noyes DH, Behrmann DL (1992) An electromechanical spinal injury device with dynamic sensitivity. J Neurotrauma 9:187–195
50. Kuhn PL, Wrathall JR (1998) A mouse model of graded contusive SCI. J Neurotrauma 15: 125–140
51. Jakeman LB, Guan Z, Wei P, Ponnappan R, Dzwonczyk R, Popovich PG, Stokes BT (2000) Traumatic SCI produced by controlled contusion in mouse. J Neurotrauma 17:299–319
52. Scheff SW, Rabchevsky AG, Fugaccia I, Main JA, Jr Lumpp JE (2003) Experimental modeling of spinal cord injury: characterization of a force-defined injury device. J Neurotrauma 20:179–193
53. Anderson KD, Sharp KG, Oswald S (2009) Bilateral cervical contusion spinal cord injury in rats. Exp Neurol 220(1):9–22
54. Zhang YP, Burke DA, Shields LB, Chekmenev SY, Dincman T, Zhang Y et al (2008) Spinal cord contusion based on precize vertebral stabilization and tissue displacement measured by combined assessment to discriminate small functional differences. J Neurotrauma 25: 1227–1240
55. Rivlin AS, Tator CH (1978) Effect of duration of acute spinal cord compression in a new acute cord injury model in rat. Surg Neurol 10(1):38–43
56. Khan M, Griebel R (1983) Acute spinal cord injury in the rat: comparison of three experimental techniques. Can J Neurol Sci 10:161–165
57. Vanický I, Urdzíková L, Saganová K, Cízková D, Gálik J (2001) A simple and reproducible model of SCI induced by epidural balloon inflation in the rat. J Neurotrauma 18:1399–1407
58. Martin D, Schoenen J, DelrCe P, Gilson V, Rogister B, Leprince P et al (1992) Experimental acute traumatic injury of the adult rat spinal cord by a subdural inflatable balloon: methodology, behavioral analysis, and histopathology. J Neurosci Res 32: 539–550

59. Rivlin AS, Tator CH (1977) Objective clinical assessment of motor function after experimental spinal cord injury in the rat. J Neurosurg 47: 577–581
60. Fraidakis MJ, Spenger C, Olson L (2004) Partial recovery after treatment of chronic paraplegia in rat. Exp Neurol 188:33–42
61. Houle JD, Tom VJ, Mayes D, Wagoner G, Phillips N, Silver J (2006) Combining an autologous peripheral nervous system "bridge" and matrix modification by chondroitinase allows robust, functional regeneration beyond a hemisection lesion of the adult rat spinal cord. J Neurosci 26:7405–7415
62. Schrimsher GW, Reier P (1992) Forelimb motor performance following cervical spinal cord contusion injury in the rat. Exp Neurol 117:287–298
63. Anderson KD, Abdul M, Steward O (2004) Quantitative assessment of deficits and recovery of forelimb motor function after cervical SCI in mice. Exp Neurol 190:184–191
64. Bregman BS, McAtee M, Dai HN, Kuhn PL (1997) Neurotrophic factors increase axonal growth after SCI and transplantation in the adult rat. Exp Neurol 148:475–494
65. Göritz C, Dias DO, Tomilin N, Barbacid M, Shupliakov O, Frisén J (2011) A pericyte origin of spinal cord scar tissue. Science 333: 238–242
66. Watson BD, Prado R, Dietrich D, Ginsberg MD, Green BA (1986) Photochemically induced spinal cord injury in the rat. Brain Res 367:296–300
67. Blakemore WF, Palmer AC (1982) Delayed infarction of spinal cord white matter following X-irradiation. J Pathol 137(4):273–280
68. Le May DR, Neal S, Neal S, Zelenock GB, D'Alecy LG (1987) Paraplegia in the rat induced by aortic cross-clamping: model characterization and glucose exacerbation of neurologic deficit. J Vasc Surg 6:383–390
69. Yang L, Jones NR, Stoodley MA, Blumbergs PC, Brown CJ (2001) Excitotoxic model of post-traumatic syringomyelia in the rat. Spine 26(17):1842–1849
70. Seki T, Fehlings MG (2008) Mechanistic insights into posttraumatic syringomyelia based on a novel in vivo animal model. J Neurosurg Spine 8(4):365–375
71. Basso DM, Beattie MS, Bresnahan JC, Anderson DK, Faden AI, Gruner JA et al (1996) MASCIS evaluation of open field locomotor scores: effects of experience and teamwork on reliability. Multicenter animal SCI study. J Neurotrauma 13:343–359
72. Gale K, Kerasidis H, Wrathall JR (1985) Spinal cord contusion in the rat: behavioral analysis of functional neurologic impairment. Exp Neurol 88:123–134
73. Basso DM, Beattie MS, Bresnahan JC (1995) A sensitive and reliable locomotor rating scale for open field testing in rats. J Neurotrauma 12(1):1–21
74. Basso DM, Beattie MS, Bresnahan JC (1996) Graded histological and locomotor outcomes after spinal cord contusion using the NYU weight-drop device versus transection. Exp Neurol 139:244–256
75. Lankhorst AJ, Verzijl MR, Hamers FPT (1999) Experimental spinal cord contusion injury: comparison of different outcome parameters. Neurosci Res Commun 24: 135–148
76. Basso DM, Fischer LC, Anderson AJ, Jakeman LB, McTigue DM, Popovich PG (2006) Basso mouse scale for locomotion detects differences in recovery after SCI in five common mouse strains. J Neurotrauma 23:635–659
77. Scheff SW, Saucier DA, Cain ME (2002) A statistical method for analyzing rating scale data: the BBB locomotor score. J Neurotrauma 19:1251–1260
78. Martinez M, Brezun JM, Bonnier L, Xerri C (2009) A new rating scale for open-field evaluation of behavioral recovery after cervical SCI in rats. J Neurotrauma 26:1043–1053
79. Gentile AM, Green S, Neilburge A, Schmelzer W, Stein DG (1978) Disruption and recovery of locomotor and manipulatory behavior following cortical lesions in rats. Behav Biol 22:417–455
80. Prakriya M, McCabe PM, Holets V (1993) A computerized grid walking system for evaluating the accuracy of locomotion in rats. J Neurosci Methods 48:15–25
81. Metz GA, Whishaw IQ (2009) The ladder rung walking test: a scoring system and its practical application. J Vis Exp 28:1204
82. Apostolova I, Irintchev A, Schachner M (2006) Tenascin-R restricts posttraumatic remodeling of motoneuron innervation and functional recovery after spinal cord injury in adult mice. J Neurosci 26:7849–7859
83. Koopmans GC, Deumens R, Honig WM, Hamers FP, Steinbusch HW, Joosten EA (2005) The assessment of locomotor function in spinal cord injured rats: the importance of objective analysis of coordination. J Neurotrauma 22:214–225
84. Hamers PT, Lankhorst AJ, van Laar TJ, Veldhuis WB, Gispen WH (2001) Automated quantitative gait analysis during overground locomotion in the rat: its application to spinal cord contusion and transection injuries. J Neurotrauma 18(2):187–208

85. Beare JE, Morehouse JR, DeVries WH, Enzmann GU, Burke DA, Magnuson DS et al (2009) Gait analysis in normal and spinal contused mice using the TreadScan system. J Neurotrauma 26:2045–2056
86. Smith RR, Burke DA, Baldini AD, Shum-Siu A, Baltzley R, Bunger M et al (2006) The Louisville swim scale: a novel assessment of hindlimb function following SCI in adult rats. J Neurotrauma 23:1654–1670
87. Liebscher T, Schnell L, Schnell D, Scholl J, Schneider R, Gullo M et al (2005) Nogo-A antibody improves regeneration and locomotion of spinal cord-injured rats. Ann Neurol 58:706–719
88. Magnuson DSK, Smith RR, Brown EH, Enzmann G, Angeli C, Quesada PM et al (2009) Swimming as a model of task-specific locomotor retraining after spinal cord injury in the rat. Neurorehabil Neural Repair 23: 535–545
89. Zörner B, Filli L, Starkey ML, Gonzenbach R, Kasper H, Röthlisberger M et al (2010) Profiling locomotor recovery: comprehensive quantification of impairments after CNS damage in rodents. Nat Methods 7:701–708
90. Hernandez TD, Schallert T (1988) Seizures and recovery from experimental brain damage. Exp Neurol 102:318–324
91. Liu Y, Kim D, Himes BT, Chow SY, Schallert T, Murray M et al (1999) Transplants of fibroblasts genetically modified to express BDNF promote regeneration of adult rat rubrospinal axons and recovery of forelimb function. J Neurosci 19:4370–4387
92. Anderson KD, Gunawan A, Steward O (2007) Spinal pathways involved in the control of forelimb motor function in rats. Exp Neurol 206:318–331
93. Webb AA, Muir GD (2005) Sensorimotor behavior following incomplete cervical SCI in the rat. Behav Brain Res 165:147–159
94. Pallier PN, Drew CJG, Morton AJ (2009) The detection and measurement of locomotor deficits in a transgenic mouse model of Huntington's disease are task- and protocol-dependent: influence of non-motor factors on locomotor function. Brain Res Bull 78: 347–355
95. Basso DM (2004) Behavioral testing after spinal cord injury: congruities, complexities, and controversies. J Neurotrauma 21:395–404
96. Bradbury EJ, Moon LDF, Popat RJ, King VR, Bennett GS, Patel PN et al (2002) Chondroitinase ABC promotes functional recovery after SCI. Nature 416:636–640
97. Merkler D, Metz GA, Raineteau O, Dietz V, Schwab ME, Fouad K (2001) Locomotor recovery in spinal cord-injured rats treated with an antibody neutralizing the myelin-associated neurite growth inhibitor Nogo-A. J Neurosci 21:3665–3673
98. Bretzner F, Liu J, Curri E, Roskams AJ, Tetzlaff W (2008) Undesired effects of a combinatorial treatment for spinal cord injury—transplantation of olfactory ensheathing cells and BDNF infusion to the red nucleus. Eur J Neurosci 28:1795–1807
99. Hofstetter CP, Holmström NA, Lilja JA, Schweinhardt P, Hao J, Spenger C et al (2005) Allodynia limits the usefulness of intraspinal neural stem cell grafts; directed differentiation improves outcome. Nat Neurosci 8:346–353
100. Chaplan S, Bach F, Pogrel J, Chung J, Yaksh T (1994) Quantitative assessment of tactile allodynia in the rat paw. J Neurosci 16:7711–7724
101. Yu W, Hao J-X, Xu X-J, Saydoff J, Haegerstrand A, Hökfelt T et al (1998) Long-term alleviation of allodynia-like behaviors by intrathecal implantation of bovine chromaffin cells in rats with spinal cord injury. Pain 74:115–122
102. Choi Y, Yoon YW, Na HS, Kim SH, Chung JM (1994) Behavioral signs of ongoing pain and cold allodynia in a rat model of neuropathic pain. Pain 59:369–376
103. Hargreaves K, Dubner R, Brown F, Flores C, Joris J (1988) A new sensitive method for measuring thermal nociception in cutaneous hyperalgesia. Pain 32:77–88
104. Kishino A, Ishige Y, Tatsuno T, Nakayama C, Noguchi H (1997) BDNF prevents and reverses adult rat motor neuron degeneration and induced axonal outgrowth. Exp Neurol 144:273–286
105. Keirstead HR, Nistor G, Bernal G, Totoiu M, Cloutier F, Sharp K et al (2005) Human embryonic stem cell-derived oligodendrocyte progenitor cell transplants remyelinate and restore locomotion after SCI. J Neurosci 25:4694–4705
106. Madhavan L, Ourednik V, Ourednik J (2008) Neural stem/progenitor cells initiate the formation of cellular networks that provide neuroprotection by growth factor-modulated antioxidant expression. Stem Cells 26: 254–265
107. Jäderstad J, Jäderstad LM, Li J, Chintawar S, Salto C, Pandolfo M et al (2010) Communication via gap junctions underlies early functional and beneficial interactions between grafted neural stem cells and the host. Proc Natl Acad Sci USA 107:5184–5189
108. Lee H, McKeon RJ, Bellamkonda RV (2010) Sustained delivery of thermostabilized chABC enhances axonal sprouting and functional

recovery after SCI. Proc Natl Acad Sci USA 107:3340–3345

109. Decherchi P, Gauthier P (2002) Regeneration of acutely and chronically injured descending respiratory pathways within post-traumatic nerve grafts. Neuroscience 112:141–152
110. Houle JD, Ye JH (1997) Changes occur in the ability to promote axonal regeneration as the post-injury period increases. Neuroreport 8:751–755
111. Houle JD, Ye JH (1999) Survival of chronically injured neurons can be prolonged by treatment with neurotrophic factors. Neuroscience 3:929–936
112. Lemons ML, Howland DR, Anderson DK (1999) Chondroitin sulfate proteoglycan immunoreactivity increases following SCI and transplantation. Exp Neurol 160:51–65
113. Morgenstern DA, Asher RA, Fawcett JW (2002) Chondroitin sulphate proteoglycans in the CNS injury response. Prog Brain Res 137:313–332
114. Irizarry-Ramirez M, Wilson CA, Cruz-Orengo L, Figueroa J, Velázquez I, Jones H et al (2005) Upregulation of EphA3 receptor after SCI. J Neurotrauma 22:929–935
115. Gonzenbach RR, Zoerner B, Schnell L, Weinmann O, Mir AK, Schwab ME (2012) Delayed anti-Nogo-A antibody application after SCI shows progressive loss of responsiveness. J Neurotrauma 29(3):567–578
116. García-Alíles G, Lin R, Akrimi SF, Story D, Bradbury EJ, Fawcett JW (2008) Time window for the application of chondroitinase ABC after spinal cord injury. Exp Neurol 210: 331–338
117. Andrade MSR, Hanania FR, Daci K, Leme RJA, Chadi G (2008) Contuse lesion of the rat spinal cord of moderate intensity leads to a higher time-dependent secondary neurodegeneration than severe one. An open-window for experimental neuroprotective interventions. Tissue Cell 40:143–156
118. Ratan RR, Noble M (2009) Novel multimodal strategies to promote brain and spinal cord injury recovery. Stroke 40(3 Suppl): S130–S132

Chapter 14

Spinal Cord Injury: Modern Clinical Management and Its Correlation to Advances in Basic Science

Anders Holtz, Leif Anderberg, Shala Ghaderi Berntsson, and Håkan Aldskogius

Abstract

This chapter reflects the yearning for discoveries within the field of spinal cord injury (SCI) and points out the difficulties, opportunities, as well as possibilities to relate and translate the results from basic science to clinical management. After a brief presentation of "spinal cord history," we provide a short overview of the cornerstones of modern management of patients with SCI. This overview will, besides introducing basic scientists in the field of clinical practice, also expose the inherent imprecision of everyday clinical care compared to the high methodological requirements of experimental SCI research. We will focus our presentation on medical management aimed at preventing the consequences of secondary injury in the acute period, i.e., neuroprotection. A brief overview of various techniques of neural regeneration in the subacute (late) and chronic phase will follow this. Finally, based on the progress in basic science within the fields of neuroprotection and regeneration, some thoughts about future clinical therapeutic avenues will be discussed.

Key words: Acute, Subacute, Secondary injury, Neuroprotection, Regeneration, Transplantation, Stem cell, Combinatorial therapy, Translation

1. Introduction

The medical history of spinal cord injuries (SCI) starts with the ancient Egyptian medical papyrus known as the Edwin Smith Surgical Papyrus (1, 2). This medical treasure was written about 2,500 years bc, i.e., at the time of great pyramid building. The construction of the pyramids made it possible for the first time to study CNS trauma, due to a high incidence of accidents. The papyrus presents 48 trauma cases, all written by the physician and architect of the Sakkara pyramids, Imhotep. It includes six descriptions of vertebral injuries ranging from injuries not involving the spinal cord to cases with severe neurological deterioration. In the well-known

Håkan Aldskogius (ed.), *Animal Models of Spinal Cord Repair*, Neuromethods, vol. 76,
DOI 10.1007/978-1-62703-197-4_14, © Springer Science+Business Media, LLC 2013

case 33, as an example, "Instructions concerning a crushed vertebra in his neck," the author states that "having a crushed vertebra in his neck" is considered as "an ailment not to be treated." Imhotep describes in that case symptoms associated with SCI such as loss of motion in arms and legs (tetraplegia), loss of sensation below the level of injury, and loss of bladder control. The Egyptians refrained from surgical interventions involving the spine but avoided urinary retention by using bronze catheters.

It was not until Hippocrates (about 460–377 bc) that vertebral injuries were correlated with paralysis. He introduced the "Hippocratic bench" to reduce spinal deformities and advocated, in the treatment of the SCI patients, a diet consisting of 4–9 pints of donkey milk combined with honey and a special mild white wine from Mendez in Egypt. Hippocrates described the difficulties for the patients with paralysis such as constipation, dysuria, skin problems, and edema. He also recommended the use of exposure to the air in combination with direct application of wine, juice, vinegar, and oils as well as aluminum salt and lead and frequent changes of bandages in the treatment of pressure sores.

The Greek Galen (130–201) was the first to perform animal experiments resulting in SCI. He described that a longitudinal incision to the spinal cord did not give any loss of function but a transverse incision at cervical level resulted in loss of both motor and sensory function below the level of injury. The surgeon Paulus (625–690) from the Greek island of Eagina described the importance of the clinical examination. He also introduced laminectomy; i.e., the removal of the vertebral arch, in order to decompress the dural sac and spinal cord. The knowledge of human anatomy of the spine and spinal cord was extensively enhanced when Andreas Vesalius in 1543, during the renaissance period, published one of the greatest medical books ever written—entitled "De humani corporis fabrica libri septem" (On the fabric of the human body in seven books). It contains seven books in which the human nervous system for the first time was illustrated in detail.

A "breakthrough" in surgical treatment following a gunshot trauma to the spine was achieved in 1762 during the battle of Amenenburg when the surgeon Andre Louis successfully removed a bullet located in the lumbar spine and the patient survived. During the nineteenth century, however, famous persons such as Lord Nelson and the US president James Garfield died following a SCI. Alexander Graham Bell made serious, but unsuccessful, attempts to localize the bullet in Garfield's spinal cord with the help of a special metal detector device. The president died after a few months due to infections and internal bleedings probably caused by non-septic practices by his physicians.

The "father of neuroscience" Santiago Ramon y Cajal was rewarded with the Nobel Prize in medicine 1906 for the description of the central nervous system structure and in 1911 Reginald

Allen introduced the experimental weight-drop technique starting the "modern era" of SCI research. Both these researchers, already at that time, were aware of the contribution of secondary injury mechanisms to the final neurological deficit after an injury to the spinal cord. During that period, including the First World War era, the mortality was still up to 80 % within the first 2 weeks following a SCI at the cervical level. This lead to a continued fatalistic attitude towards treatment of SCI during the inter world war period. However, the modern era of rehabilitation started in 1943 when Sir Ludwig Guttmann opened the Stoke-Mandeville National Spinal Centre. The establishment of rehabilitation clinics focused on special programs for SCI together with the introduction of antibiotics meant a big step forward for this group of patients. The subsequent overall improved medical management has led to a substantial improvement in quality of life for SCI patients and a nearly equal life-long expectancy for these patients compared to the uninjured population.

2. Early Medical Management After Acute Traumatic SCI

The overall goal in the early management following acute SCI is besides the life-saving objective to counteract an ongoing neurological deterioration, to initiate posttraumatic mobilization, as well as to commence the rehabilitation process as soon as possible. For reviews in the concept of pre-hospital management and treatment in the acute stage at hospitals, see (3–6). Our knowledge about the pathophysiological process following an injury to the spinal cord has increased in parallel with the improved medical management. The concept of primary and secondary mechanisms has been accepted both experimentally and in the acute care of these patients (7–9). The primary events, caused by the moment of initial spinal cord compression by fractured bone, disc material, or blood, are morphologically defined as changes appearing within 15 min. These changes consist of mechanical disruption of spinal cord vasculature resulting in hemorrhage, loss of spinal cord pathway conduction, and ionic derangement. The secondary injury processes are initiated by these initial mechanically induced irreversible changes and are enhanced by remaining static compression of bone and/or disc material and epidurally located blood. Vascular damage (loss of auto-regulation, microvascular damage, and reduced spinal cord blood flow), free radical formation, and lipid peroxidation generate axonal damage, disturbed calcium homeostasis, alterations in opiate receptor mechanisms, apoptosis, inflammation, and excitotoxicity. All these mechanisms contribute to subsequent ischemia, hypoxia, reduced energy metabolism, and finally cell death.

To prevent secondary damage and self-destruction of neurons in order to rescue tissue is the core of the concept "neuroprotection." Neuroprotection research entails characterization of injury mechanisms and testing the influence of different substances in animal experiments, as well as clinical studies aimed at minimizing neurological deficits after trauma. The major goal of neuroprotective therapy is to preserve neurological function by prevention of the secondary injury mechanisms, i.e., to correct hypoxia and/or hypotension, to treat associated injuries, and to monitor the degree and possible spreading of the neurological deterioration. This is in the acute phase achieved by an active intensive medical treatment focusing on the following six cornerstones:

- Respiratory management
- Cardiovascular management
- Neurological examinations
- Pharmacological treatment
- Radiological investigations
- Surgical treatment

2.1. Medical Treatment

Earlier studies in which respiratory and circulatory parameters were monitored in patients with SCI revealed insufficient ventilation resulting in hypoxia, hemodynamic instability, and heart rhythm disturbances. The identification and treatment of these complications is of crucial importance in order to minimize morbidity and mortality following SCI. Tator and co-workers presented in 1984 the results of an active treatment towards the consequences of the secondary injury mechanisms (10). The effect of what was at that time a new treatment was related to a comparable group of patients not receiving the same therapy. The authors introduced administration of volume expanders (using crystalloid solutions, plasma, and whole blood) in order to avoid hypotension, the use of adequate ventilation including endotracheal intubation and assisted ventilation if required and routines to avoid sepsis and urological complications. These measures resulted in reduced mortality, morbidity, and shorter in-hospital stay and finally reduced costs to the society. The spinal cord injured patients were during the three National Acute Spinal Cord Injury Studies (NASCIS) that started the same decade upgraded to intensive care status. This intensive care attitude remains and patients with SCI are treated in intensive care units during the acute stage. The possibility to avoid neurological deterioration as well as prevent failure of other organ systems may hopefully increase from an aggressive intensive care treatment in the acute phase. This medical management should be tailored with regard to the results from previous experimental knowledge (11, 12). Future phase III clinical trials should enable the development of highly effective and safe neuroprotective therapies for human SCI (13).

2.1.1. Respiratory Management

Preservation of adequate oxygenation is of utmost importance in the management of the acute spinal cord injured patients (14, 15). Paralysis of the diaphragm and the respiratory accessory muscles results in an insufficient breathing ability and patients exhibit a significantly reduced inspiratory capacity. This results in a relative hypoxia, which might lead to impaired oxygen supply to the spinal cord. The resulting hypoxia speeds up and augments the destructive effects of the secondary injury mechanisms. It is thus of greatest importance to establish an adequate oxygenation in order to counteract the consequences of secondary pathophysiological events. Clinical symptoms such as agitation, disorientation, anxiety, and decreased breathing capacity (verified through arterial blood gases) determine the need for intubation and assisted ventilation. Arterial blood gases are measured frequently, in order to guarantee the patient optimal ventilation. The oxygenation is adequate if the PO_2 is 100 mmHg (12 kPa) or higher and PCO_2 is less than 45 mmHg or between 4.5 and 5.5 kPa. Some authors advocate prophylactic intubation and assisted ventilation of patients with high cervical lesions. Intubation is usually required if the PO_2 is less than 70 mmHg measured on room air and/or if the PCO_2 rises above 45 mmHg. Optimal oxygenation and ventilation is probably the most important therapeutic measure in treating the consequences following an injury to the spinal cord. An adequate supply of oxygen not only guarantees the spinal cord a sufficient oxygenation but also prevents additional organ failure.

2.1.2. Cardiovascular Management

The initial mechanical impact and the subsequent vasospasm result in local changes in the spinal cord blood flow (SCBF) and loss of SCBF auto-regulation. An injury located at or above the level of ThI results in an almost unopposed parasympathetic influence due to the loss of sympathetic impact. A neurogenic shock develops at systemic level as a consequence of this "autosympathectomy." This results among other things in a systemic hypotension and in a reduced spinal cord blood flow. This decrease in SCBF will lead to spinal cord ischemia and further destruction of nervous tissue. In this situation of ongoing secondary injury processes, the spinal cord is dependent on the systolic blood pressure in order to receive sufficient blood supply and oxygen to the area of injury.

Hypotension in patients with an isolated injury to the spinal cord is thus not due to hypovolemia (hypovolemic shock) but rather to a redistribution of blood from the central to the peripheral system (neurogenic shock). It has been shown that a systolic blood pressure below approximately 90 mmHg increases morbidity and mortality following an injury to the spinal cord. The blood pressure must therefore be maintained aggressively in order to assure an adequate blood flow to the spinal cord during the first hours to days following injury. Some authors recommend a mean arterial blood pressure (MABP) of 80 mmHg or a systolic blood pressure of 130 mmHg in order to maintain a satisfactory perfusion of the spinal cord. The principles of fluid replacement follows those for brain injury and the treatment to counteract the consequences of the secondary injury mechanisms starts as soon as the neurogenic shock has been diagnosed and the hypovolemic shock excluded. A maximum of 2.5 L fluid delivery during the first 24 h are accepted. Volume expanders (colloid solutions) and vasopressors (dopamine) are administrated if needed and bradycardia is treated with atropine (16, 17). Finally, special prophylactic considerations are taken to counteract the formation of venous tromboembolism (18) and avoid fever (19).

2.1.3. Neurological Examinations

It is of great importance to perform a detailed neurological examination following the initial resuscitation. The neurological evaluation should be performed repeatedly in the early posttraumatic phase, in order to detect alterations in neurological status. The ASIA (American Spinal Injury Association) "International Standards for Neurological and Functional Classification of SCI" is the most accepted standard for assessment of neurological function in spinal cord injured patients. Examination according to the principles of ASIA enables a description of sensorimotor function and level of injury, for details see the ASIA Web site and references (20, 21). Many experimental assessment methods have been presented during the last decades. The Basso, Beattie, and Bresnahan (BBB) locomotor rating scale relates to interlimb coordination and it is the most frequent used scale to evaluate different kinds of neurological dysfunction (or usually deterioration) in rats (22). The aim of neurological investigation, regardless of human or experimental use, is to diagnose the severity of the SCI and detect ongoing changes in neurological function (prior to or after various treatments) such as prior to surgical decision making. Both the experimental and clinical mode of assessment are easy to repeat, enables prognosis and the results are easy to communicate and thus of great value for future research.

2.1.4. Pharmacological Treatment

Drug therapy in the treatment of the spinal cord injured patients has a three to four decade old history. Increased knowledge regarding secondary injury mechanisms provides a basis for pharmacotherapy in the acute stage following injury. Various drugs have been tried based on the results of animal experiments. Such, so-called, neuroprotective agents, counteract the consequences of the secondary pathophysiological events. Two types of corticosteroids (methylprednisolone and tirilazad mesylate), naloxone and the GM-1-ganglioside, showed experimentally promising results regarding the possibility to counteract the neurological deterioration in the acute stage and were subsequently used in clinical trials. The results from these clinical trials did not meet the expectations from the experimental studies. Thus, so far, pharmacological neuroprotection in the treatment of acute SCI has fallen short and the only remaining drug currently in use is methylprednisolone (Solu-Medrol®), which according to the overall opinion is a "possible optional treatment" (23) and a recent example of the use of the drug in a European country is given in (24).

2.1.5. Radiological Investigations

The development within the field of radiology is explosive and the possibility to visualize posttraumatic changes in the spinal column and spinal cord has improved tremendously. The available techniques today are plain radiographs, CT (computer tomography), and MRI (magnetic resonance imaging). The introduction of MRI technology made it possible for the first time to detect small lesions in the spinal cord tissue. The main advantage of MRI is the ability to provide morphological information of the spinal cord and to visualize hemorrhages and edema, as well as other changes within spinal cord tissue. The extent of bleeding and edema also provides some indication of the severity of injury and therefore information of prognostic value (25, 26).

2.1.6. Surgical Treatment

The sixth and last cornerstone in the management of the spinal cord injured patients is the surgical decision making. The option between surgical intervention versus conservative treatment (nonsurgical treatment) in the management of fractures and/or ligament injuries must be evaluated carefully. There is, thus, still controversies regarding the benefits of decompressing the spinal cord and restoring the spinal column alignment following trauma as well as regarding the timing of surgery (27, 28). It is known from animal experimental studies that remaining compression by artificial weights increases the severity of the neurological deterioration and that this increase in severity is time dependent. It has also been shown that early removal of foreign material, i.e., decompression of the spinal cord, results in less neurological deficit. Similar circumstances should therefore, logically, be valid among humans. However, it has not yet been strictly scientifically proven that early decompression and stabilization is to prefer in comparison to nonsurgical treatment among humans. On the other hand, neither has it been strictly scientifically proven that conservative treatment is favorable compared to surgical intervention. Studies fulfilling all scientific requirements are extremely difficult to perform since the treatment today tends to involve an increased surgical activity and less interest for the nonsurgical option. However, a prospective, randomized, large scale multicentre study, Surgical Timing in Acute Spinal Cord Injury Study (STASCIS), addressing this problem, has now been conducted (29). One year results from the STASCIS study show that patients who underwent surgical decompression within 24 h regained greater improvement in neurological function accompanied by a lower complication rate compared to individuals operated after 24 h. In the STASCIS study patients who underwent early surgery had a 2.8 time higher chance of improving two grades on the ASIA Impairment Scale compared to the delayed surgery group.

In addition, a Phase II study evaluating surgical management of spinal cord injuries in the neck is conducted in patients with central cord syndrome. The hypothesis tested is if surgical decompression within 5 days following traumatic central cord syndrome will result in a more rapid recovery than decompression 6 weeks following injury (30). This rather wide time frame in inclusion criteria between the early and late surgery groups for traumatic central cord syndrome reflects the difficulties in the clinical management of this most often elderly patient group and the result of this study will be of great interest.

2.2. Summary of Clinical Strategies to Provide Neuroprotection

Neuroprotection is defined as measures to counteract secondary injury mechanisms and/or limit the extent of damage caused by self-destructive cellular and tissue processes. Based on extensive experimental research, a variety of therapeutic agents targeting one or more of those mechanisms have been proposed for possible clinical use to achieve neuroprotection and prevent further tissue damage. The results of translating this research to the clinical setting have been very discouraging and pharmacological neuroprotection for patients with SCI has fallen short of the expectations. The knowledge and experiences that have been accumulated during the neuroprotective journey from animal experiments to clinical studies has led to improved medical management, summarized as the six cornerstones above, but not to any new therapeutic principle that is able to even partially restore the neurological function in patients. The outcome of this previous extensive research on spinal cord neuroprotection has obviously dampened the optimism for a therapeutic breakthrough in clinical neuroprotection within a foreseeable future and the focus of translational research in spinal cord repair is now largely transferred to the field of neural regeneration and plasticity.

3. Neural Regeneration, Plasticity, and Their Clinical Prospects

Ever since the ground breaking studies by Aguayo and collaborators (31) extensive experimental research has encouraged expectations that lost neurological functions in patients with SCI can at least be in part restored by regeneration of injured axons and/or by formation of new synapses from uninjured axons. Here, we will highlight selected lines of experimental spinal cord regeneration and plasticity research and, in the end, consider their possible implications for human research and the bedside perspective. The reader

Table 1
Regeneration obstacles

Nerve cell disability	Environmental factors
Nerve cell body response	Presence of growth inhibiting factors Scar tissue formation Formation of cavities and cysts at, below, and above the level of injury

is referred to Chap. 6 for an update on recent developments in pharmacological treatment aimed at activating the central pattern generator after complete SCI. Regeneration and plasticity are hampered by the combination of a deficient neuronal response to CNS injury (nerve cell disability) and the emergence of a non-permissive environment (environmental factors; Table 1).

4. Regeneration Obstacles

4.1. Nerve Cell Disability

Following axon injury, a shift in gene expression from a "transmitting mode" to a "growing (regenerating) mode" among damaged neurons is seen as a sign of nerve cell adaptability. Neurons with axons in the central nervous system are able to maintain a sustained expression of regeneration-associated genes (RAGs) only for a limited period. The injured neurons gradually enter an atrophic state and may, eventually, degenerate and die as a result of the absence of sufficient intrinsic and extrinsic growth stimulating factors.

4.2. Environmental Hostility

Neurons which attempt to regenerate are exposed to a growth-inhibitory environment which is created like a three-step rocket after the injury. Firstly, myelin-associated growth inhibitors such as Nogo-A, myelin-associated glycoprotein (MAG), and oligodendrocyte myelin glycoprotein (OMgp) produced by oligodendrocytes inhibit axonal growth (32). Secondly, a chemical barrier composed of chondroitin sulfate proteoglycans (CSPGs) and collagens is produced that together with components from astrocytes, microglial cells, oligodendrocytes, and meningeal cells contribute to the formation of a "scar tissue." Finally, necrotic and apoptotic cell death in the injured area enhances the development of cavities and cysts in the spinal cord itself after many weeks. The following part of this presentation will describe obstacles of interest and address regeneration strategies to prevent those mechanisms (33–37).

5, Regeneration Strategies

Successful regeneration in the spinal cord requires the combination of several approaches divided as suggested below in order to fulfill the aims of bridging the lesion gap at the site of the SCI, replacing dead cells and to contributing to an non-hostile environment for axon regeneration:

Nerve cell strategy

- Overcoming the intrinsic growth deficient neuronal capacity

Environmental strategies

- Inhibiting the early inhibitory mechanisms for axon growth in the environment of the injured neurons
- Counteracting the late inhibitory mechanisms on axon growth in the environment of the injured neurons

5.1. Nerve Cell Strategy

5.1.1. Overcoming the Intrinsic Growth Deficient Neuronal Capacity

The nerve cell targeted strategy aims at increasing the *intrinsic neuron capacity for regeneration*. Following injury, the normally inactive regeneration "program" becomes activated in order to rebuild the lost part of the axon and restore lost connections (38, 39). This program includes activation of RAGs and is significantly stimulated by treatment with specific growth factors, also referred to as neurotrophic factors (NTFs; Table 2). Their activities in general promote neuron survival and axonal outgrowth, as well as the maintenance of neuron–target interactions. In the context of SCI they stimulate survival of injured neurons, regulate the expression of RAGs in a way that increases the regeneration capacity of the injured neurons. However, as a result of this stimulation, NTFs may also cause undesired axonal growth leading to, e.g., neuropathic pain, or new functionally inappropriate, or unproductive connections. Furthermore, since NTFs act on multiple neuronal populations in the nervous system they may cause side effects, due to their action on uninjured neurons. Future research must

Table 2
Examples of growth factors (neurotrophic factors)

Neurotrophic factors	Abbreviation
Nerve growth factor	NGF
Neurotrophin 3 and 4	NT-3/NT-4
Brain-derived neurotrophic factor	BDNF
Fibroblast growth factor	FGF
Glia cell line derived neurotrophic factor	GDNF

Table 3
Strategies stimulating the neuronal intrinsic regenerating capacity

Target/mechanism	Delivery	Effect
CNS-tissue (neurons)	Intrathecal injections	Regrowth of damaged axons
Stimulating regeneration-associated genes (RAGs)	Local deposition and others	Functional improvement

determine which growth factors are optimal for promoting survival and regeneration of different types of neuron populations, at what time point they should be administered, and in which fashion the delivery should be performed in order to guarantee an efficient and controlled treatment. Until now, intrathecal delivery through osmotic pumps containing genetically modified neurotrophin releasing cells, e.g., fibroblasts, the implantation of slow-release cell-free systems, or gene therapy, has been possible mainly experimental options (Table 3). Unfortunately clinical trials using delivery of growth factors for various disorders have failed due to lack of efficacy or unacceptable side effects or both. However, at least one phase III study regarding efficacy and safety of growth hormone treatment in SCI is performed with the aim to study whether growth hormone can improve motor function. The growth hormone is given subcutaneously daily for 1 year in a double blinded fashion with a follow-up after 1 year (40). In the future, combining the delivery of promising growth factors with other therapeutic approaches seems to be the mode by which growth factors will be administered.

5.2. Environmental Strategies

5.2.1. Inhibiting the Early Inhibitory Mechanisms for Axon Growth in the Environment of the Injured Neurons

Strategies that obstruct the early inhibitory mechanisms on axon growth in the environment of the injured neurons involve measures that can be taken as long a formation of scar tissue is estimated to take place at the level of injury. Here, we discuss some of the main strategies that have been adopted in the early post-lesion period to counteract the accumulation of myelin-associated inhibitors and by factors that decrease the formation of scar tissue in the environment surrounding the nerve cell (34, 35, 37, 41) (Table 4).

Blocking Axonal Growth Inhibitory Molecules

The uninjured spinal cord contains powerful inhibitory substances that prevent neurite growth. Following trauma, the same factors create a major obstacle by delaying and/or preventing the onset of the regenerating process (42, 43). The myelin growth inhibiting factor NOGO-A, a myelin protein produced by oligodendrocytes, exerts its inhibitory effect by blocking specific Nogo-A-receptors on the surface of the axons resulting in a receptor-mediated inhibition of axon growth (Table 4). Administration intrathecally of neutralizing antibodies (IN-1) that either binds to the Nogo-A itself

Table 4
Strategies to counteract the early inhibitory mechanisms of the CNS environment

Factor	Target	Delivery	Effect
Antibodies (IN-1)	Nogo-A receptors on regenerating axons	Pumps Pills?	Blocked inhibition of axonal outgrowth Immunization
4-Amino-pyridine	Demyelinated axons	Intravenous Intrathecal	Restored signal transmission
Monoclonal antibody	ICAM	Immunization	Reduced edema increased spinal cord blood flow (SCBF)
	Interleukins	Systemically Intraperitoneally	Switches off the inflammatory response
Activated macrophages Chondroitinase ABC Erythropoietin (EPO) Rho	Proteoglycans Actin cytoskeleton	Injury epicenter Local infusion Systemic Injury site	Improved motor function Reduces cyst formation Digestion of scar Functional improvement Retain neuroprotective properties Improves SCBF and locomotion

or act as antagonists to the Nogo-A receptor blocks the effect of Nogo-A and thereby indirectly stimulates axon regeneration. Results from a Phase I trial with intrathecal administration of human anti-Nogo-A antibody in spinal cord injured patients indicate that this treatment is feasible and safe. Based on these observations, a Phase II study is underway (44).

The myelin components MAG and OMgp exert their effect in a similar receptor-mediated fashion. Anti-MAG is a monoclonal antibody directed against MAG which is a protein that inhibits axonal regeneration. The first anti-MAG administration to human has been completed (45). It is a Phase I study and the anti-MAG is given as an intravenous infusion.

Combination therapies targeting MAIS (myelin-associated inhibitors) have recently been reported and combination treatments are the future approach when using these antibodies (46).

Restoring Impulse Propagation by Modulating Axon Membrane Properties

After contusion injuries, the most frequent cause of SCI, many axons survive both primary and secondary events. However, degeneration of oligodendrocytes, mainly by apoptosis, leads to partial or total destruction of the myelin that covers the axons. As a result of this demyelination impulse propagation is severely disturbed or completely blocked. In addition, propagated electrical impulses will be spread between the demyelinated axons like a short circuit in an electrical cable when the outer insulation is damaged. An increased number of potassium channels are exposed when an axon is demyelinated and potassium (K^+) leaks into the extracellular space resulting in conduction failure. Infusion of the fast voltage-sensitive potassium channel blocker Fampridine-SR (4-amino-pyridine, 4-AP) blocks the potassium channels at those gaps which results in increased action potentials, thereby facilitating propagation of action potentials in demyelinated axons (Table 4). A Phase II study published in 2007 showed an increased effect of Fampridine on spasticity when given in a tolerable dose (47). A Phase III clinical trial showed a trend for improvement only in spasticity and the therapeutic use of this drug is now focused on multiple sclerosis.

Modulating the Immune System

The inflammatory response after SCI includes degenerative and reparative processes summarized into three consecutive periods: The cellular debris and breakdown products are eliminated during the initiation period (the phase of disintegration), thereby promoting the restoration of a beneficial environment for surviving cells. This period is followed by the maintenance period (the scar forming phase) during which the trauma area is isolated from surrounding healthy tissue. Finally, the phase of shutting off macrophages, transformed from mononuclear cells in the phagocyte system, plays a key role in inflammation and an inadequate macrophage response may be an important factor behind the failure of axon regeneration in the spinal cord (48, 49). Most of the macrophages originate from microglia although the astrocytes exert a tight control of the activation of microglia to fully developed macrophages, thereby preventing rapid removal of, e.g., growth inhibitors that are present in degenerating myelin. Thus, an optimal phagocytic response may benefit from a facilitated entry of monocytes into the degenerating spinal cord white matter and their subsequent differentiation to competent macrophages, provided that this will not result in a fully developed inflammation. Recruitment into the spinal cord of circulating monocytes requires that they first attach to the endothelial cells in capillaries before entering the spinal cord tissue. This attachment and the subsequent entry are regulated by cell adhesion molecules (CAMs) located at the cell surface (50).

Intercellular adhesion molecule (ICAM) at the surface of endothelial cells is a ligand for macrophages and plays a key role in the initiation phase of the inflammatory response (Table 4). It might be possible to regulate the monocyte-mediated components of inflammation following SCI by modulating ICAM expression. A sufficient level of macrophage activity in the phase of initiation, followed by a downregulation of this activity in the phase of maintenance, would first promote rapid removal of cellular debris and reduce the impact of macrophage in the scar forming phase, respectively.

The interleukin (IL) family of cytokines is important modulators of immune and inflammatory responses following neural trauma. Whereas IL-1 and IL-6 are considered proinflammatory, IL-10 plays a role in switching off the inflammatory response (Table 4). Intravenous or intraperitoneal administration of IL-10 attenuates the inflammatory response and mitigates the secondary injury process. Interleukins have been shown to reduce edema formation, reduce the period of secondary damage, and improve local blood flow and neurological recovery following trauma to the spinal cord. In conclusion, ICAM and interleukins are molecules potentially useful for pharmacological treatment in SCI (51).

Activated Autologous Macrophages

Macrophages clean up cell debris promoting a controlled inflammatory reaction that forms the first phase of the wound healing process. This does not occur in CNS because their activities are prevented by a mechanism known as immune privilege. In addition, CNS macrophages are significantly less activated compared to macrophages in the peripheral nervous system following injury. This may also contribute to less efficient removal of myelin-associated inhibitors and suboptimal production and release of growth factors. However, it appears that incubated microphages circumvent the immune privilege and thereby support regrowth of axons through the injury site, enabling some recovery of neurological function (52, 53). Implantation of activated macrophages derived from peripheral nerves has been shown experimentally to have neuroprotective effects and stimulate axonal regeneration (Table 4). These findings were the basis for the initiation of a Phase I clinical trial in patients with SCI, using activated macrophages isolated from the patient's own blood (ProCord). The following phase II study has however been suspended (54) and no clinical study is ongoing.

Chondroitinase ABC: A Molecular Machete

Proteoglycans, collagens, and adhesive proteins in the extracellular matrix participate in the scar formation which is hostile towards axonal regeneration but a natural response after injury (55). Chondroitin sulfate proteoglycans (CSPGs) are the main contributors to the scar formation and the major culprit of the failure of injured axons to grow beyond the level of injury. A bacterial enzyme was discovered in 2002 with the capability to digest chondroitin sulfate (56). Bacterial chondroitinase ABC (ChABC) given intrathecally following a trauma in the posterior horn in adult rats reduced scar formation by acting as a molecular machete (Table 4). As a result, regeneration of sensory and corticospinal axons was supported, the level of regeneration-associated proteins was increased, and functional recovery of movements and proprioception was observed. The use of implanted channels containing Schwann cells together with ChABC and olfactory ensheathing cells (OECs), as an example of combined treatment, promoted regeneration of serotonergic axons beyond the grafts (57, 58). Finally, in a clinical Phase III setup, ChABC has been tested following herniated lumbar discs resulting in less leg pain in the treatment group compared to controls (59).

Erythropoietin

Erythropoietin (EPO), the prime stimulator of erythroid progenitor cell proliferation, has been shown to exert neuroprotective effects by reducing lipid peroxidation and inflammation as well as by counteracting apoptosis and stimulating oligodendrogenesis (60) (Table 4). Administration of carbamyl erythropoietin (CEPO) within the first 24 h following experimental SCI reduces the neurological deficits compared to controls (61). However, two clinical trials using EPO following spinal shock and spinal cord compression have been suspended and no further clinical studies along this line are ongoing.

C-AMP and RHO Pathway Antagonists

Altering the intracellular signaling events following inhibition of neurons has been investigated as a therapeutic strategy for SCI. By increasing the intracellular levels of cyclic adenosine monophosphate (c-AMP) and protein kinase A activity, neurons were shown to overcome myelin-based inhibition. Cyclic AMP is able to turn on growth factor genes in nerve cells, thereby stimulating axon growth and helping to overcome signals that normally inhibit regeneration. Post-injury administration to zebrafish of Rolipram, an agent that prevents the hydrolysis of c-AMP and thereby increases c-AMP levels, lead to regeneration of severed CNS axons and restored function. Increasing c-AMP levels after SCI has also been tried in combination with other treatments such as Rolipram and Schwann cell transplants resulting in improved function in paralyzed rats (62). Before any human clinical test can begin, the therapeutic windows of c-AMP delivery must be defined as well as the role of the substance in various combination treatments.

The GTPase-associated signaling protein Rho influences cell motility by transducing extracellular signals to alterations in the actin cytoskeleton (63) (Table 14.4). SCI increases Rho-activity resulting in growth cone collapse and neurite retraction. Promoted neurite extension, improved spinal cord blood flow, and improved locomotion have been seen after administration of Rho-associated kinase inhibitors early after experimental SCI (64). A cell permeable variant of Rho inhibitor known as Cethrin has been developed and used in a Phase I/IIa study (65). Cethrin is an engineered version of C3 transferase derived from clostridium botulinum. The drug blocks the common pathway, called RHO, through which myelin-related inhibitory proteins affect the spinal cord nerve cells. Cethrin was administered epidurally within 72 h after SCI simultaneously with decompression surgery. The drug seems to be safe and individuals with cervical injuries improved better than expected both in sensory and motor scores. Hopefully, a future prospective randomized clinical study will bring more knowledge about the therapeutic effect of this drug.

5.2.2. Counteracting the Late Inhibitory Mechanisms on Axon Growth in the Environment of the Injured Neurons

This period starts once the scar formation is established. Morphologically the upper and lower stumps of the injured area are separated by a gap consisting of intra- and extramedullary scar tissue and/or cyst formations, which creates a mechanical barrier and prevents injured axons to navigate this area. The growing axons usually fail to re-enter the spinal cord above or below the lesion and they need to be guided through or around the mechanical obstacle by biological or biosynthetic bridges or supportive structures with or without the combination of growth factors. The term "filling the gap" includes a variety of bridging strategies to overcome this mechanical obstacle and the combination of supportive bridging structures and specific cell implantations may be a promising way to support and maintain axon elongation across the injury site. Cellular transplantation after SCI has in that respect several aims: to bridge "the gap," to replace dead cells, and to contribute to an affirmative environment for axon regeneration and transplants and can be more or less used as a complement to the following supportive structures.

Supportive Structures

Supportive structures promote axonal growth by acting as a scaffold for growth factors and/or as a substrate for growth interactions with regeneration axons. Peripheral nerve tissue or its components are the most frequently used scaffold due to the ability of injured axons to grow long distances in a transplanted peripheral nerve graft.

Peripheral Nerves

Peripheral nerve grafts with different combinations of therapies have been utilized. An elaborate example of the use of peripheral nerve grafts to bridge SCI is the studies by Olson and collaborators (66) (Table 5). In their initial experiments, multiple intercostal nerve grafts were inserted between the stumps of a complete spinal cord transection to provide conduits across the injury site from gray matter caudally to white matter rostrally for ascending pathways and from white to gray matter for descending pathways. The nerve grafts were stabilized with fibrin glue allowing slow release of acidic fibroblast growth factor (aFGF). Morphologically, axonal regrowth across the lesion site was seen as well as recovery of hindlimb sensorimotor function. Similar results were subsequently obtained even after a chronic lesion (67). Furthermore, clear evidence for at least a partial causal connection was established in a term experiment in which functional recovery was completely reversed after sectioning the grafts (67). However, much work remains to be done to determine whether therapies that involve peripheral nerve bridge grafting can safely and effectively improve outcome after human SCI.

Schwann Cells

Schwann cells are the essential growth promoting cellular element of peripheral nerves. Isolated Schwann cells or growth promoting Schwann cell molecules would therefore be suitable as guidance channel for injured spinal cord axons. This approach was used by Bunge and co-workers who filled "guidance channels" with Schwann cells and growth factors (68) (Table 5). Increased axonal growth reduced secondary degeneration of axons

Table 5
Strategies to overcome the late inhibitory mechanisms of the CNS environment: supportive structures

Factor	Target	Delivery	Effect
Peripheral nerve + aFGF	Axotomized axons	Locally with glue	Structural and functional recovery
Schwann cells (SC) + growth factors	Axotomized axons	Placement of SC transplants in guidance channels	Reduced secondary degeneration Functional improvement
Olfactory ensheathing cells (OEC)	Axotomized axons	Local deposition	Reduced secondary degeneration Functional improvement
Non-cellular elements + stem cells	Cut axons Scar forming glia	Placement of stem cells in transplanted polymers	Improved tissue sparing Reduced scar formation Functional improvement

and functional recovery was observed in the animal models (69). The hostile astrocytic reaction towards Schwann cells hampers, however, the ability of these cells to produce myelin. The first remaining part towards human clinical trials is to conduct safety and efficiency evaluations of transplanting autologous Schwann cells into non-human primates after contusive SCI. There are no reports of ongoing clinical trials involving transplantation of Schwann cells after SCI.

Olfactory Ensheathing Cells

Olfactory ensheathing cells (OECs) associate with olfactory sensory axons along their way to the target neurons in the olfactory bulb. OECs are continuously supporting the growth of axons into the olfactory bulb from newly formed olfactory sensory neurons in the olfactory mucosa. OECs are fully integrated into the adult CNS; they release growth factors and other growth promoting molecules and are able to produce myelin around demyelinated or regenerating CNS axons. OECs stimulate regrowth of damaged axons in the spinal cord when locally implanted as well as growth of axons through areas with scar-tissue formation resulting in signs of functional recovery of sensory and postural functions (Table 5). In addition to their supportive and guiding properties, OECs have the ability to promote synaptic plasticity and to contribute to structural and functional repair when stem cells from olfactory mucosa are present in the transplants. The hostile astrocytic reactions which are seen with Schwann cell implants are essentially absent after implants of OECs making those cells more favorable candidates than Schwann cells for supporting axonal regeneration in injured

CNS issue. OECs for transplanting purposes can be harvested under local anesthesia from human olfactory mucosa, grown in vitro and thereafter used for transplantation.

Among other studies a Phase I/IIa design study whose main aim was to test the feasibility and safety of transplantation of autologous olfactory ensheathing cells into the injured spinal cord in human paraplegia was published in 2008 (70). The authors concluded that transplantation of autologous olfactory ensheathing cells into the injured spinal cord is feasible and safe up to 3 years of post-implantation. However, this conclusion should be considered preliminary because of the small number of trial patients. For an extended review see (71). In summary, issues like optimal source of cells, their age, and also graft strategy are not yet fully answered but studies in humans have in any case not shown any adverse side effects.

Artificial Supportive Structures

A variety of artificial biocompatible materials have been used to bridge SCI (72). As an example Rochkind and colleagues placed a multi-component polymer filled with stem cells coated with a suitable substrate for axonal growth into the gap of the hemisected spinal cord (73). Improved hindlimb functional recovery was seen in the grafted animals compared to the controls (Table 5).

Transplantation of Embryonic/Fetal Neural Tissue for Spinal Cord Repair

The use of embryonic/fetal neural tissue for repair of CNS disorders including SCI has played an important role in establishing the principle that immature neural tissue/cells are able to survive, differentiate, and promote functional recovery in experimental animal studies. Embryonic/fetal neural tissue has a number of properties, which makes it attractive in SCI, although ethical aspects complicate the use of this tissue both for experimental and clinical purposes (Table 6). The tissue must have the potential to survive transplantation into the mature and injured host spinal cord and further to thereafter differentiate to mature neurons with the ability to make functionally useful synaptic connections. Embryonic/fetal neural tissue also contains stem cells and non-neural cells with trophic and substrate supporting ability for transplanted immature neurons as well as for damaged neurons in the host spinal cord (Table 6).

The ideal cellular candidate for cell-based repair strategy should therefore be able to:

- Support axonal elongation and regrowth through the adult CNS.
- Restore myelin around spared axons.
- Migrate within the CNS to appropriate place(s) and coexist with astrocytes.
- Fully integrate within the lesion microenvironment.

Table 6
Properties of embryonic tissue

1	Rapid growth and cell division
2	Ability to develop to similar cells as in surrounding tissue
3	Less often rejected
4	Contain factors stimulating in growth of vessels as well as a higher proportion of growth factors promoting survival when transplanted
5	Less sensitive to ischemia and thus able to survive in surroundings with lowered oxygen levels

Table 7
Strategies to overcome the late inhibitory mechanisms of the CNS environment: transplantation of embryonic/fetal tissue

Factor	Target	Delivery	Effect
Solid embryonic/fetal tissue	Red nucleus axons	Transplantation to cavity	Rescue axotomized neurons
Brain stem fragments	Denervated terminal field	Injection	Local reinnervation Functional restitution
Solid embryonic/fetal tissue	Axotomized axons	Transplantation to cavity	Axonal elongation into and out of transplant
Human embryonic/fetal tissue	Axotomized axons	Transplantation	Feasibility Cyst obliteration

Different strategies have been used in transplanting embryonic/fetal neural tissue in experimental spinal cord injuries (Table 7) and the transplantation procedure has five primary objectives:

- To replace specific cells in order to restore lost functions.
- To minimize neuron degeneration and scar formation.
- To facilitate regeneration and plasticity by providing a scaffold and trophic support for axonal growth.
- To serve as a relay station in which descending or ascending impulses can terminate and subsequently be transferred via axons from transplanted cells to neurons caudal/rostral to the lesion.
- To serve as harbor for growth factors.

Animal Tissue to Animal Recipients and Human Tissue to Animal Recipients

Transplantation of solid rodent embryonic spinal cord tissue into experimentally induced posttraumatic cysts was shown to counteract retrograde cell death of, e.g., rubrospinal tract neurons. Transplants of embryonic rodent brainstem tissue into the spinal cord below the level of injury induced local re-innervations and even some functional recovery. Furthermore, spinal cord tissue harvested from early abortions in humans (5–8 weeks) and then transplanted to animal spinal cord promoted tissue repair by increasing survival of injured neurons and axons, reducing cyst formation, and providing bridges for axonal growth (74). Fetal spinal cord tissue in combination with, for instance, neurotrophin transplanted to animal recipients has also been shown to improve functional recovery (75) (see Table 7).

Human Tissue to Patients with SCI

Embryonic/fetal tissue was initially used in humans in the treatment of, for example, Parkinson's disease and blindness due to macular degeneration. In the light of the promising results with transplantation of human embryonic tissue to animals the question arose whether implanting such tissue into human spinal cord could counteract further cyst expansion in a similar way as following experimental SCI. The transplants consisted of solid tissue or suspension of cells from human embryonic/fetal spinal cord; both methods of transplantation have their advantages and disadvantages. It should be emphasized that at the current stage these transplantations do not aim for functional recovery, but the experience from them has provided information mainly regarding feasibility, how the transplant fills the cavity, and if rejection is provoked (76). However, because of ethical issues, the limited access of suitable tissue, and the emergence of the exciting options provided by stem cell research, the use of embryonic tissue for transplantation to patients with SCI is currently on halt.

Stem Cells for Spinal Cord Repair

The presence of stem cells in the subventricular zone of the adult human brain offers a potential source for replacement of lost neurons and glial cells after, e.g., brain trauma or stroke. However, since neurogenesis in the adult spinal cord does not appear to occur or is at least very limited, current stem cell-based approaches aimed at SCI repair therefore rely on transplantation of stem/progenitor cells. In many instances the beneficial effects of stem/progenitor cell transplants are likely to depend on their trophic effects rather than on replacement of lost neurons or glial cells. There is an extensive literature documenting beneficial effects of such transplantations in spinal cord injured animals (77, 78). To direct differentiation of transplanted stem/progenitor cells to desired neurons or glial cells and their appropriate integration into host neural network(s) remains a major challenge in the use of stem cells for replacement therapy, since many different kinds of cells are affected by the injury.

The primary objective in SCI is to repair diseased white matter, often referred to as "partial repair" of SCI. The gray matter damage is of less interest since it only produces limited loss of function within the injured segment. A major cause for the loss of sensorimotor and autonomic functions after partial SCI is focal demyelination as a result of oligodendrocyte degeneration, leading to permanent loss of impulse propagation through the lesion site. Moreover, demyelinated axons become themselves more vulnerable than normally and are therefore likely to undergo delayed degeneration. To repopulate damaged white matter with oligodendrocytes, which are competent to form myelin around demyelinated axons and thereby restore impulse propagation, would be an important goal and also probably more realistic, than to achieve more than limited axon growth from transplanted stem cell-derived neurons. Therefore, a prime goal in the use of stem cells for partial SCI is to learn how to replace lost oligodendrocytes with myelin producing cells in order to restore impulse propagation. With complete SCI, the objectives with stem cell transplants will focus on forming functionally useful connections across the injury site. Stem cells could provide support which allows axons of injured endogenous neurons to grow across the injury and restore connectivity. Alternatively, stem cell-derived neurons could themselves make these connections and thereby convey information across the injury site. In this situation, stem cell-derived neurons also have to be integrated into intrinsic neural networks in a functionally meaningful manner.

Neuroprotection and neurotrophic support with stem cell transplants can be achieved with stem cells from a wide range of non-neural sources which are potentially relevant for clinical translation, e.g., bone marrow-derived hematogenous or mesenchymal stem cells (79), adipose tissue-derived mesenchymal stem cells (80), and umbilical cord blood stem cells (81). The possibility for autologous transplantation makes several of these sources attractive for clinical use (see below). Although neural cells have been possible to generate from these sources, the yield is insufficient for the requirements of cell replacement purposes (see, e.g., (82)). For these purposes stem cells are therefore isolated either from adult (83), or embryonic neural tissues, or are pre-differentiated to neural progenitors from embryonic stem cells (ESCs) (84), or induced pluripotent stem cells (iPSCs) (85). ESCs and iPSCs are attractive since they can differentiate into any type of neural cells needed for replacement therapy. Moreover, iPSCs can be used for autologous transplantation, although the time to generate desired patient-derived type of cell(s) for transplantation may allow stem cell-based treatment only in the subacute or chronic phase. The use of embryonic or adult stem cells from the central nervous system for clinical purposes is complicated by ethical issues. However, adult neural stem cells, which are potentially useful for autologous, clinical transplantation, can be isolated from hair follicles (86, 87).

The most straightforward way to deliver stem cells in SCI is to inject them directly into the lesion site. However, stem cells have a unique ability to home to sites of injury, making it possible to deliver them intrathecally or even systemically. By using these routes the risk of inducing any additional damage at the lesion site is avoided, but at the same time the control of where the majority of stem cells will be located is obviously much more limited.

Several Phase I/II clinical trials have been carried out with implantation of autologous bone marrow-derived stem cells to patients with acute, subacute, or chronic cervical or thoracic SCI. Cells have either been injected directly into the lesion site (88–91) delivered intrathecally (92–96) or systemically (97, 98). The transplantation procedures were feasible and without serious complications and in some patients minor-moderate improvements were reported.

Umbilical cord blood stem cells have attracted increasing interest as a source for transplantation in neurological disorders, including SCI (99, 100). In a patient who had become paraplegic after a low thoracic SCI, multipotential stem cells were isolated from umbilical cord blood and injected into the subarachnoid space just rostral to the injury site and also deposited in the subdural and epidural space around the injury (101). Remarkably, signs of some functional and morphological improvements were reported after this treatment which was done more than 19 years after the injury. Except for this case report, no reported outcomes of umbilical cord stem cell transplants have been published, but several phase I/II clinical trials are ongoing (102–104).

Two clinical trials are ongoing or underway with the ultimate goals of obtaining neural specific cells in patients with SCI. A limited clinical trial to test the safety of human ESC-derived oligodendrocytes progenitor transplants to the injury site in patients with thoracic SCI is ongoing in the USA since 2010 (105) and a study with delivery of purified human neural stem cells (HuCNS-SC®) into the lesion site of patients with thoracic SCI is underway in Switzerland (106).

Stem cell transplantation to patients with SCI is obviously only in an early phase, focusing on feasibility and safety issues. Based on the data from clinical trials which have been evaluated, stem cell transplantation appears to be safe and still holds promises as a potentially valuable therapeutic strategy for patients with SCI. However, the potential adverse effects of stem cell transplants will continue to require careful attention. The neurotrophic effects of stem cells may induce aberrant axonal growth which could contribute to or exacerbate neuropathic pain and/or spasticity. The tumorigenic potential of stem cells, particularly when derived from ESCs, has to be completely eliminated prior to transplantation. Thus, much basic and translational research is needed before the possible place of this treatment utility can be defined, including its combination with other neuroprotective and neuroregenerative approaches.

5.3. Gene Therapy

Implantation of genetically modified cells to promote axonal regeneration and plasticity is a relatively new approach in SCI (107, 108). Genetically modified autologous or syngenic cells by viral vectors (ex vivo gene therapy) can serve as biological mini pumps for localized targeted delivery of growth factors into a lesion site leading to the activation of the intrinsic growth capacity which will provide a potential substrate for axonal extension. The ability to turn off gene expression after axons have entered the lesion site might further enhance the efficacy and safety of growth factor gene delivery.

Novel ways of vector delivery could enhance the feasibility of therapeutic gene delivery and reduce the invasiveness of intraparenchymal injections. The ability of viral vectors to be retrogradely transported along axons from the injection site is of high interest in the spinal cord. Adenovirus and lentivirus injected in the skin, muscle, or peripheral nerve can be taken up by nerve terminals and transported to motor neurons in the spinal cord, sensory neurons in the dorsal root ganglia, across these ganglia into the spinal cord (transganglionic transport), and even trans-synaptically.

Gene therapy for neurological disorders has moved to phase I and II clinical trials in neurodegenerative diseases. The translation of animal studies will therefore advance with improved vector and promoter systems for minimally invasive administration. Gene transfer to supply factors stimulating axonal growth within and beyond the lesion site and to activate a neuron intrinsic growth capacity for axonal regeneration may be a useful component in the future treatment of SCI.

6. Future Challenges in SCI Research and Clinical Management

The intensive care approach in the acute management of SCI has led to a remarkable improvement in the health state of patients with SCI by minimizing complications from the respiratory and circulatory systems and also from other organ systems, for example, urinary tracts. Life expectancy is now nearly the same for spinal cord injured patients as for uninjured individuals but the challenge to cure sensory loss, paralysis, and autonomic dysfunctions still remains.

The current state of art in the treatment of patients with SCI has its translational scientific background in the vastly increased knowledge and understanding of the pathophysiology of primary and secondary injury which has accumulated during recent decades. In addition, clinical research which has led to modern intensive care for minimizing long-term sequelae after, e.g., brain trauma has had an impact on medical management of patients with SCI. The main challenge for the future will be to explore and implement the translational potential of the progress in basic science.

This will require a dedicated multidisciplinary approach based on the realization that success will most likely need the implementation of multiple, complementary approaches.

No single therapy will reduce the secondary injury mechanisms or increase the capacity for regeneration of damaged axons. A combined treatment using updated knowledge from the fields of basic and clinical research covering the neuroprotection efforts and neural regeneration must be used. At the same time, risk–benefit analyses must be made since treatment in patients with incomplete injuries can result in deterioration rather than gained function. The "golden drug" or "treatment" has not yet been invented and new questions arise as soon as new pieces of knowledge are acquired. "Future challenges" include initially the quest to further increase knowledge within the fields of SCI research and treatment. These include experimental basic sciences with hypothesis relevant to clinical practice, possible pharmacological interventions, additional developments within all cornerstones of medical management including the field of radiology, for instance, functional MRI. The need for multidisciplinary cooperation seems to be a logical starting point in order to bring together newly obtained scientific findings. We have in this chapter summarized approaches to "fill" and/or "close" the gap within the injured spinal cord. The need for closing/filling the gap and build bridges between scientists in the field of basic science and clinical practice cannot be overemphasized. Highlighting various hypotheses from a broad angle is crucial for further improvements in the clinical management. The interpretation and transmission of new findings to "common interests" require good relations between all scientific fields and to create optimal conditions for multidisciplinary research which incorporates basic science, translational research and clinical management is perhaps the most important future challenge.

Early treatment according to basic principles described initially in this review is accepted in most of the wealthy part of the world. SCI patients are treated at intensive care units receiving all modern high-tech treatment based on the principles of counteracting the effects of secondary injury mechanisms. The objectives of the treatment are to minimize both the neurological deficits and other organ failure and to facilitate early mobilization. However, additional questions need multidisciplinary approaches to be properly addressed and solved for the future. This includes questions as to whether the current treatment given today uses all knowledge explored within the fields of basic science, for example:

- Is the importance of treating injuries caused by "secondary mechanisms" overestimated in order to minimize (avoid) future neurological deterioration and/or does our knowledge regarding secondary injury mechanisms indicate the possibility of a regeneration oriented approach, even in the acute stage?

- Does our knowledge within the field of regeneration advocate changes or modifications in the acute management such as surgical intervention and what is the optimal overall tissue treatment in the acute stage in order to facilitate later regeneration treatment? This also includes how do we optimize the chemical environment surrounding the nerve cells in order to maintain the nerve tissue that survived in the acute stage in as good a condition as possible?
- Is the surgical removal of mechanical pressure on the neural sac followed by fixation of the spinal column the best surgical treatment for all patients despite degree of neurological deterioration and level of injury, i.e., should the initial procedure be identical from the spinal cord point of view if patients achieve an immediate complete injury to the mid-thoracic level with poor prognosis or a partial injury with signs of initial recovery at the cervical level with relatively good possibility to regaining useful function?
- Could, as an alternative in some cases, the dural sac be opened and the subarachnoid space be exposed and flushed in order to eliminate blood accumulated subarachnoidally and is that a therapeutic measure in order to decrease or eliminate the inflammatory response and further minimize the regeneration obstacles developing over time?
- Should we in the future use, for example, embryonic/fetal tissue, stem cells, OECs, or scaffold structures with or without neural growth factors already in the acute surgical procedure as a complement or treatment option in some cases or should we delay surgical procedures until we are able to artificially start the regeneration procedure at a later stage and, if so, when is the most appropriate time for this combined treatment?
- Hypothetically, is, in addition to the overall medical management, a two-step surgical procedure a future solution for both addressing the compression of the spinal cord and resulting secondary damage and regeneration process most relevant from a spinal cord perspective?

Let us present a future patient with a SCI at the cervical level caused by a burst fracture with remaining bony fragments that compresses the spinal cord. The surgical procedure starts with a corpectomy (removal of the damaged vertebra alleviating the spinal cord from compression) followed by a stabilizing procedure. The patient is then placed in a prone position and the dural sac is exposed from the posterior view using a laminectomy (removal of the arches) and then opened through a midline incision and the spinal cord is exposed. The blood products and inflammatory debris which covers the spinal cord (i.e., the components required for scar formation and subsequent obstacle for the regeneration procedure) are removed through careful rinsing. An artificial dural sac is then applied in our hypothetically future treated patient in such a fashion that it later counteracts compression from postoperative extradural blood. The artificial dura is perforated with a needle after which an intradural/subarachnoid infusion of factors that obstruct scar formation and promote regeneration is administered. Neurotrophic growth factors, antibodies against Nogo-A receptors, 4-aminopyridine, monoclonal antibodies against ICAM-1, as well as chondroitinase ABC are examples of substances included in this infusion cocktail. Theoretically, this infusion can start immediately following the application of the artificial dural sac. A percutaneous intradural/subarachnoid insertion of the needle followed by a similar infusion will be the alternative in conservatively treated patients.

The treatment period could be extended until the estimated scar formation period has ended. The length of this may hypothetically vary between 2 and 6 weeks depending on the amount of blood products and inflammatory debris but also on the severity of the damage to the spinal cord seen at the exposure. In the meantime, scaffold structures of different sources could now be prepared for subsequent implantation at the injury site. The choice of scaffold depends on the type of injury seen at the exposure. Transection injuries may need one type of bridge while contusion injuries will be bridged by other types of scaffolds. Schwann cells, NTFs, and stem cells may be attached to the scaffold structures by means of different types of glues leading to a gradual release of the substances. OECs could be administered either through infusion or as a cell suspension and activated macrophages from the injured patients, itself counteracting the inflammatory response, and Roliprane acting as c-AMP preserver will be administered at different time points during the posttraumatic period. Stem/progenitor cell research is now being translated to clinical feasibility trials. Besides its biological and medical issues, this research also requires detailed analyses of the associated ethical, legal, and donor issues. Will stem cell or fetal tissue be our source or will purified adult stem cells, collected from the injured patient, be used as replacement tissue? Will it be possible to use T-cells already at the time of the first surgical procedure? Those and other medical “science fiction” thoughts need to be clarified during the route to the goal of giving the patients the best treatment following a SCI. Accumulating the knowledge attained during the latest two to three decades and looking forward in the same time perspective may indicate some hope to solve one of the most pressing medical challenges, i.e., how to regenerate transected and/or damaged axons in the spinal cord and thereby create opportunities for regain of function, although limited, below the level of injury.

References

1. Lifshutz J, Colohan A (2004) A brief history of therapy for traumatic spinal cord injury. Neurosurg Focus 16(1):E5
2. Wilkins R (1964) The Edwin Smith surgical papyrus (Neurosurgical classic XVII). J Neurosurg 21:240–244
3. Ahn H, Singh J, Nathens A, MacDonald RD, Travers A, Tallon J et al (2011) Pre-hospital care management of a potential spinal cord injured patient: a systematic review of the literature and evidence-based guidelines. J Neurotrauma 28(8):1341–1361
4. Fehlings MG, Cadotte DW, Fehlings LN (2011) A series of systematic reviews on the treatment of acute spinal cord injury: a foundation for best medical practice. J Neurotrauma 28(8):1329–1333
5. Markandaya M, Stein DM, Menaker J. Acute treatment options for spinal cord injury. Curr Treat Options Neurol 2012 (Epub ahead of print)
6. Werndle MC, Zoumprouli A, Sedgwick P, Papadopoulos MC (2012) Variability in the treatment of acute spinal cord injury in the United Kingdom: results of a national survey. J Neurotrauma 29(5):880–888
7. Anderberg L, Aldskogius H, Holtz A (2007) Spinal cord injury—scientific challenges for the unknown future. Ups J Med Sci 112(3): 259–288
8. Balentine JD (1978) Pathology of experimental spinal cord trauma. I. The necrotic lesion as a function of vascular injury. Lab Invest 39(3):236–253
9. Tator CH, Fehlings MG (1991) Review of the secondary injury theory of acute spinal cord trauma with emphasis on vascular mechanisms. J Neurosurg 75(1):15–26
10. Tator CH, Rowed DW, Schwartz ML, Gertzbein SD, Bharatwal N, Barkin M et al (1984) Management of acute spinal cord injuries. Can J Surg 27(3):289–293, 296
11. Gupta R, Bathen ME, Smith JS, Levi AD, Bhatia NN, Steward O (2010) Advances in the management of spinal cord injury. J Am Acad Orthop Surg 18(4):210–222
12. Oyinbo CA (2011) Secondary injury mechanisms in traumatic spinal cord injury: a nugget of this multiply cascade. Acta Neurobiol Exp (Wars) 71(2):281–299
13. Anderson DK, Beattie M, Blesch A, Bresnahan J, Bunge M, Dietrich D et al (2005) Recommended guidelines for studies of human subjects with spinal cord injury. Spinal Cord 43(8):453–458
14. Berney S, Bragge P, Granger C, Opdam H, Denehy L (2011) The acute respiratory management of cervical spinal cord injury in the first 6 weeks after injury: a systematic review. Spinal Cord 49(1):17–29
15. Winslow C, Rozovsky J (2003) Effect of spinal cord injury on the respiratory system. Am J Phys Med Rehabil 82(10):803–814
16. Lazaridis C (2012) Advanced hemodynamic monitoring: principles and practice in neurocritical care. Neurocrit Care 16(1):163–169
17. Ploumis A, Yadlapalli N, Fehlings MG, Kwon BK, Vaccaro AR (2010) A systematic review of the evidence supporting a role for vasopressor support in acute SCI. Spinal Cord 48(5):356–362
18. Raslan AM, Fields JD, Bhardwaj A (2010) Prophylaxis for venous thrombo-embolism in neurocritical care: a critical appraisal. Neurocrit Care 12(2):297–309
19. McKinley W, McNamee S, Meade M, Kandra K, Abdul N (2006) Incidence, etiology, and risk factors for fever following acute spinal cord injury. J Spinal Cord Med 29(5):501–506
20. Fehlings MG, Wilson JR (2010) Spine trauma: the challenges in assessing outcomes. J Neurosurg Spine 13(5):636–637, discussion 7
21. Furlan JC, Noonan V, Singh A, Fehlings MG (2011) Assessment of impairment in patients with acute traumatic spinal cord injury: a systematic review of the literature. J Neurotrauma 28(8):1445–1477
22. Basso DM, Beattie MS, Bresnahan JC (1995) A sensitive and reliable locomotor rating scale for open field testing in rats. J Neurotrauma 12(1):1–21
23. Apuzzo M (2002) Pharmacological therapy after acute cervical spinal cord injury. Neurosurgery 50(3 Suppl):S63–S72
24. Felleiter P, Muller N, Schumann F, Felix O, Lierz P (2012) Changes in the use of the methylprednisolone protocol for traumatic spinal cord injury in Switzerland. Spine (Phila Pa 1976) 37(11):953–956
25. Lammertse D, Dungan D, Dreisbach J, Falci S, Flanders A, Marino R et al (2007) Neuroimaging in traumatic spinal cord injury: an evidence-based review for clinical practice and research. J Spinal Cord Med 30(3): 205–214
26. Miyanji F, Furlan JC, Aarabi B, Arnold PM, Fehlings MG (2007) Acute cervical traumatic spinal cord injury: MR imaging findings correlated with neurologic outcome—prospective

study with 100 consecutive patients. Radiology 243(3):820–827
27. Bagnall AM, Jones L, Duffy S, Riemsma RP. Spinal fixation surgery for acute traumatic spinal cord injury. Cochrane Database Syst Rev 2008(1):CD004725
28. Furlan JC, Noonan V, Cadotte DW, Fehlings MG (2011) Timing of decompressive surgery of spinal cord after traumatic spinal cord injury: an evidence-based examination of preclinical and clinical studies. J Neurotrauma 28(8):1371–1399
29. Fehlings MG, Vaccaro A, Wilson JR, Singh A, Cadotte DW, Harrop JS et al (2012) Early versus delayed decompression for traumatic cervical spinal cord injury: results of the Surgical Timing in Acute Spinal Cord Injury Study (STASCIS). PLoS One 7(2):e32037
30. ClinicalTrials.gov (Internet). Araabi (MD): National Library of Medicine (US). 2009-(cited 2012 Mar 13). Available from: http://clinicaltrials.gov/show/NCT00475748 NLM Identifier:NCT00475748. (database on the Internet)
31. David S, Aguayo AJ (1981) Axonal elongation into peripheral nervous system "bridges" after central nervous system injury in adult rats. Science 214(4523):931–933
32. Goldshmit Y, Spanevello MD, Tajouri S, Li L, Rogers F, Pearse M et al (2011) EphA4 blockers promote axonal regeneration and functional recovery following spinal cord injury in mice. PLoS One 6(9):e24636
33. Fawcett JW (2006) The glial response to injury and its role in the inhibition of CNS repair. Adv Exp Med Biol 557:11–24
34. Kuffler DP (2011) Are we approaching research into the repair of spinal cord trauma from the best perspective? J Neurodegen Regen 3(1):55–77
35. Reier PJ (2004) Cellular transplantation strategies for spinal cord injury and translational neurobiology. NeuroRx 1(4):424–451
36. Silver J, Miller JH (2004) Regeneration beyond the glial scar. Nat Rev Neurosci 5(2):146–156
37. Thuret S, Moon LD, Gage FH (2006) Therapeutic interventions after spinal cord injury. Nat Rev Neurosci 7(8):628–643
38. Schmitt AB, Breuer S, Liman J, Buss A, Schlangen C, Pech K et al (2003) Identification of regeneration-associated genes after central and peripheral nerve injury in the adult rat. BMC Neurosci 4:8
39. Zhou FQ, Snider WD (2006) Intracellular control of developmental and regenerative axon growth. Philos Trans R Soc Lond B Biol Sci 361(1473):1575–1592
40. ClinicalTrials.gov (Internet). Oliviero (MD): National Library of Medicine (US). 2011-(cited 2012 May 03). Available from: http://clinicaltrials.gov/show/NCT01329757 NLM Identifier: NCT01329757 (database on the Internet)
41. Fawcett JW (2006) Overcoming inhibition in the damaged spinal cord. J Neurotrauma 23(3–4):371–383
42. Schwab ME (2004) Nogo and axon regeneration. Curr Opin Neurobiol 14(1):118–124
43. Yiu G, He Z (2006) Glial inhibition of CNS axon regeneration. Nat Rev Neurosci 7(8):617–627
44. Zorner B, Schwab ME (2010) Anti-Nogo on the go: from animal models to a clinical trial. Ann N Y Acad Sci 1198(Suppl 1):E22–E34
45. ClinicalTrials.gov (Internet). GSK Clinical Trials: National Library of Medicine (US). 2011- (cited 2012 May 03). Available from: http://clinicaltrials.gov/show/NCT00622609 NLM Identifier: NCT00622609. (database on the Internet)2011
46. Lee JK, Zheng B (2012) Role of myelin-associated inhibitors in axonal repair after spinal cord injury. Exp Neurol 235(1):33–42
47. Cardenas DD, Ditunno J, Graziani V, Jackson AB, Lammertse D, Potter P et al (2007) Phase 2 trial of sustained-release fampridine in chronic spinal cord injury. Spinal Cord 45(2):158–168
48. Jones TB, McDaniel EE, Popovich PG (2005) Inflammatory-mediated injury and repair in the traumatically injured spinal cord. Curr Pharm Des 11(10):1223–1236
49. Rice T, Larsen J, Rivest S, Yong VW (2007) Characterization of the early neuroinflammation after spinal cord injury in mice. J Neuropathol Exp Neurol 66(3):184–195
50. Kubasak MD, Hedlund E, Roy RR, Carpenter EM, Edgerton VR, Phelps PE (2005) L1 CAM expression is increased surrounding the lesion site in rats with complete spinal cord transection as neonates. Exp Neurol 194(2):363–375
51. Segal JL (2005) Immunoactivation and altered intercellular communication mediate the pathophysiology of spinal cord injury. Pharmacotherapy 25(2):145–156
52. Knoller N, Auerbach G, Fulga V, Zelig G, Attias J, Bakimer R et al (2005) Clinical experience using incubated autologous macrophages as a treatment for complete spinal cord injury: phase I study results. J Neurosurg Spine 3(3):173–181
53. Schwartz M, Yoles E (2006) Immune-based therapy for spinal cord repair: autologous

macrophages and beyond. J Neurotrauma 23(3–4):360–370
54. ClinicalTrials.gov (Internet). National Library of medicine (US). 2003- (cited 2012 May 25). Available from: http://clinicltrials.gov/show/NCT00073853 NLM Identifier: NCT00073853 (database on the Internet)
55. Busch SA, Silver J (2007) The role of extracellular matrix in CNS regeneration. Curr Opin Neurobiol 17(1):120–127
56. Bradbury EJ, Moon LD, Popat RJ, King VR, Bennett GS, Patel PN et al (2002) Chondroitinase ABC promotes functional recovery after spinal cord injury. Nature 416(6881):636–640
57. Barritt AW, Davies M, Marchand F, Hartley R, Grist J, Yip P et al (2006) Chondroitinase ABC promotes sprouting of intact and injured spinal systems after spinal cord injury. J Neurosci 26(42):10856–10867
58. Fouad K, Schnell L, Bunge MB, Schwab ME, Liebscher T, Pearse DD (2005) Combining Schwann cell bridges and olfactory-ensheathing glia grafts with chondroitinase promotes locomotor recovery after complete transection of the spinal cord. J Neurosci 25(5): 1169–1178
59. http://www.seikagaku.co.jp/english/pdf/50.pdf (database on the Internet)
60. Hong Z, Hong H, Chen H, Wang Z, Hong D (2012) Protective effects of erythropoietin in experimental spinal cord injury by reducing the C/EBP-homologous protein expression. Neurol Res 34(1):85–90
61. Matis GK, Birbilis TA (2009) Erythropoietin in spinal cord injury. Eur Spine J 18(3): 314–323
62. Pearse DD, Pereira FC, Marcillo AE, Bates ML, Berrocal YA, Filbin MT et al (2004) cAMP and Schwann cells promote axonal growth and functional recovery after spinal cord injury. Nat Med 10(6):610–616
63. McKerracher L, Higuchi H (2006) Targeting Rho to stimulate repair after spinal cord injury. J Neurotrauma 23(3–4):309–317
64. Dergham P, Ellezam B, Essagian C, Avedissian H, Lubell WD, McKerracher L (2002) Rho signaling pathway targeted to promote spinal cord repair. J Neurosci 22(15):6570–6577
65. Fehlings MG, Theodore N, Harrop J, Maurais G, Kuntz C, Shaffrey CI et al (2011) A phase I/IIa clinical trial of a recombinant Rho protein antagonist in acute spinal cord injury. J Neurotrauma 28(5):787–796
66. Olson L, Cheng H, Zetterstrom RH, Solomin L, Jansson L, Gimenez-Llort L et al (1998) On CNS repair and protection strategies: novel approaches with implications for spinal cord injury and Parkinson's disease. Brain Res Brain Res Rev 26(2–3):302–305
67. Fraidakis MJ, Spenger C, Olson L (2004) Partial recovery after treatment of chronic paraplegia in rat. Exp Neurol 188(1):33–42
68. Xu XM, Chen A, Guenard V, Kleitman N, Bunge MB (1997) Bridging Schwann cell transplants promote axonal regeneration from both the rostral and caudal stumps of transected adult rat spinal cord. J Neurocytol 26(1):1–16
69. Takami T, Oudega M, Bates ML, Wood PM, Kleitman N, Bunge MB (2002) Schwann cell but not olfactory ensheathing glia transplants improve hindlimb locomotor performance in the moderately contused adult rat thoracic spinal cord. J Neurosci 22(15):6670–6681
70. Mackay-Sim A, Feron F, Cochrane J, Bassingthwaighte L, Bayliss C, Davies W et al (2008) Autologous olfactory ensheathing cell transplantation in human paraplegia: a 3-year clinical trial. Brain 131(Pt 9):2376–2386
71. Chuah MI, Hale DM, West AK (2011) Interaction of olfactory ensheathing cells with other cell types in vitro and after transplantation: glial scars and inflammation. Exp Neurol 229(1):46–53
72. Nomura H, Tator CH, Shoichet MS (2006) Bioengineered strategies for spinal cord repair. J Neurotrauma 23(3–4):496–507
73. Rochkind S, Shahar A, Fliss D, El-Ani D, Astachov L, Hayon T et al (2006) Development of a tissue-engineered composite implant for treating traumatic paraplegia in rats. Eur Spine J 15(2):234–245
74. Akesson E, Holmberg L, Jonhagen ME, Kjaeldgaard A, Falci S, Sundstrom E et al (2001) Solid human embryonic spinal cord xenografts in acute and chronic spinal cord cavities: a morphological and functional study. Exp Neurol 170(2):305–316
75. Ogawa Y, Sawamoto K, Miyata T, Miyao S, Watanabe M, Nakamura M et al (2002) Transplantation of in vitro-expanded fetal neural progenitor cells results in neurogenesis and functional recovery after spinal cord contusion injury in adult rats. J Neurosci Res 69(6):925–933
76. Falci S, Holtz A, Akesson E, Azizi M, Ertzgaard P, Hultling C et al (1997) Obliteration of a posttraumatic spinal cord cyst with solid human embryonic spinal cord grafts: first clinical attempt. J Neurotrauma 14(11):875–884
77. Hawryluk GW, Mothe A, Wang J, Wang S, Tator C, Fehlings MG (2012) An in vivo characterization of trophic factor production

following neural precursor cell or bone marrow stromal cell transplantation for spinal cord injury. Stem Cells Dev 21(12): 2222–2238
78. Ruff CA, Wilcox JT, Fehlings MG (2012) Cell-based transplantation strategies to promote plasticity following spinal cord injury. Exp Neurol 235(1):78–90
79. Wright KT, El Masri W, Osman A, Chowdhury J, Johnson WE (2011) Concise review: bone marrow for the treatment of spinal cord injury: mechanisms and clinical applications. Stem Cells 29(2):169–178
80. Chung JY, Kim W, Im W, Yoo DY, Choi JH, Hwang IK et al (2012) Neuroprotective effects of adipose-derived stem cells against ischemic neuronal damage in the rabbit spinal cord. J Neurol Sci 317(1–2):40–46
81. Park DH, Lee JH, Borlongan CV, Sanberg PR, Chung YG, Cho TH (2011) Transplantation of umbilical cord blood stem cells for treating spinal cord injury. Stem Cell Rev 7(1):181–194
82. Lindvall O, Kokaia Z (2010) Stem cells in human neurodegenerative disorders—time for clinical translation? J Clin Invest 120(1):29–40
83. Hernandez J, Torres-Espin A, Navarro X (2011) Adult stem cell transplants for spinal cord injury repair: current state in preclinical research. Curr Stem Cell Res Ther 6(3):273–287
84. Peljto M, Wichterle H (2011) Programming embryonic stem cells to neuronal subtypes. Curr Opin Neurobiol 21(1):43–51
85. Tsuji O, Miura K, Fujiyoshi K, Momoshima S, Nakamura M, Okano H (2011) Cell therapy for spinal cord injury by neural stem/progenitor cells derived from iPS/ES cells. Neurotherapeutics 8(4):668–676
86. Amoh Y, Katsuoka K, Hoffman RM (2010) The advantages of hair follicle pluripotent stem cells over embryonic stem cells and induced pluripotent stem cells for regenerative medicine. J Dermatol Sci 60(3):131–137
87. Sieber-Blum M (2010) Epidermal neural crest stem cells and their use in mouse models of spinal cord injury. Brain Res Bull 83(5): 189–193
88. Bhanot Y, Rao S, Ghosh D, Balaraju S, Radhika CR, Satish Kumar KV (2011) Autologous mesenchymal stem cells in chronic spinal cord injury. Br J Neurosurg 25(4): 516–522
89. Deda H, Inci MC, Kurekci AE, Kayihan K, Ozgun E, Ustunsoy GE et al (2008) Treatment of chronic spinal cord injured patients with autologous bone marrow-derived hematopoietic stem cell transplantation: 1-year follow-up. Cytotherapy 10(6):565–574
90. Park HC, Shim YS, Ha Y, Yoon SH, Park SR, Choi BH et al (2005) Treatment of complete spinal cord injury patients by autologous bone marrow cell transplantation and administration of granulocyte-macrophage colony stimulating factor. Tissue Eng 11(5–6):913–922
91. ClinicalTrials.gov (Internet). Sabaawy (M, PhD): Natrional Library of Medicine (US). 2009- (cited 2012 Mar 13). Available from http://clinicaltrialss.gov/show/NCT00816803 NLM Identifier: NCT00816803 (database on the Internet)
92. Callera F, do Nascimento RX (2006) Delivery of autologous bone marrow precursor cells into the spinal cord via lumbar puncture technique in patients with spinal cord injury: a preliminary safety study. Exp Hematol 34(2): 130–131
93. Geffner LF, Santacruz P, Izurieta M, Flor L, Maldonado B, Auad AH et al (2008) Administration of autologous bone marrow stem cells into spinal cord injury patients via multiple routes is safe and improves their quality of life: comprehensive case studies. Cell Transplant 17(12):1277–1293
94. Pal R, Venkataramana NK, Bansal A, Balaraju S, Jan M, Chandra R et al (2009) *Ex vivo*-expanded autologous bone marrow-derived mesenchymal stromal cells in human spinal cord injury/paraplegia: a pilot clinical study. Cytotherapy 11(7):897–911
95. Saito F, Nakatani T, Iwase M, Maeda Y, Hirakawa A, Murao Y et al (2008) Spinal cord injury treatment with intrathecal autologous bone marrow stromal cell transplantation: the first clinical trial case report. J Trauma 64(1):53–59
96. Yoon SH, Shim YS, Park YH, Chung JK, Nam JH, Kim MO et al (2007) Complete spinal cord injury treatment using autologous bone marrow cell transplantation and bone marrow stimulation with granulocyte macrophage-colony stimulating factor: phase I/II clinical trial. Stem Cells 25(8):2066–2073
97. Cristante AF, Barros-Filho TE, Tatsui N, Mendrone A, Caldas JG, Camargo A et al (2009) Stem cells in the treatment of chronic spinal cord injury: evaluation of somatosensitive evoked potentials in 39 patients. Spinal Cord 47(10):733–738
98. Sykova E, Homola A, Mazanec R, Lachmann H, Konradova SL, Kobylka P et al (2006) Autologous bone marrow transplantation in patients with subacute and chronic spinal cord injury. Cell Transplant 15(8–9):675–687

99. Malgieri A, Kantzari E, Patrizi MP, Gambardella S (2010) Bone marrow and umbilical cord blood human mesenchymal stem cells: state of the art. Int J Clin Exp Med 3(4):248–269
100. Momin EN, Mohyeldin A, Zaidi HA, Vela G, Quinones-Hinojosa A (2010) Mesenchymal stem cells: new approaches for the treatment of neurological diseases. Curr Stem Cell Res Ther 5(4):326–344
101. Kang KS, Kim SW, Oh YH, Yu JW, Kim KY, Park HK et al (2005) A 37-year-old spinal cord-injured female patient, transplanted of multipotent stem cells from human UC blood, with improved sensory perception and mobility, both functionally and morphologically: a case study. Cytotherapy 7(4): 368–373
102. ClinicalTrials.gov (Internet). Yihua (MD): National Library of Medicine (US). 2011- (cited 2012 Mar 17). Available from: http://clinicaltrials.gov/show/NCT01393977 NLMIdentifier: NCT01393977. (database on the Internet)
103. ClinicalTrials.gov (Internet). Zhu (MD): National Library of Medicine (US). 2011- (cited 2012 Mar 17). Available from: http://clinicaltrials.gov/show/NCT01354483 NLM Identifier: NCT01354483. (database on the Internet)
104. ClinicalTrials.gov (Internet).Sang Poon (MD): National Library of Medicine (US). 2011- (cited 2012 May 03). Available from: http://clinicaltrials.gov/show/NCT01046786 NLM Identifier: NCT01046786 (database on the Internet)
105. Mayor S (2010) First patient enters trial to test safety of stem cells in spinal injury. BMJ 341:c5724
106. Stem Cells IUHB, Switzerland. A phase I/II study of the safety and preliminary efficacy of intramedullary spinal cord transplantation of human central nervous system (CNS) stem cells (HuCNS-SC) in subjects with thoracic (T2-T11) spinal cord trauma. In: ClinicalTrials.gov (Internet). Stemcells.Inc: National Library of Medicine (US). 2011- (2012 May 25). Avaiable from: http://clinicaltrials.gov/show/NCT01321333 NLM Identifier: NCT01321333
107. Huang WC, Kuo HS, Tsai MJ, Ma H, Chiu CW, Huang MC et al (2011) Adeno-associated virus-mediated human acidic fibroblast growth factor expression promotes functional recovery of spinal cord-contused rats. J Gene Med 13(5):283–289
108. Taha MF (2010) Cell based-gene delivery approaches for the treatment of spinal cord injury and neurodegenerative disorders. Curr Stem Cell Res Ther 5(1):23–36

INDEX

Håkan Aldskogius (ed.), *Animal Models of Spinal Cord Repair*, Neuromethods, vol. 76,
DOI 10.1007/978-1-62703-197-4, © Springer Science+Business Media, LLC 2013

D

E

MIX
Papier aus verantwortungsvollen Quellen
Paper from responsible sources
FSC® C105338

If you have any concerns about our products,
you can contact us on
ProductSafety@springernature.com

In case Publisher is established outside the EU,
the EU authorized representative is:
Springer Nature Customer Service Center GmbH
Europaplatz 3, 69115 Heidelberg, Germany

Printed by Libri Plureos GmbH
in Hamburg, Germany